Frederick W. Menk
and Colin L. Waters

Magnetoseismology

Related Titles

Guest, G.

Electron Cyclotron Heating of Plasmas

264 pages with approx. 40 figures
2009
Hardcover
ISBN: 978-3-527-40916-7

Blaunstein, N., Christodoulou, C.

Radio Propagation and Adaptive Antennas for Wireless Communication Links
Terrestrial, Atmospheric and Ionospheric

614 pages
2006
Hardcover
ISBN: 978-0-471-25121-7

Bohren, C. F., Clothiaux, E. E.

Fundamentals of Atmospheric Radiation
An Introduction with 400 Problems

490 pages with 184 figures
2006
Softcover
ISBN: 978-3-527-40503-9

Hippler, R., Pfau, S., Schmidt, M.

Low Temperature Plasma Physics
Fundamental Aspects and Applications

523 pages with 244 figures and 23 tables
2001
Hardcover
ISBN: 978-3-527-28887-8

Smirnov, B. M.

Physics of Ionized Gases

398 pages
2001
Hardcover
ISBN: 978-0-471-17594-0

Frederick W. Menk and Colin L. Waters

Magnetoseismology

Ground-Based Remote Sensing of Earth's Magnetosphere

WILEY-VCH Verlag GmbH & Co. KGaA

The Authors

Prof. Frederick W. Menk and
Prof. Colin L. Waters
School of Mathematical and Physical Sciences
The University of Newcastle
University Drive
Callaghan NSW 2308
Australia

Cover

Artist's depiction of geomagnetic field lines mapping from Earth's surface into the magnetosphere, where colors represent mass density in the equatorial plane, which in turn determines the resonant frequency of these field lines. Kindly provided by Matthew Waters.

Library of Congress Card No.: applied for

British Library Cataloguing-in-Publication Data
A catalogue record for this book is available from the British Library.

Bibliographic information published by the Deutsche Nationalbibliothek
The Deutsche Nationalbibliothek lists this publication in the Deutsche Nationalbibliografie; detailed bibliographic data are available on the Internet at <http://dnb.d-nb.de>.

Print ISBN: 978-3-527-41027-9
ePDF ISBN: 978-3-527-65208-2
ePub ISBN: 978-3-527-65207-5
mobi ISBN: 978-3-527-65206-8
oBook ISBN: 978-3-527-65205-1

Typesetting Thomson Digital, Noida, India
Printing and Binding Markono Print Media Pte Ltd, Singapore

Cover Design Adam Design, Weinheim

Printed on acid-free paper

Contents

Preface

One of the joys as a student was building and using relatively simple equipment – magnetometers and ionospheric sounders – to probe the region of space around Earth and gain insight into processes there. This is the essence of this book: remote sensing, mostly using ground-based instruments and techniques, to understand our space environment, the magnetosphere. This region dynamically links interplanetary space with Earth's atmosphere, and is where satellites orbit.

The agents involved are ultralow-frequency plasma waves, since they propagate from the solar wind through the magnetosphere and atmosphere to the ground. These waves transfer energy and momentum and are not only involved in many types of instabilities and interactions but can also be used as a diagnostic monitor of these processes. This book focuses on the second aspect through understanding of the first.

With the move to online data access, undergraduate students can conduct original research using observations from ground arrays, radar networks, and satellites. The magnetosphere is there for everyone to explore. This in turn provides wonderful insight into all the relevant physics, from the cycles of the Sun to the nature of the geomagnetic field and the atmosphere, and exploring other planets.

This book focuses on the underlying principles and their interconnectedness. We do not assume familiarity with physics or mathematics concepts beyond undergraduate level.

Many people have guided our personal journeys. Our scientific mentors include Brian Fraser, John Samson, Keith Cole, and Valerie Troitskaya. Other colleagues include Sean Ables, Brian Anderson, Mark Clilverd, Bob Lysak, Ian Mann, Pasha Ponomarenko, Murray Sciffer, Peter Sutcliffe, and Tim Yeoman. Many students taught us at least as much as we taught them. The development of this monograph was patiently and enthusiastically guided by our editors at Wiley, Nina Stadthaus and Christoph Friedenburg. Of course, this book would not have been possible without the continual support of our families and wives, who suffered in silence a great many evenings while we disappeared into offices to pursue our arcane endeavors.

Newcastle, July 2012

Frederick W. Menk
Colin L. Waters

Color Plates

Figure 1.1 Comet Hale–Bopp, showing a white dust tail and a blue ion tail, resulting from the effect of the solar wind and entrained magnetic field. *Source:* Alessandro Dimai and Davide Ghirardo (Associazione Astronomica Cortina) at Passo Giau (2230 m), Cortina d'Amprezzo, Italy, March 16, 1997, 03:42 UT. e-mail: info@cortinastella.it, web: www.cortinastelle.it - www.skyontheweb.org. (This figure also appears on page 2.)

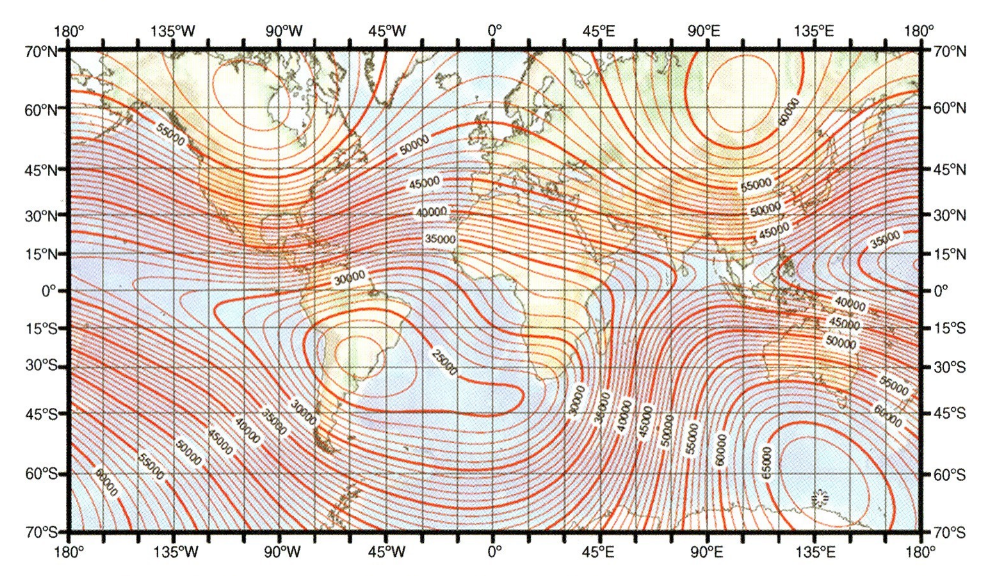

Figure 2.3 Mercator projection of the total intensity F of the main geomagnetic field computed using the 2010 World Magnetic Model. Contour interval is 1000 nT. From Maus *et al.* (2010). (This figure also appears on page 17.)

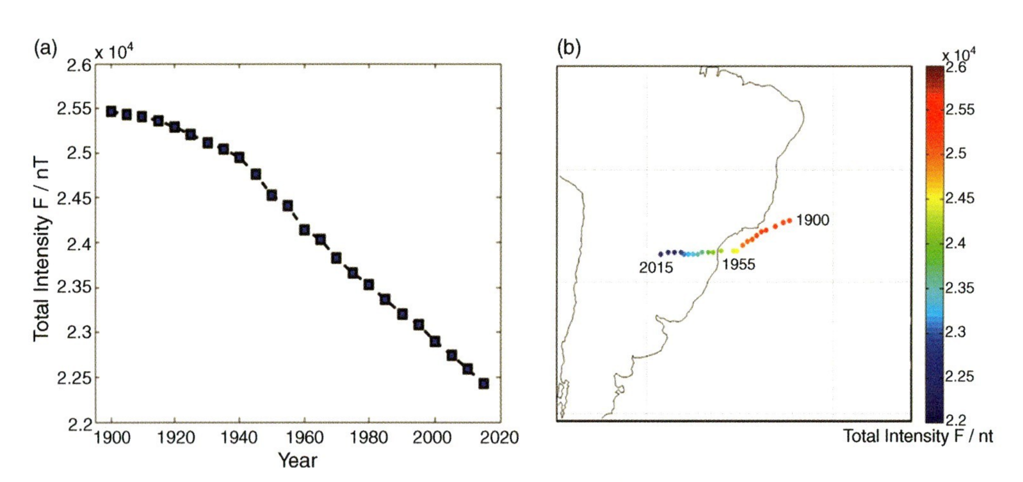

Figure 2.5 Variation in (a) strength and (b) location of the minimum in the total field intensity F in the South Atlantic Anomaly region during the past century. From Finlay *et al.* (2010). (This figure also appears on page 18.)

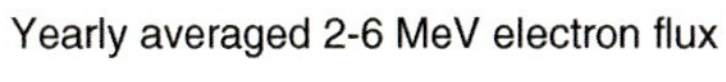

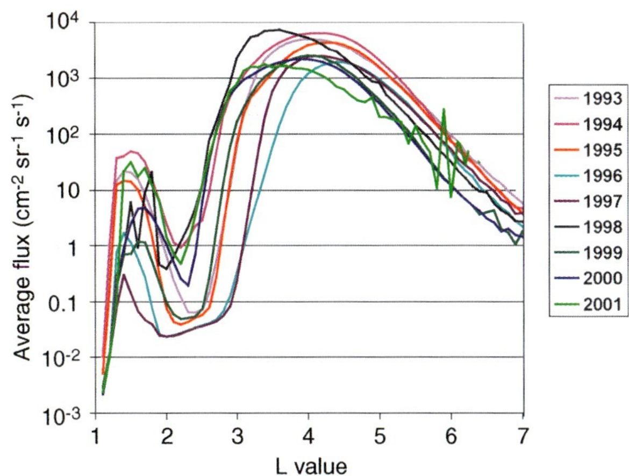

Figure 2.9 Yearly averaged 2–6 MeV electron flux measured at low altitudes by the SAMPEX spacecraft during 1993–2001, showing location and intensity of the radiation belts. Data courtesy of Shri Kanekal and SAMPEX Data Center staff. (This figure also appears on page 28.)

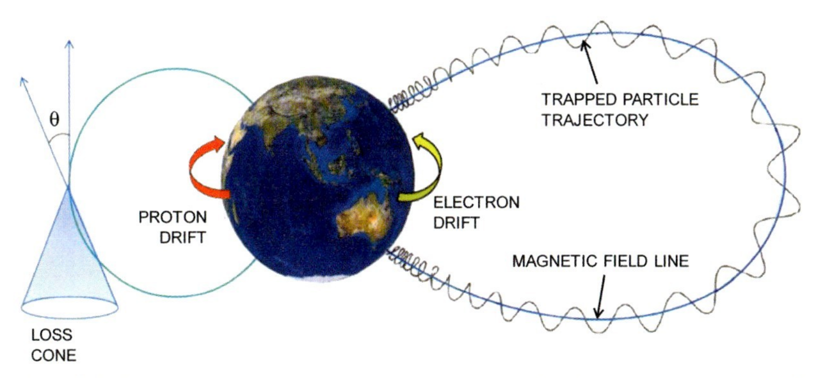

Figure 3.1 Charged particle motions in the magnetosphere, showing gyration around a field line, bouncing between mirror points, and azimuthal drift along L shells. Pitch angle is θ. (This figure also appears on page 47.)

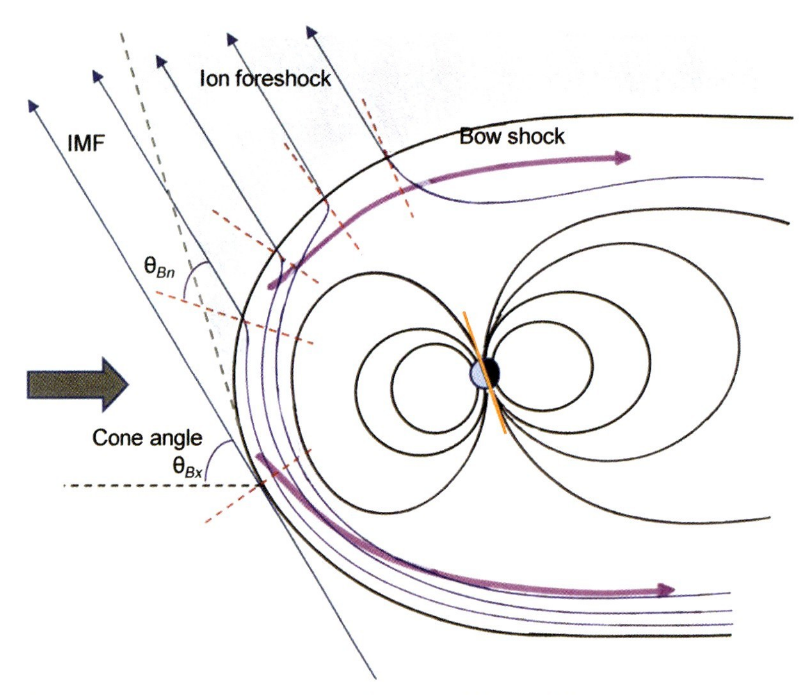

Figure 4.3 Schematic representation of the magnetopause, bow shock, and the ion foreshock region (shaded) where ULF waves are likely generated. The IMF is shown northward, and thick arrows represent plasma streamlines. Field lines (solid) map around the magnetopause and the plasma convects antisunward. (This figure also appears on page 66.)

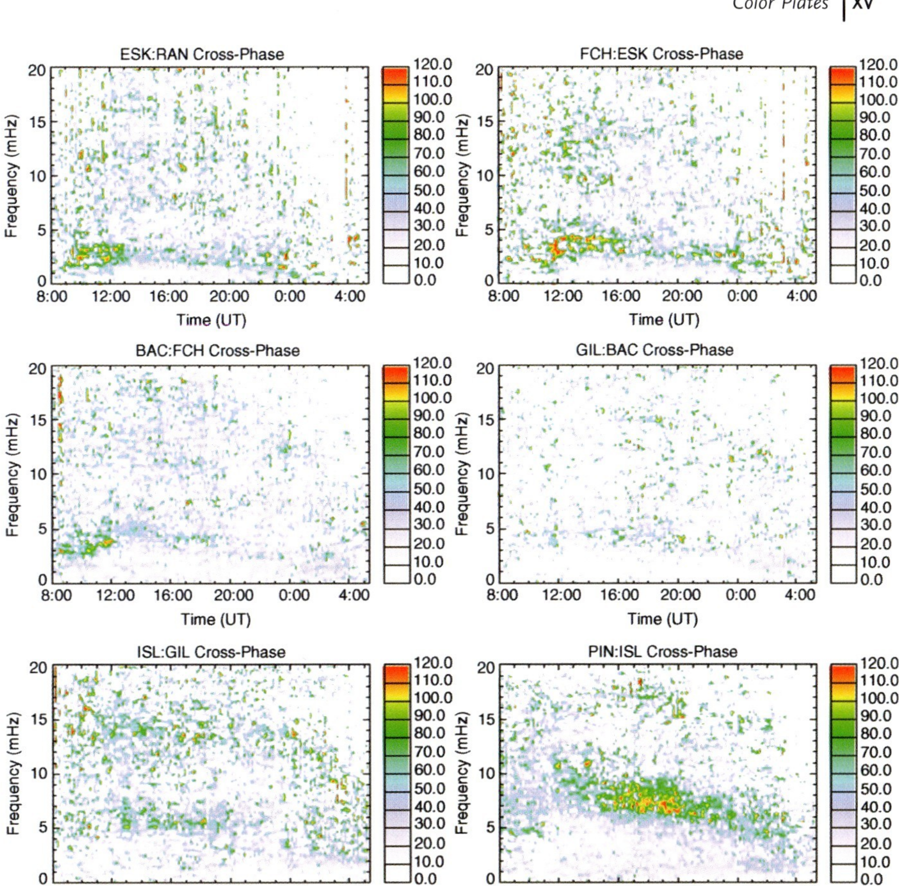

Figure 5.9 Dynamic cross-phase spectra data recorded on February 9, 1995 by the Churchill line of magnetometers of the Canadian array. Time axis is from 0800 to 0530 UT and local noon is at 1800 UT. Cross-phase scale is from 0° to 120°. (This figure also appears on page 99.)

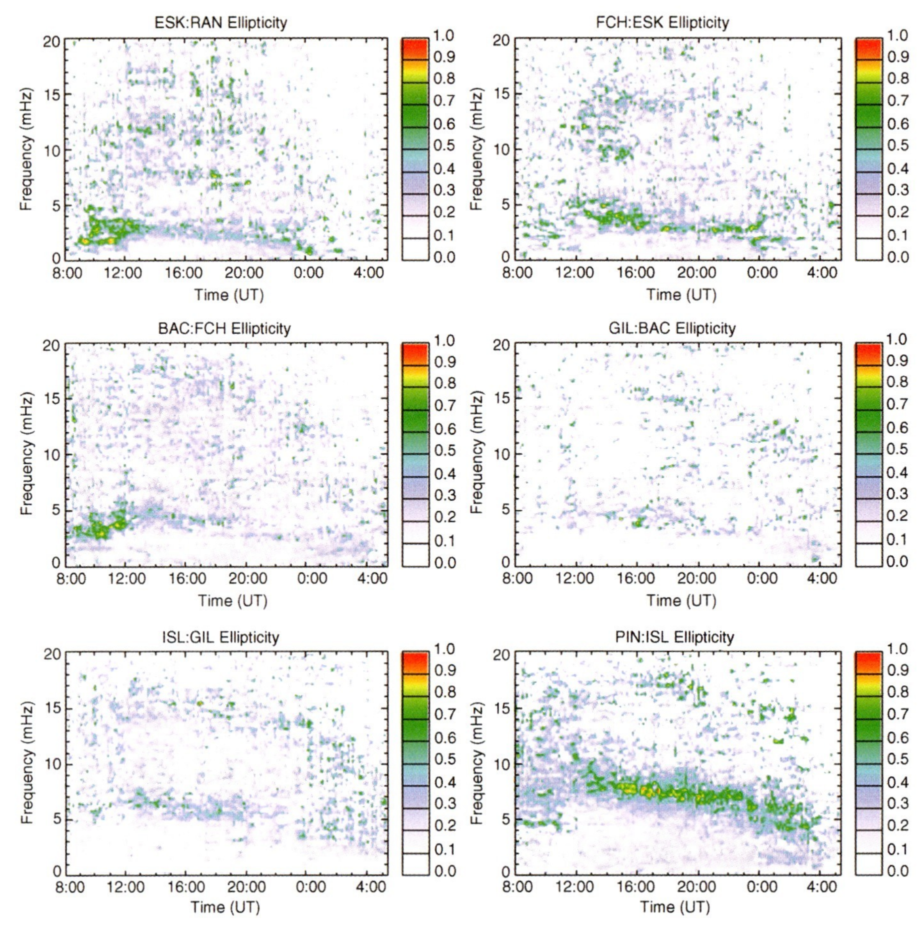

Figure 5.10 The "ellipticity" spectra computed from the north–south component magnetic field data from pairs of latitudinal spaced stations of the Canadian Churchill line. The processing used the same time series as Figure 5.9. (This figure also appears on page 102.)

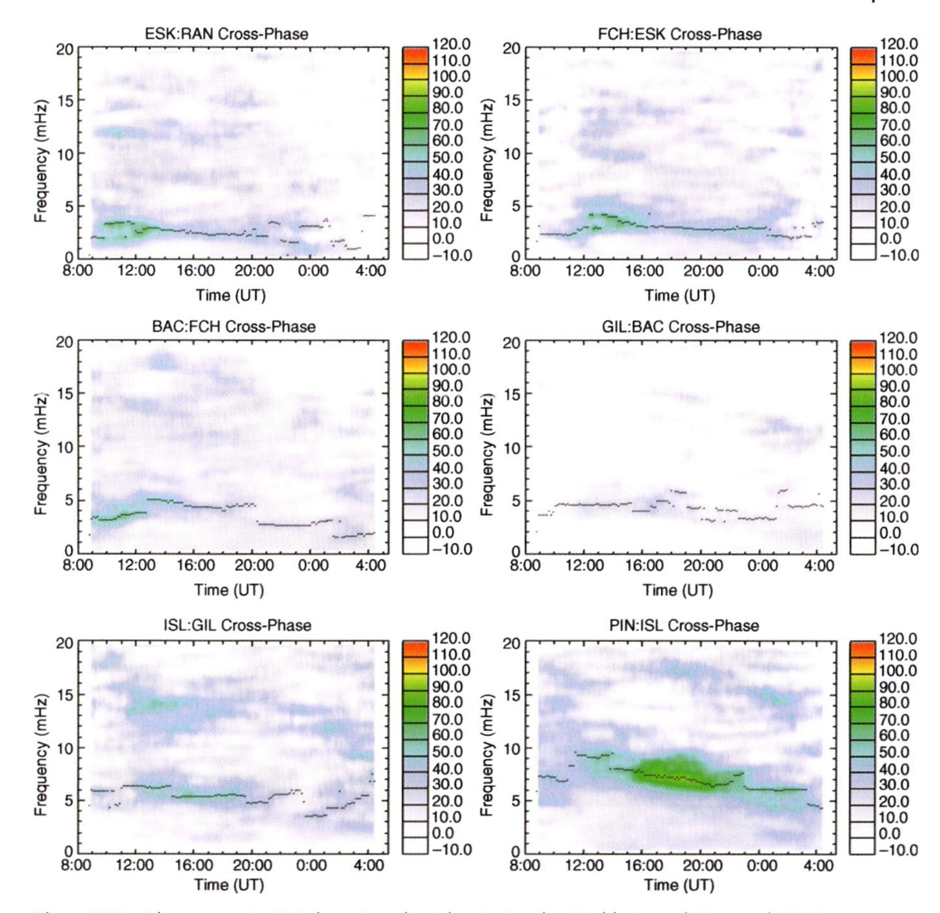

Figure 5.11 The automatic FLR detection algorithm in Berube, Moldwin, and Weygand (2003) applied to the cross-phase data in Figure 5.9. (This figure also appears on page 104.)

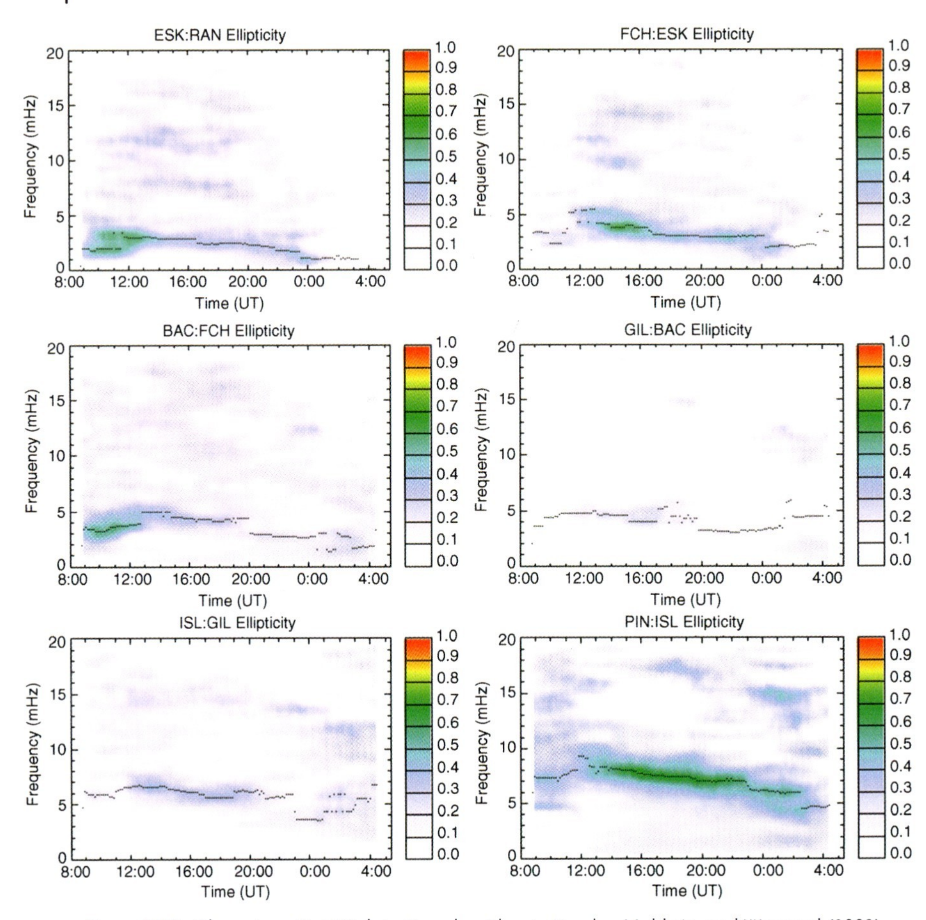

Figure 5.12 The automatic FLR detection algorithm in Berube, Moldwin, and Weygand (2003) applied to the ellipticity data in Figure 5.10. (This figure also appears on page 105.)

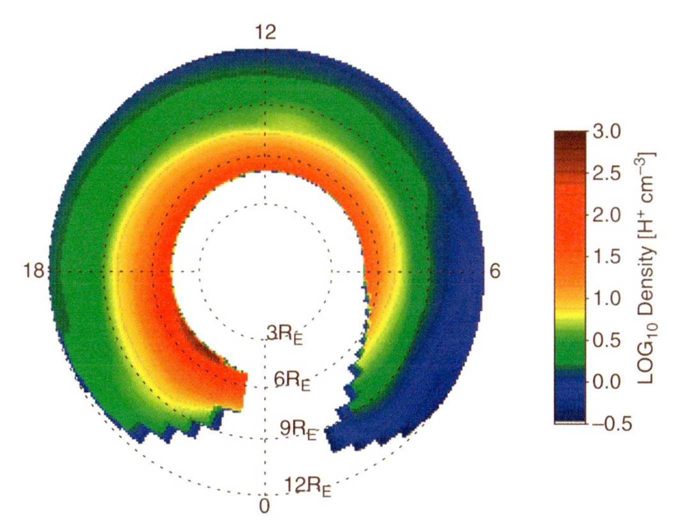

Figure 6.1 Logarithm of magnetosphere plasma mass density in units of H^+ cm^{-3} as a function of radial distance and MLT, derived from FLRs detected with the CANOPUS magnetometer array on February 9, 1995. From Waters *et al.*, (2006). (This figure also appears on page 109.)

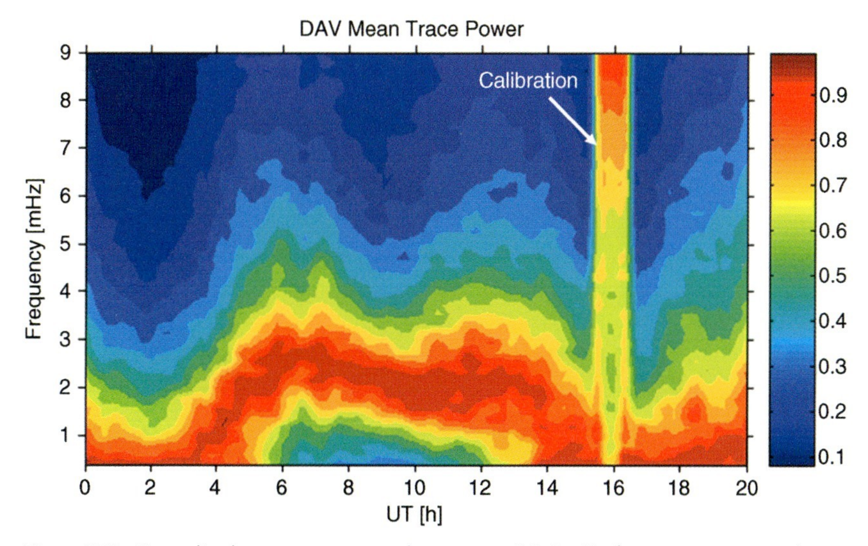

Figure 6.14 Normalized, mean trace spectral power over 0.1–9 mHz from magnetometer data recorded at Davis, Antarctica. The data are for the full year 1996 and local magnetic noon is near 0940 UT. (This figure also appears on page 129.)

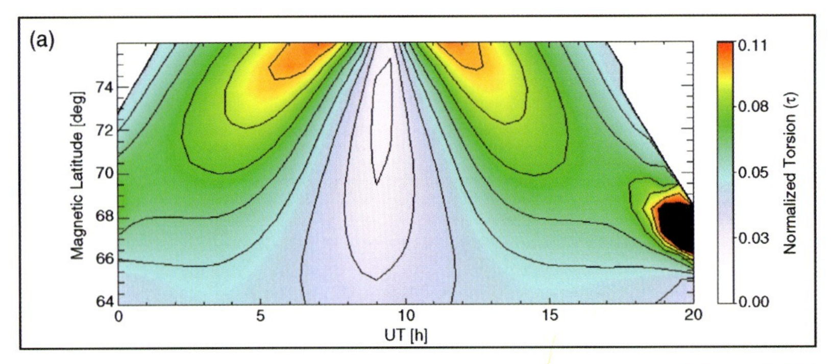

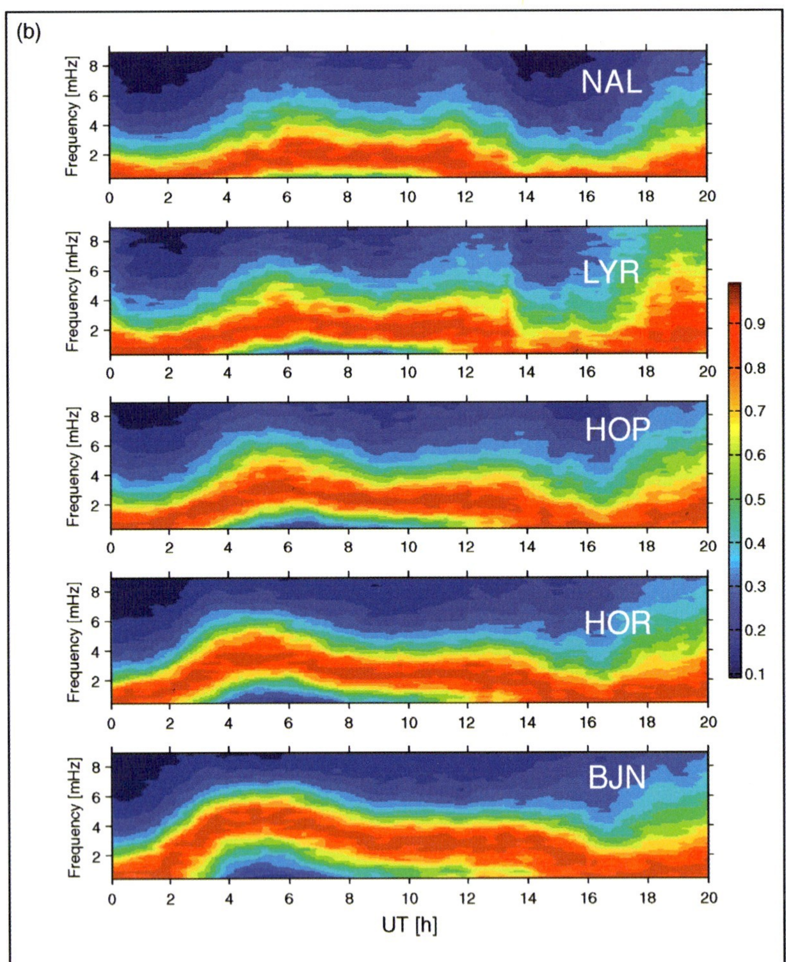

Figure 6.16 Extent in latitude of field line tension and torsion that affects FLR frequencies. (a) Estimates using the Tsyganenko 1996 model. (b) Normalized trace spectra of the horizontal components of magnetometer data from various stations in the Scandinavian IMAGE magnetometer array for the year 1996. (This figure also appears on page 131.)

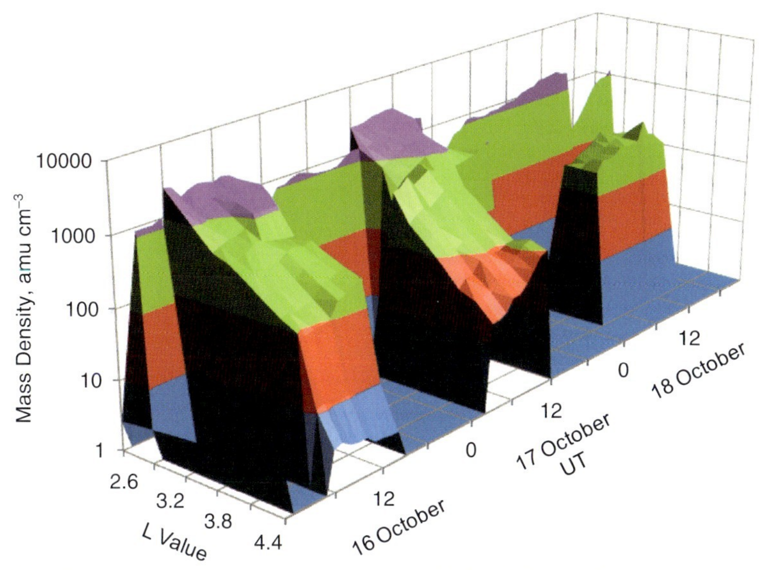

Figure 7.2 Plasma mass density map for October 16–18, 1990, based on observations from two magnetometer arrays separated by 10 h in local time. Adapted from Menk *et al.* (1999). (This figure also appears on page 135.)

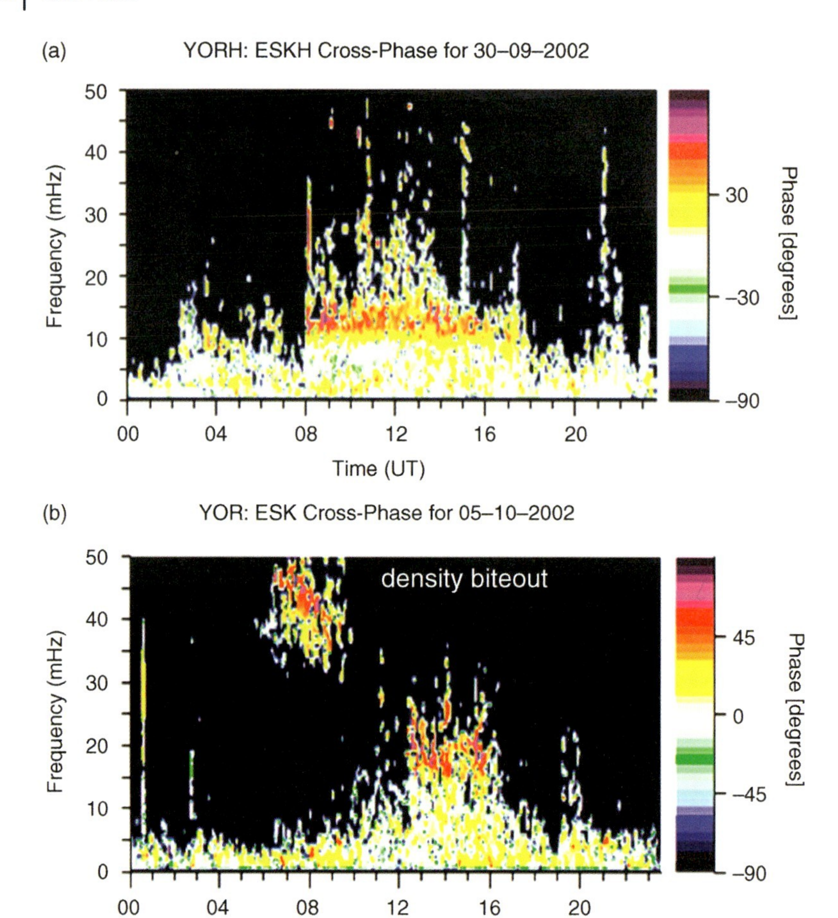

Figure 7.7 Whole-day $L = 2.67$ cross-phase frequency–time spectra for (a) September 30, 2002 and (b) October 5, 2002 when a density biteout occurred. (This figure also appears on page 140.)

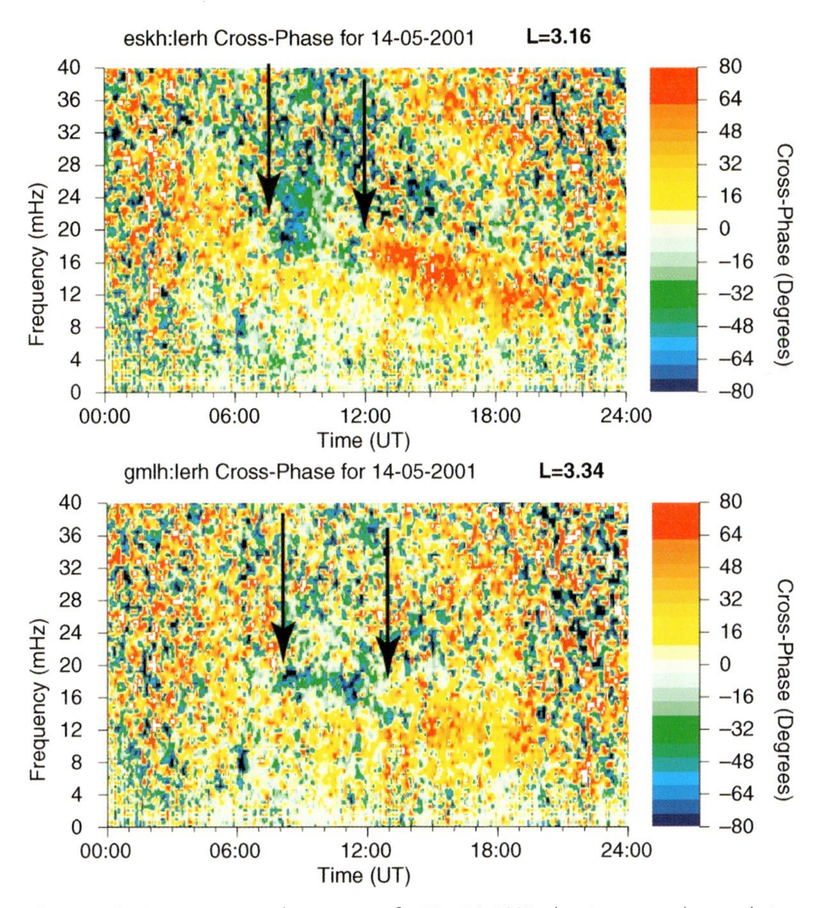

Figure 7.10 Dynamic cross-phase spectra for May 14, 2001, showing cross-phase polarity reversals, arrowed, at 0730 and 1200–1230 UT. From Kale *et al.* (2007). (This figure also appears on page 143.)

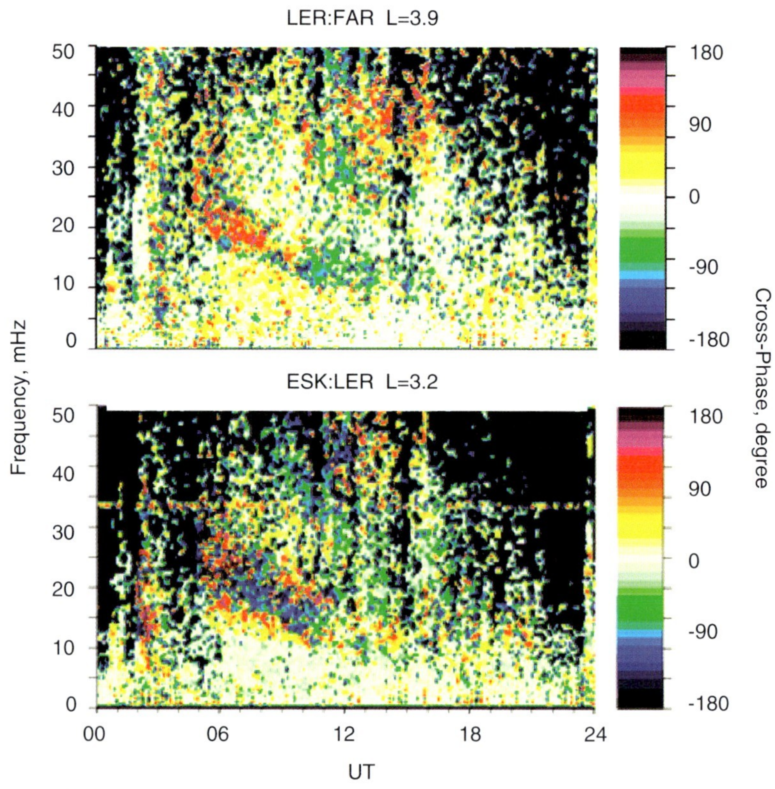

Figure 7.11 Dynamic cross-phase spectra for station pairs centered on $L = 3.9$ and $L = 3.2$ on June 11, 2001, a day after a $K_p = 6$ storm. A cross-phase reversal with time appears in the upper plot, and a reversal with frequency in the lower plot. (This figure also appears on page 144.)

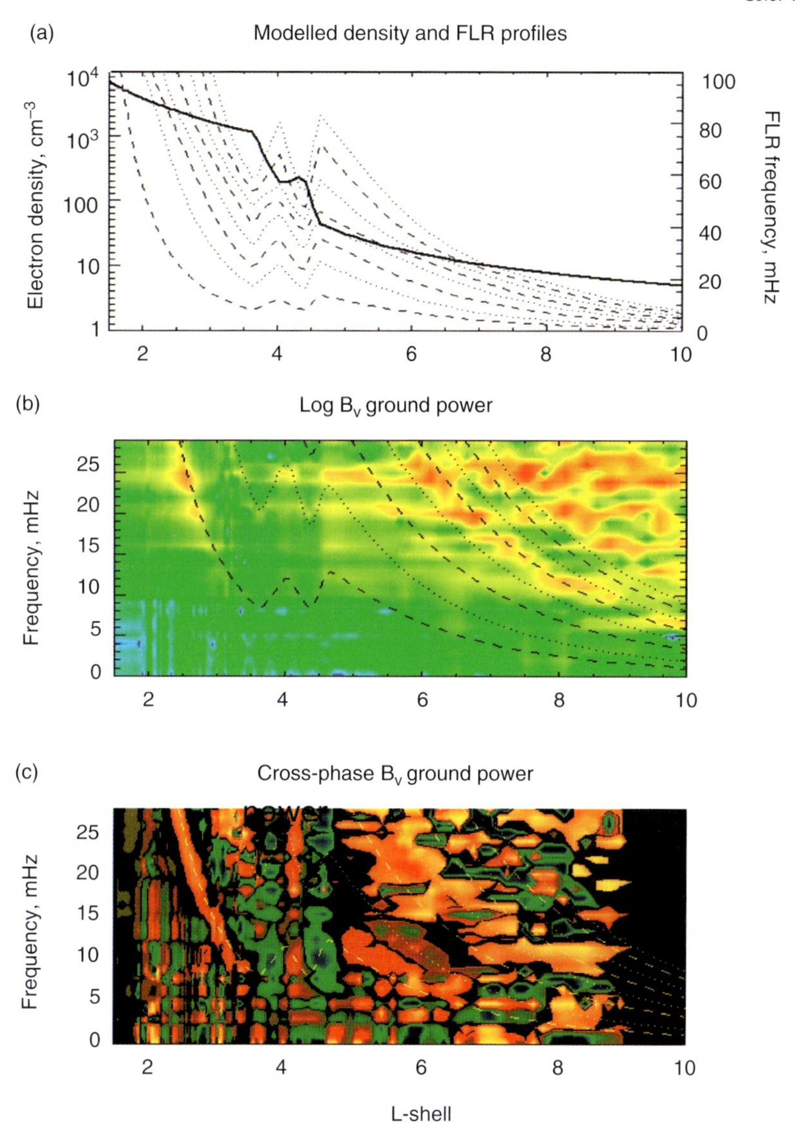

Figure 7.13 (a) Equatorial electron density (solid line) and resultant resonant frequency profiles for 19–20 UT on June 11, 2001 from a 2.5D numerical model. (b) Corresponding predicted power spectral density for the north–south ground-level magnetic perturbation. (c) Predicted ground cross-phase profile for an interstation spacing of 2°. (This figure also appears on page 145.)

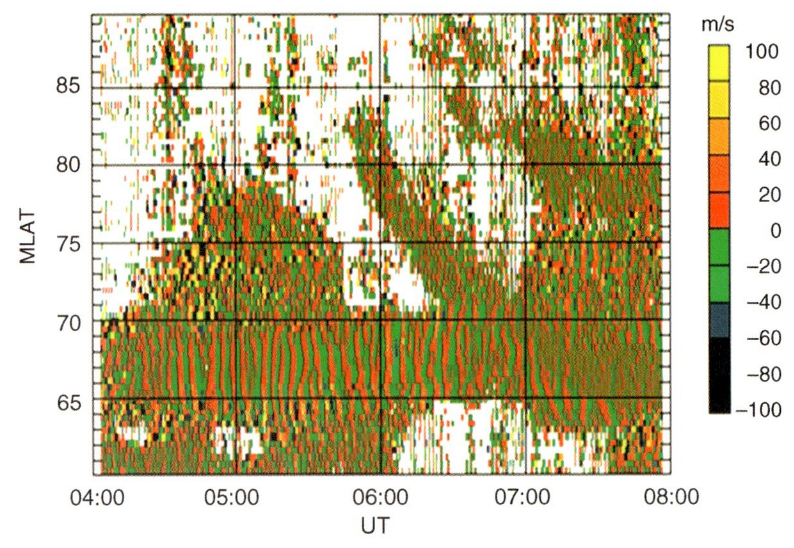

Figure 7.21 Doppler velocity oscillations in beam 5 of the Finland (Hankasalmi) HF radar from 0400–0800 UT on January 6, 1998. (This figure also appears on page 156.)

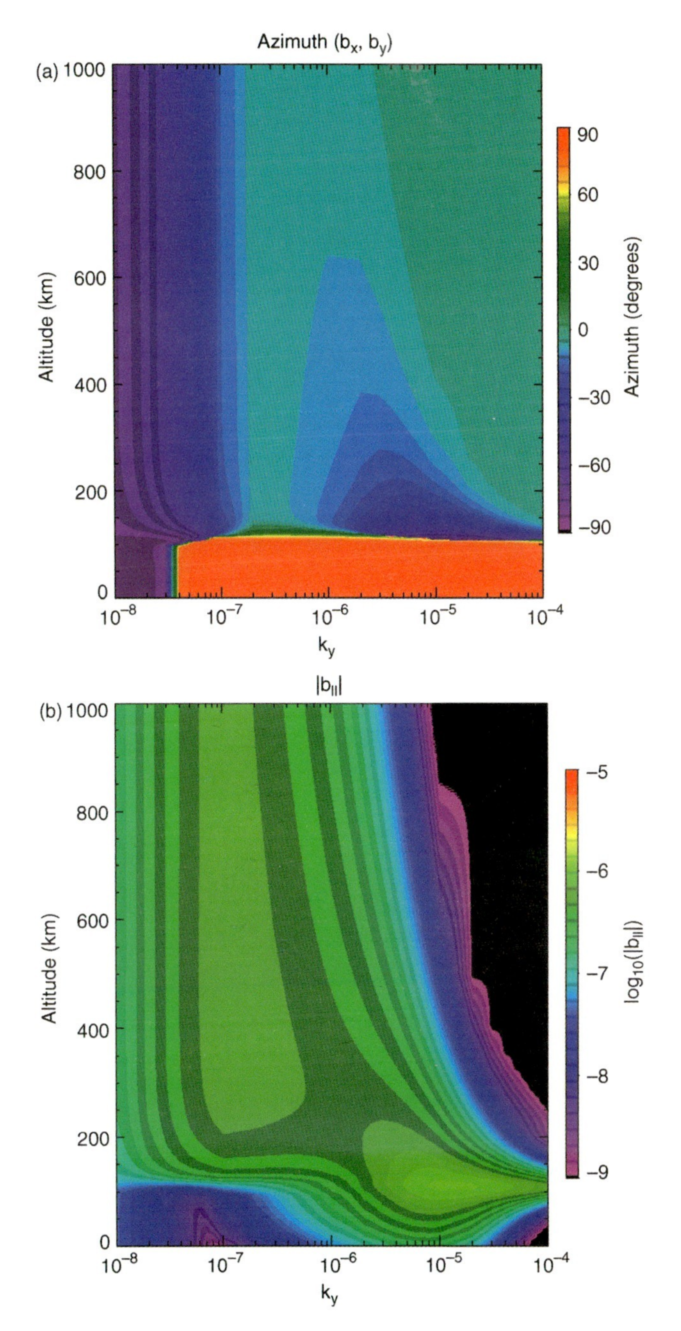

Figure 8.4 (a) The polarization azimuth computed from ULF b_x and b_y for a frequency of 16 mHz and dip angle of $I = 70^\circ$ at solar maximum ionosphere conditions. The wave numbers are $k_x = 10^{-10}\,\mathrm{m}^{-1}$ and k_y varying between 10^{-8} and $10^{-4}\,\mathrm{m}^{-1}$. (b) The amplitude of the field-aligned (compressional) component of the ULF wave magnetic field for the parameters used in panel (a). From Sciffer, Waters, and Menk (2005). (This figure also appears on page 175.)

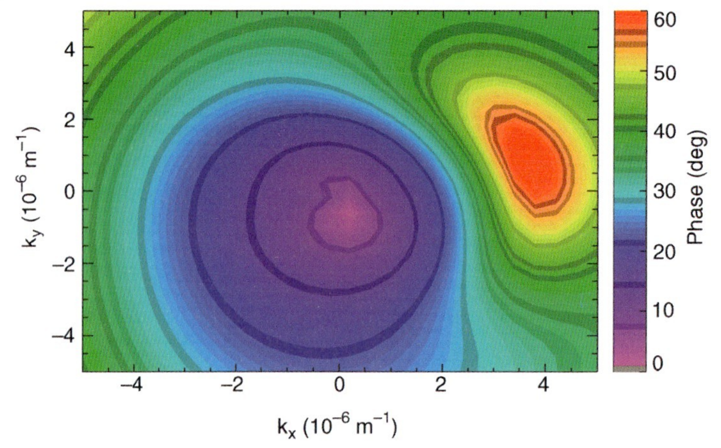

Figure 8.16 The variation in differential phase for a 70 MHz signal due to changes in TEC from a 15 mHz ULF wave with ULF wave mix of 80% shear Alfvén mode at 1000 km altitude, as a function of the ULF wave spatial scale size. Conditions were for local noon using the divergence term (last term in Equation 8.42) only. From Waters and Cox (2009). (This figure also appears on page 193.)

1
Introduction

1.1
Purpose of This Book

This book describes how measurements of naturally occurring variations of Earth's magnetic field can be used to provide information on the near-Earth space environment. This is a complex and highly dynamic region, the home of space weather that affects orbiting spacecraft and technological systems on the ground. The measurements come mostly from ground-based magnetometers but also from high-frequency radars, very low-frequency radio propagation circuits, and satellite platforms. Such remote sensing is possible because magnetic field lines originating in Earth extend through the atmosphere into space and respond to perturbations in the solar wind, which are transmitted earthward by periodic magnetic and electric field perturbations called plasma waves.

This area of research is called magnetoseismology. Its study and use for remote sensing require knowledge of and provide information on the solar wind, the interface between the solar wind and Earth's (geo)magnetic field, the near-Earth plasma environment and its variable particle populations, the ionized region of the atmosphere, and to some extent the subsurface structure of the ground.

The book does not assume familiarity with concepts in space physics and plasma physics. However, there is a strong emphasis on understanding of the core concepts and the consequent science applications. This is a new and exciting field, which greatly extends the utility of ground and *in situ* observations and mathematical descriptions of the observed phenomena.

1.2
The Solar Wind

Our planet Earth is immersed in the Sun's outer atmosphere. Particles streaming outward from the Sun exert pressure upon interplanetary matter, evident from observations of comet tails. As seen in Figure 1.1, comets may form two tails: a dust tail arising from the combined effects of radiation pressure on the low-mass dust particles and inertia of the heavier grains, and an ion tail due to the pressure exerted on gas in the

Magnetoseismology: Ground-based remote sensing of Earth's magnetosphere, First Edition. Frederick W. Menk and Colin L. Waters.

Figure 1.1 Comet Hale–Bopp, showing a white dust tail and a blue ion tail, shown here as gray, resulting from the effect of the solar wind and entrained magnetic field. *Source:* Alessandro Dimai and Davide Ghirardo (Associazione Astronomica Cortina) at Passo Giau (2230 m), Cortina d'Amprezzo, Italy, March 16, 1997, 03:42 UT. e-mail: info@cortinastella.it, web: www.cortinastelle.it-www.skyontheweb.org. (For a color version of this figure, please see the color plate at the beginning of this book.)

comet's coma by the streaming solar particles and an embedded magnetic field. Biermann (1951) thus deduced that particles flow continuously outward from the Sun with velocities of order $10^3\,\mathrm{km\,s^{-1}}$. Parker (1959) called this stream the solar wind and showed that it arises from the supersonic expansion of the solar corona into space along magnetic lines of force originating in the Sun and due to the pressure gradient between the coronal gas ($\sim 10^{-3}$ Pa) and interplanetary space ($\sim 10^{-13}$ Pa). Further details appear in a number of reviews (e.g. Aschwanden, 2005; Goldstein *et al.*, 2005; Hundhausen, 1995; Watermann *et al.*, 2009). The solar wind energy flux reaching Earth's magnetosphere boundary is around 10^{12} W, imparting a force of order 4×10^7 N.

The solar wind carries with it an embedded magnetic field that extends from the Sun into interplanetary space. This is called the interplanetary magnetic field (IMF), and because of the high electrical conductivity of the solar wind plasma, the IMF is "frozen in" to the solar wind. This means that the magnetic flux through a volume of solar wind plasma remains constant, so the magnetic field is transported with the plasma and the field lines may be regarded as streamlines of the flow. Due to the conservation of angular momentum of the plasma, the field lines trace out Archimedean spiral patterns (called the Parker spiral) with distance from the Sun. This is an example of a rotating "garden hose" effect, so at Earth's orbit the IMF makes an azimuthal angle $\varphi \approx 45^\circ$ with respect to the Sun–Earth line, given by

$$\tan\phi = R_{\mathrm{SE}}\,\Omega_{\mathrm{S}}/V_{\mathrm{sw}}, \tag{1.1}$$

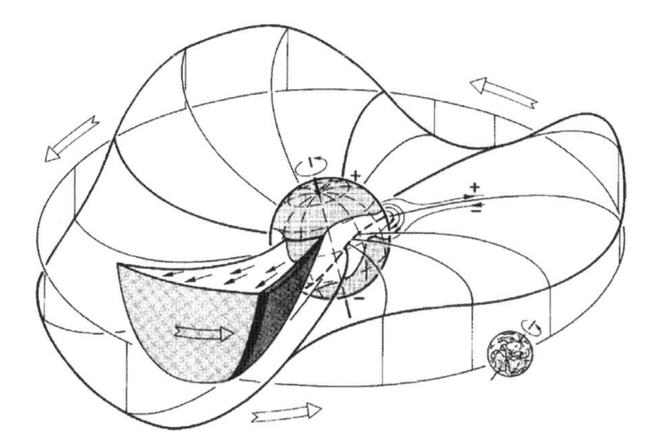

Figure 1.2 Artistic representation of IMF structure in the inner heliosphere during the declining phase of the solar cycle. Coronal holes near the Sun's poles are the sources of open field lines (here, positive or outward from the north pole), with bright active regions near the equator. The separatrix between the positive and negative solar magnetic fields forms a warped three-dimensional structure. Due to the tilt of the coronal magnetic field, high-speed streams are embedded among slow flows in the solar wind, resulting in corotating interaction regions (CIRs) on Earth. From Schwenn (2001).

where R_{SE} is the Sun–Earth distance, 1 AU (Astronomical Unit) $= 1.50 \times 10^9$ m, Ω_S is the angular rotation speed of the Sun $\approx 2.87 \times 10^{-6}$ rad s^{-1}, and V_{sw} is the solar wind speed.

The direction of the IMF is usually described in terms of vector components B_x, B_y, and B_z with respect to the Geocentric Solar Magnetospheric (GSM) coordinate system, in which the X-axis is directed from Earth to the Sun, the Z-axis points roughly northward such that Earth's magnetic dipole axis is in the X–Z plane, and the Y-axis completes the right-handed system. Since the polarity of solar magnetic fields is usually of one sign or the other over much of a hemisphere of the Sun, the IMF B_x component at Earth tends to be directed sunward or antisunward for extended periods of time. The boundary between these toward/away regions is referred to as the heliospheric current sheet, and in three dimensions the structure resembles a fluted ballerina skirt. This is illustrated in Figure 1.2. Under quiet conditions distinct crossings of these "sector boundaries" are observed during each solar rotation. The angle between the IMF and the Sun–Earth line is often called the cone angle, defined as

$$\theta_{B_x} = \cos^{-1}\frac{B_x}{B}, \tag{1.2}$$

where B is the IMF magnitude. In reality, the cone angle fluctuates considerably, although it is often described as being small ($\leq 40°$) or large.

Properties of the solar wind are now well known (e.g. Hundhausen, 1995; Schwenn, 2001) and are monitored in real time by satellites upstream from Earth. The solar wind is highly variable over a range of timescales, but at Earth's orbit comprises a neutral, fully ionized plasma with typical ion composition of 95–96%

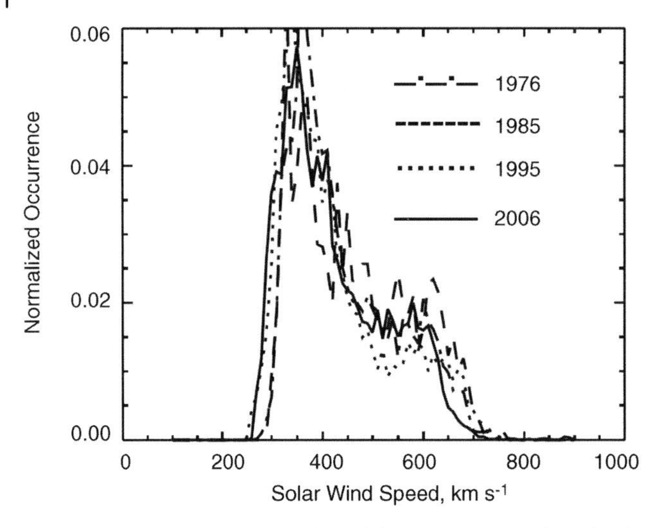

Figure 1.3 Normalized solar wind speed distributions at Earth's orbit during the solar minimum years 2006 (solid curves), 1995 (dotted curves), 1985 (dashed curves), and 1976 (dash-dotted curves). Adapted from McGregor *et al.* (2011).

protons, 4–5% He, and a minor proportion of heavy ions. Proton temperature is $T_p \approx 4\text{–}10 \times 10^4$ K and electron temperature $T_e \approx 1.5 \times 10^5$ K. The average flow speed is $V_{SW} \approx 350\text{–}400\,\mathrm{km\,s^{-1}}$, the proton density is typically in the range $n_p = 3\text{–}10\,\mathrm{cm^{-3}}$, the resultant dynamic pressure (i.e., momentum flux, $D_P = n_p v_p^2$) is around 3 nPa, and the magnetic field strength is $B \approx 5$ nT. The solar wind energy is dominated by the flow, $(1/2)(\rho V_{SW}^2)$, which has energy density $\sim 7 \times 10^{-10}\,\mathrm{J\,m^{-3}}$ at Earth. The speed V_S at which pressure disturbances propagate in the solar wind is about $50\,\mathrm{km\,s^{-1}}$ at 1 AU, so the solar wind is supersonic, with Mach number $M_s = V_{SW}/V_S$.

Various features of the solar corona determine the characteristics of the solar wind. Figure 1.3 shows normalized yearly solar wind speed distributions during solar minimum years derived from the NASA OMNIWeb database. A bimodal structure is evident, suggesting that there are two types of solar wind streams (McGregor *et al.*, 2011; Schwenn, 2001). The high-speed, low-density solar wind streams originate from coronal holes, well-defined low-temperature regions from which coronal magnetic field lines are open to, and expand superradially into, interplanetary space. These are represented in Figure 1.2, and result in the formation of corotating interaction regions (CIRs) due to the tilt of the coronal magnetic field with respect to the Sun's rotation axis during the declining phase of the solar cycle (Gosling and Pizzo, 1999). Coronal holes are regarded as features of the quiet Sun, when they extend to lower latitudes on the solar surface, near the ecliptic plane. The helium abundance in such streams is higher and more constant than that under slow wind conditions, and the ionization state of heavier ions is lower, suggesting that the high-speed streams originate from lower and cooler altitudes in the corona.

The relatively slow component of the quiet solar wind typically has higher proton density, lower helium density, and higher ionization temperatures. It appears to originate from hotter magnetic loops that intermittently reconnect with open flux tubes at coronal hole boundaries.

During the active phase of the solar cycle, large transient events associated with active centers and closed loop-like magnetic structures are common. These include coronal mass ejections (CMEs), large eruptions of plasma and magnetic field into interplanetary space, causing interplanetary shocks and triggering geomagnetic storms at Earth when magnetic reconnection occurs between the CME and the geomagnetic field. The frequency of CME eruptions peaks near solar maximum, although a slow solar wind component, with high He^+ concentration, is also present during this phase. Howard (2011) has provided a detailed review of current understanding of CMEs.

Multisatellite observations reveal the presence of complex features in the solar wind, including rotational and tangential discontinuities, magnetic clouds, flux ropes and holes, and isolated electrostatic structures indicative of the presence of double layers (Goldstein *et al.*, 2005).

An extremely quiet solar wind condition occurred during May 11–12, 2009, when the solar wind density dropped to $\leq 0.2\,cm^{-3}$ and the velocity to $\sim 300\,km\,s^{-1}$ for an extended period. This event was associated with highly nonradial solar wind flows and unipolar IMF and has been traced to a small coronal hole lying adjacent to a large active region on the solar disk (Janardhan *et al.*, 2008). In contrast, a series of violent solar flares and CMEs during late October to early November 2003 triggered major geomagnetic storms at Earth resulting in extreme space weather events with significant impacts on Earth-orbiting spacecraft, the ionosphere and atmosphere, and terrestrial systems (Gopalswamy *et al.*, 2005). During this time, solar wind gusts exceeding $2000\,km\,s^{-1}$ were accompanied by increases in the >10 MeV proton flux at geostationary orbit of over five orders of magnitude relative to nominal quiet conditions.

1.3 Fluctuations in the Solar Wind

The region of interplanetary space dominated by the solar wind is the heliosphere. At some large distance from the Sun, the solar wind speed (which depends on density) becomes subsonic, with large changes in plasma flow direction and magnetic field orientation. This is the termination shock. Beyond this is a shock region and the heliopause, where solar wind particles encounter and are stopped by particles of the interstellar medium. The Voyager 1 and Voyager 2 spacecraft, launched 2 weeks apart in 1977, crossed the termination shock in December 2004 and August 2007, respectively, at about 94.0 AU in the northern hemisphere and 83.6 AU in the southern hemisphere.

Plasma waves are a characteristic feature of shocks, where they dissipate energy and drive particle motions, and have been observed in bow shocks upstream of the magnetized planets and also at the solar wind termination shock (e.g. Gurnett and Kurth, 2008). Since the termination shock may move due to variations in the solar wind dynamic pressure or waves on the shock front, Voyager 2 has been seen to cross the shock multiple times (Li, Wang, and Richardson, 2008). By December 2010, the solar wind recorded by Voyager 1 at 116 AU had zero radial component and was purely parallel to the heliosphere surface.

The solar wind is full of fluctuations, although whether this is due to turbulence or waves is not always evident from single point measurements (Goldstein *et al.*, 2005; Tu and Marsch, 1995). A correlation often exists between magnetic field and velocity fluctuations. If the perturbed magnetic field $\boldsymbol{b} = \delta \boldsymbol{B}$ is related to the velocity perturbation $\boldsymbol{v} = \delta \boldsymbol{V}$ and number density ρ in the form

$$\boldsymbol{b} = \pm \boldsymbol{v}\sqrt{\mu_0 \rho}, \tag{1.3}$$

then the perturbation is said to be Alfvénic and may be related to the propagation of transverse Alfvén plasma waves. These waves are described in detail in Chapter 3. High-speed solar wind streams contain a complex spectrum of wave-like oscillations in velocity, density, and magnetic field strength, with periods ranging from seconds to days (Gurnett, 2001), while large-amplitude, nonsinusoidal Alfvén waves that probably originate in coronal holes and propagate outward from the Sun dominate the microscale structure of the solar wind (Belcher and Davis, 1971).

Power spectra from near-Earth orbit of solar wind magnetic field fluctuations in the millihertz frequency range often follow a power law similar to that for uncorrelated turbulence, of the form $f^{-5/3}$. Monthly averaged spectra of solar wind speed and density from upstream satellites exhibit no statistically significant peaks, as shown in Figure 1.4, although magnetic fluctuations may occur specifically near the local proton gyrofrequency (Jian *et al.*, 2010; Tsurutani *et al.*, 1994). Plasma waves play a fundamental role in determining the properties of the solar wind, and can also be used as a diagnostic of energetic processes. Such plasma waves are an important focus of this book.

Oscillations exist within the Sun in the form of pressure and gravity waves. The latter are trapped in the deep interior of the Sun below the convection zone, while the former can propagate throughout the Sun's interior. Acoustic waves can thus propagate around the Sun and arrive in phase at the original point to form global standing modes, called normal modes, producing surface oscillations as the wave energy couples with the boundary of the cavity. This boundary is usually the photosphere where the density decreases rapidly.

Study of these surface oscillations provides insights into the internal structure and dynamics of the Sun (Christensen-Dalsgaard, 2002). This area of science is called helioseismology, while the study of the internal structure of other stars using similar techniques is called asteroseismology. Chapter 9 provides a brief introduction to these interesting topics. The period of the solar normal mode oscillations is about 5 min (3.33 mHz), although long-duration ground-based observations reveal a multitude of Doppler shift frequency peaks between about 2.5 and 4.5 mHz in power spectra of the Sun's full disk (Claverie *et al.*, 1979; Grec, Fossat, and Pomerantz, 1980). The precise frequencies vary with solar cycle. Plasma waves

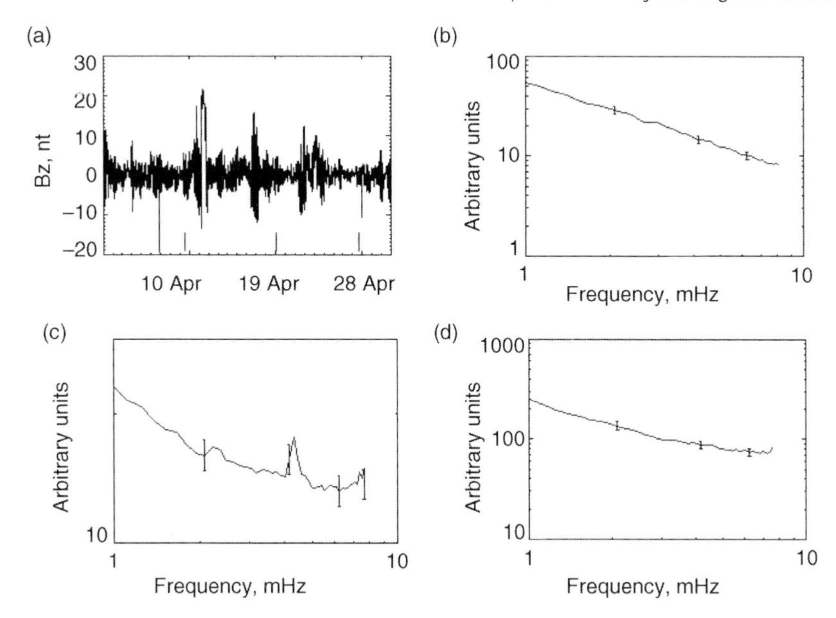

Figure 1.4 (a) Time series and (b) power spectrum of the IMF north–south component magnitude measured upstream of Earth by the WIND spacecraft during April 1997. Spectral resolution is 0.14 mHz. Error bars denote 99% confidence intervals. (c) Power spectrum of proton density from upstream ACE spacecraft during March 2003. (d) Power spectrum of solar wind velocity from ACE spacecraft during June 1999.

are believed to play an important role in heating the corona, and 5 min waves have been observed in the photosphere. Near Earth, there are many reports of geomagnetic field oscillations with discrete frequencies in the range 1.4–4.2 mHz (e.g. Mathie *et al.*, 1999a; Viall, Kepko, and Spence, 2009). It has been suggested that discrete frequencies similar to those observed at Earth in the millihertz range are present in the solar wind density (Walker, 2002) and may originate from the corona (Viall, Spence, and Kasper, 2009). The sources of millihertz frequency plasma waves are discussed in more detail in Chapter 4.

1.4
Early Observations of Geomagnetic Variations

Oscillatory variations of suspended magnetic needles were noted by several workers during the mid-nineteenth century (Schröder and Wiederkehr, 2000), including Johann von Lamont in 1841 and Balfour Stewart in 1859. The development of magnetometers, in London in 1857, using a light beam to record small variations on photosensitive waxed sheets permitted the continuous registration of geomagnetic (micro)pulsations at a number of observatories (Stewart, 1859). This allowed Stewart (1861) to note that great auroral displays seen on August 28 and September 2, 1859 at latitudes as low as Cuba and Australia were accompanied by rapid pulsatory magnetic variations with periods ranging from 0.5 min (the smallest observable interval) to 4–5 min, similar to "the nature of its action on telegraphic wires" in addition to longer variations with periods of 40–50 min and about 6 h. Remarkably, Stewart considered that an intense magnetic disturbance commencing at 05 LT on September 2, 1859 was related to solar flare eruptions, observed for the first time at 1115 LT on the previous day by Carrington. He also noted that at the instant of the flare observation, there was an abrupt magnetic disturbance lasting about 7 min. This phenomenon is now called a solar flare effect (SFE).

We now know that solar flares and CMEs trigger magnetic storms, resulting in disturbances in the geomagnetic field with periods from fractions of a second to many hours, intense auroral activity at high latitudes, and a host of other effects including space weather impacts on satellites, radio networks, and terrestrial systems. Ground-based measurements of geomagnetic variations may be used to remote sense the regions where these effects are produced. The remote sensing techniques are described in Chapter 6 and examples of applications are given in Chapter 7.

In 1896, Eschenhagen recorded geomagnetic perturbations using rapid run magnetographs with tiny suspended magnets, finding small-amplitude oscillations with a period of 12 s. Such phenomena were called geomagnetic pulsations by van Bemmelen in 1899. Using measurements at 15 observatories worldwide, Eschenhagen (1897) concluded that these fluctuations can originate in the uppermost levels of the atmosphere "unless they are first caused by special activity on the Sun itself."

Since these first observations, many thousands of papers have appeared in the refereed literature on the properties and origin of what are now called geomagnetic pulsations. The region of near-Earth space in which they are found is termed the magnetosphere, since it is enclosed and dominated by the geomagnetic field. The development of knowledge on geomagnetic pulsations may be traced through detailed reviews (Allan and Poulter, 1992; Chapman and Bartels, 1940; Kato and Watanabe, 1957; Menk, 2011; Orr, 1973; Saito, 1969).

Advances in instrumentation and data analysis techniques have stimulated this growth in understanding. These include the development of induction coil and fluxgate sensors and digital data loggers (Serson, 1973), the use of spectrum analysis techniques to display whole day frequency–time spectra (Duncan, 1961) and cross-phase spectra between closely spaced ground magnetometers (Waters, Menk, and Fraser, 1991), and the detection with high-frequency radars of perturbations in the ionosphere accompanying geomagnetic variations (Harang, 1939). These techniques are described in Chapter 5.

1.5 Properties of Geomagnetic Variations

Intense work during the International Geophysical Year, (1957–1958) established that small, periodic oscillations of the geomagnetic field occur virtually at all times and across the globe, although their amplitudes peak at high latitudes. The pulsations are generally of two forms in ground-based magnetometer records. Regular, sinusoidal oscillations with periods between about 0.1 and 1000 s occur mostly during local daytime, and are called continuous pulsations, Pc. They last from minutes to hours, often exhibiting a characteristic modulated waveform similar to the beating pattern produced by signals with closely spaced frequencies. The amplitude of these pulsations increases with increasing period and is of order 1 nT at 30 s (33 mHz). For comparison, the main background geomagnetic field on the ground has intensity of $\sim 5 \times 10^4$ nT. The amplitude of Pc signals also exhibits a diurnal variation, typically peaking near noon, varies with solar wind conditions, and is strongly dependent on latitude.

Figure 1.5a shows an early measurement of the relation between amplitude and period of Pc pulsation activity at a midlatitude station, while Figure 1.5b represents the variation with latitude of Pc occurrence frequency at solar maximum.

At local nighttime, pulsations have the appearance of impulsive damped trains of oscillations, called Pi, and are often associated with auroral activity and other geomagnetic disturbances.

To lend order to descriptions of magnetic pulsations, Jacobs *et al.* (1964) proposed a classification system based on the frequency bands in which oscillations tend to be observed. This is summarized in Table 1.1.

Although this classification did not attempt to describe the essential physics of the phenomena, it is still widely used today. Descriptions of the appearance of these signals in frequency–time spectra have been provided by several workers (e.g. Fukunishi *et al.*, 1981; Menk, 1988; Saito, 1969). As seen in Figure 1.6, daytime pulsation activity is often present in one or more well-defined spectral bands lasting up to several hours. This suggests the pulsations are generated by mechanisms that select specific frequencies for a given location.

A landmark in the understanding of geomagnetic pulsations was the realization by Hannes Alfvén in 1942 that the motion of a conducting fluid in a magnetic field gives rise to electric currents, which in turn produce mechanical forces, resulting in the formation of hydrodynamic waves propagating through space at what is today called the Alfvén speed. These ideas were developed in detail by Alfvén (1948) with the formulation of the theory of hydromagnetic waves (now called Alfvén waves) and the field of magnetohydrodynamics (MHD). Hannes Alfvén was awarded the 1970 Nobel Prize in Physics for "fundamental work and discoveries in magneto-hydrodynamics with fruitful applications in different parts of plasma physics." The theory and properties of plasma waves in the magnetosphere are described in Chapter 3.

It is now known that geomagnetic pulsations are the signature of ultralow frequency (ULF) (frequency in the range 1 mHz–10 Hz) waves that propagate through Earth's magnetosphere. There is growing evidence that Pc pulsations are connected with ULF waves, which enter the magnetosphere from the upstream solar wind or are generated at the magnetopause boundary when it undergoes rapid deformation under the action of the solar wind. These waves may have dimensions comparable to the size of the entire magnetosphere, and may therefore establish oscillations along geomagnetic field lines stretching from Earth into space. Some Pc pulsations with low spatial coherence lengths are associated with waves generated

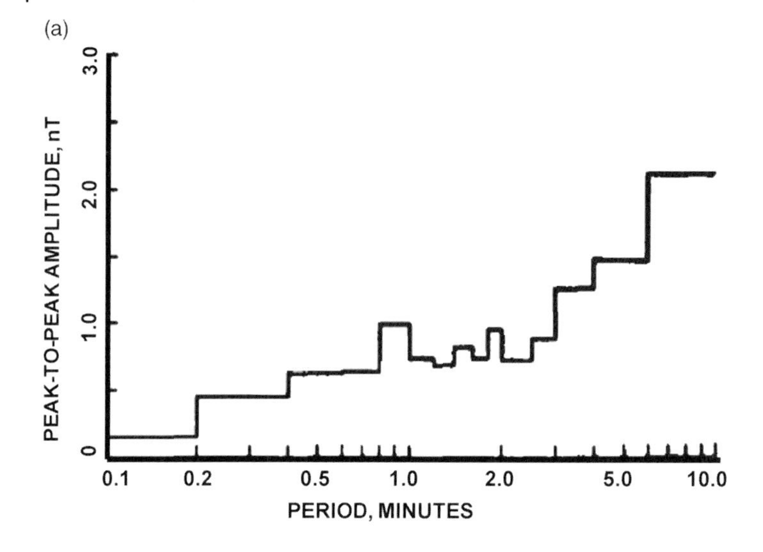

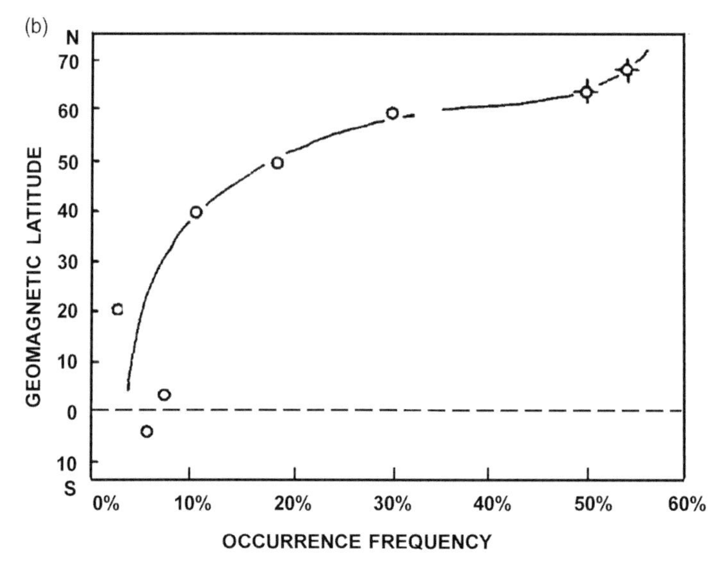

Figure 1.5 (a) Relation between amplitude and period of Pc activity at a midlatitude station. After Duffus and Shand (1958). (b) Variation with latitude of Pc occurrence frequency during October 1957 to September 1958 at some American observatories. After Jacobs and Sinno (1960).

Table 1.1 Classification system for geomagnetic pulsations.

Continuous pulsations	Pc1 $T = 0.2$–5 s	Pc2 $T = 5$–10 s	Pc3 $T = 10$–45 s	Pc4 $T = 45$–150 s	Pc5 $T = 150$–600 s
Irregular pulsations	Pi1 $T = 1$–10 s	Pi2 $T = 10$–150 s			

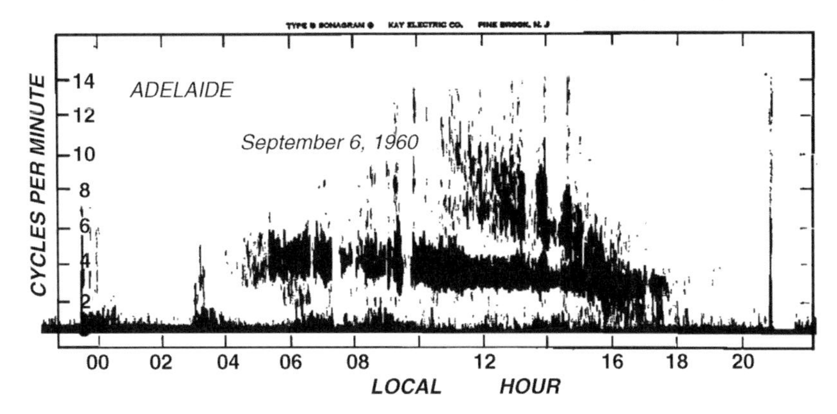

Figure 1.6 Dynamic power spectrum showing pulsation activity recorded over a whole day at a low-latitude station. Pc activity occurs mostly in a band near 50–60 mHz during local daytime, while a Pi2 event is seen near 22 LT. From Duncan (1961).

locally within the magnetosphere, and gain energy from interactions with charged particles orbiting Earth in the radiation belts. Pi pulsations are manifestations of irregular ULF waves associated with the transient magnetic fields and precipitating energetic particles responsible for auroras.

ULF waves carry energy throughout the magnetosphere, but can also be used to monitor processes in space. Obayashi and Jacobs (1958) suggested that long-period magnetic pulsations are due to standing Alfvén waves on geomagnetic field lines, called field line resonances (FLRs). The concept of standing oscillations, assisted by MHD theory, is often described in terms of two limiting types of standing oscillations, toroidal and poloidal modes, as represented in Figure 1.7. These relate respectively to azimuthal and radial motions of field lines in space. The oscillation

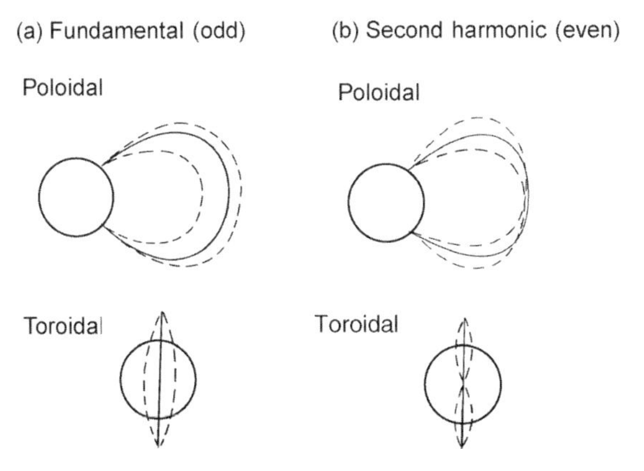

Figure 1.7 Schematic representation of (a) fundamental (odd mode) and (b) second harmonic (even mode) standing oscillations of geomagnetic field lines. Decoupled toroidal and poloidal modes are shown, with dashed lines depicting the displaced field lines.

frequency of the fundamental mode, f_R, depends on the field line length and the Alfvén speed V_A along the resonant field line:

$$f_R^{-1} \approx \int \frac{ds}{V_A(s)}, \tag{1.4}$$

where

$$V_A = \sqrt{B^2/\mu_0 \rho}, \tag{1.5}$$

$\rho = n_i m_i$ is the plasma mass density, μ_0 is the permeability of space, and the integration is carried out between conjugate ionospheres. Since field lines originating at different latitudes on the ground have different lengths and hence map out to different radial distances in space, measurement of the FLR frequency as a function of latitude provides an estimate of magnetospheric plasma mass density distribution provided the magnetic field strength is known. This is the basis for remote sensing near-Earth space described in this book.

2 The Magnetosphere and Ionosphere

2.1 The Geomagnetic Field

The origin and structure of Earth's (geo)magnetic field is described in numerous publications (e.g., Campbell, 2003; Glassmeier, 2009; Jacobs, 1987, 1991; Menk, 2007; Merrill, McElhinny, and McFadden, 1998). The first clear description of the geomagnetic field was by William Gilbert in 1600, through a series of experiments that showed that Earth is magnetized and possesses magnetic poles. We now know that the total strength of this field is around 0.066 mT near the poles and 0.024 mT near the equator. By comparison, a small bar magnet produces a field of order 10 mT. Note that Tesla, T, is actually the unit of magnetic flux density or magnetic induction, **B**, with SI units $\mathrm{T = Wb\,m^{-2} = kg\,s^{-2}\,A^{-1}}$.

The needle of a magnetic compass points at an angle to true (geographic) north. This angle of deviation is called the magnetic declination (positive when east of true north) and varies with location, as reported by Christopher Columbus in 1492. The first declination chart of Earth was produced by Edmund Halley (1702). At low latitudes, the declination is typically a few degrees, but may approach 90° in the polar regions. For example, at Davis station in Antarctica, the declination in January 2010 was 79°29′W, decreasing at $0.14^\circ\ \mathrm{yr}^{-1}$.

The geomagnetic field vector $\boldsymbol{F}$ is usually described in terms of total field intensity F, the horizontal intensity H (in nT), the inclination (or dip angle) I of $\boldsymbol{F}$ with respect to the horizontal plane, and the declination D (in degrees). At Davis in Antarctica, $I = -71^\circ 59'$, decreasing at $0.01^\circ\ \mathrm{yr}^{-1}$, and $F = 54\,459$ nT, increasing at $2\ \mathrm{nT\,yr^{-1}}$, as of January 2010. Note that in practice, $\boldsymbol{H}$ is often used to describe the magnetic field. An orthogonal X, Y, Z coordinate system (representing field intensity in the north, east, and vertical directions) is also widely used, especially at high latitudes. The relationship between these coordinates is illustrated in Figure 2.1. We see that $H = F\cos I$, $Z = F\sin I$, $X = H\cos D$ and $Y = H\sin D$.

The geomagnetic field measured at the surface of Earth includes contributions from sources within Earth's core, the crust and upper mantle, and a disturbance field representing electric currents flowing in the ionized part of the atmosphere (the ionosphere) and in the magnetosphere:

$$B(r,t) = B_{\mathrm{core}}(r,t) + B_{\mathrm{crust}}(r) + B_{\mathrm{dist}}(r,t). \tag{2.1}$$

Magnetoseismology: Ground-based remote sensing of Earth's magnetosphere, First Edition. Frederick W. Menk and Colin L. Waters.

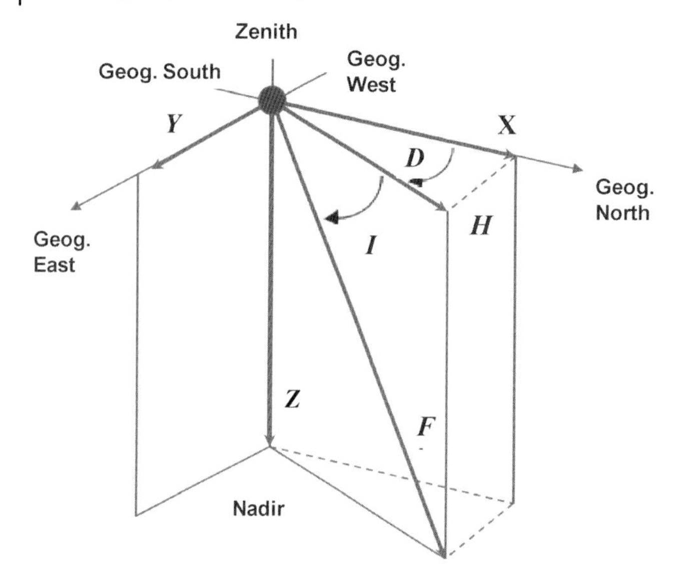

Figure 2.1 Components of the geomagnetic field, where $H = F \cos I$, $Z = F \sin I$, $X = H \cos D$, $Y = H \sin D$, and $F = \sqrt{X^2 + Y^2 + Z^2} = \sqrt{H^2 + Z^2}$. From Fraser-Smith (1987).

The core field is produced by self-exciting dynamo action due to differential rotation between the inner and outer regions of the conducting outer core of Earth. The resultant field is approximately dipolar and the axis of this centered dipole is tilted by about 10° with respect to Earth's geographic rotation axis. Such a dipole accounts for about 90% of the geomagnetic field at the surface. The core field changes slowly with time, called secular variation, and includes large-scale anomalies associated with eddy circulations in the outer core.

The crustal field arises from relatively small-scale sources of induced or remnant magnetization of rock and may significantly affect local magnetic measurements. This field varies widely with location and is mapped with ground-based, marine, aeromagnetic, and high-precision satellite surveys. Such surveys provide vital information to the geophysics exploration community that is concerned with locating and exploiting mineral deposits.

The disturbance field due to external sources varies with location and time and is of particular interest. It is discussed in more detail later.

The dipole axis cuts Earth's surface at the geomagnetic poles. Vostok station in Antarctica is located near the south geomagnetic pole, which at epoch 2010.0 was at 80°02′S, 107°8′E. Strictly speaking, this is a north pole since it attracts the south pole of magnets. The regions of peak auroral activity, the auroral ovals, are approximately centered on the geomagnetic poles.

The positions of the actual magnetic poles, where a compass needle stands vertically, depend on local effects in the crust and do not coincide with the geomagnetic poles. Determining the locations of these poles was an important goal for early explorers. At epoch 2010.0, the north magnetic pole was located at

84°97′N, 132°35′W and the south magnetic pole at 64°42′S, 137°34′E. When Edgeworth David, Mawson, and Mackay reached the south magnetic pole on January 16, 1909, it was located at 71°36′S, 152°0′E. More recently, the south magnetic pole has been accurately located and observed to move 360 km over the course of a relatively quiet day, at about 17 km h^{-1} (Barton, 2002).

The magnetic field due to a centered dipole is represented in spherical coordinates by

$$\begin{aligned} B_r &= -\frac{M}{r^3} 2\cos\theta, \\ B_\theta &= -\frac{M}{r^3}\sin\theta, \\ B_\varphi &= 0. \end{aligned} \tag{2.2}$$

The total intensity is given by

$$B = -\frac{M}{r^3}[3\cos\theta + 1]^{1/2}, \tag{2.3}$$

where r is measured from the center of Earth and θ is the colatitude measured from the dipole axis. $M \approx 7.8 \times 10^{22}$ A m^2 is the dipole moment of Earth. This has been decreasing with time, as shown in Figure 2.2 (Fraser-Smith, 1987). The equation describing a dipole field line is then given by

$$r = r_0 \sin^2\theta, \tag{2.4}$$

where r_0 is the geocentric distance at which the field line crosses the equator.

The magnetic shell parameter L is commonly used to describe the equivalent distance in R_E by which dipole field lines extend into space at the equator, mapping out magnetic shells with $LR_E = r_0$.

A centered dipole is a poor approximation to the actual field. An improvement is gained by using an eccentric axis dipole, based on a tilted dipole source displaced about 560 km from Earth's center, toward the Mariana Islands, east of the Philippines.

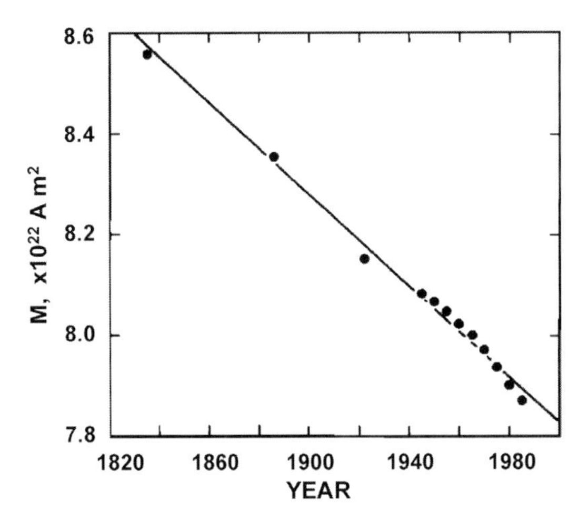

Figure 2.2 Variation of Earth's dipole magnetic moment M over 150 years. From Fraser-Smith (1987).

The first mathematical description of the geomagnetic field was provided by Carl Gauss in 1839, through spherical harmonic analysis. This is the basis for the most widely used model of the geomagnetic field, the International Geomagnetic Reference Field (IGRF) (Finlay *et al.*, 2010). The magnetic field $\boldsymbol{B}(r, \theta, \varphi, t)$ produced by internal sources is represented by the gradient of a scalar potential $V(r, \theta, \varphi, t)$, giving

$$\boldsymbol{B} = -\nabla V. \tag{2.5}$$

Since

$$\nabla \cdot \boldsymbol{B} = 0, \tag{2.6}$$

$$\nabla^2 V = 0, \tag{2.7}$$

where V is written as a spherical harmonic expansion:

$$V = R_{\mathrm{E}} \sum_{n=1}^{N} \sum_{m=0}^{n} \left(\frac{R_{\mathrm{E}}}{r} \right)^{n+1} \left[g_n^m(t) \cos m\phi + h_n^m(t) \sin m\phi \right] P_n^m(\cos\theta). \tag{2.8}$$

The g_n^m and h_n^m terms are determined from experimental measurements of the geomagnetic field, and are called the Gauss coefficients. The $P_n^m(\cos\theta)$ terms are the associated Legendre functions and r is the radial distance from the center of Earth with radius R_{E}. For the IGRF-11 model, $N = 10$ (i.e., 120 coefficients) up to the year 2000, but extended to degree 13 (195 coefficients) for later years to take advantage of Oersted and CHAMP spacecraft data. The World Magnetic Model (WMM) has been developed jointly by government agencies in the United States and United Kingdom and is based on an expansion of the magnetic potential into spherical harmonic functions to degree and order 12 (Maus *et al.*, 2010). This represents internal magnetic fields with spatial wavelengths exceeding 30° in arc length, excluding sources within Earth's crust and upper mantle, the ionosphere and magnetosphere. Consequently, local, regional, and temporal magnetic declination anomalies exceeding 10° may occur. Ground observations are required to describe accurately the actual field and its spatial and temporal variation.

Figure 2.3 shows a global map of the main total field intensity F on a Mercator projection produced using the 2010 WMM and distributed by the US National Geophysical Data Center. Contours represent 1000 nT increments. A perfectly uniform geocentric dipole field would result in contours parallel to lines of latitude. The maxima are at the dip poles and the minimum is near Brazil. The latter is called the South Atlantic Magnetic Anomaly (SAA) and arises because of the offset of the dipole field. Since the motion of charged particles arriving from space is influenced by the direction and strength of the magnetic field, such particles reach lowest altitudes near the SAA. This can result in high radiation doses that may affect humans and damage spacecraft in near-Earth orbits traversing the SAA. Such spacecraft suffer the greatest number of radiation-induced operational anomalies near the SAA (Hastings and Garrett, 1996). This is illustrated in Figure 2.4, which shows the location of single-event upsets (SEUs) in NMOS DRAM computer chips on board the UoSat-2 spacecraft over 1 year at 700 km orbit. The increase in events near the SAA is obvious.

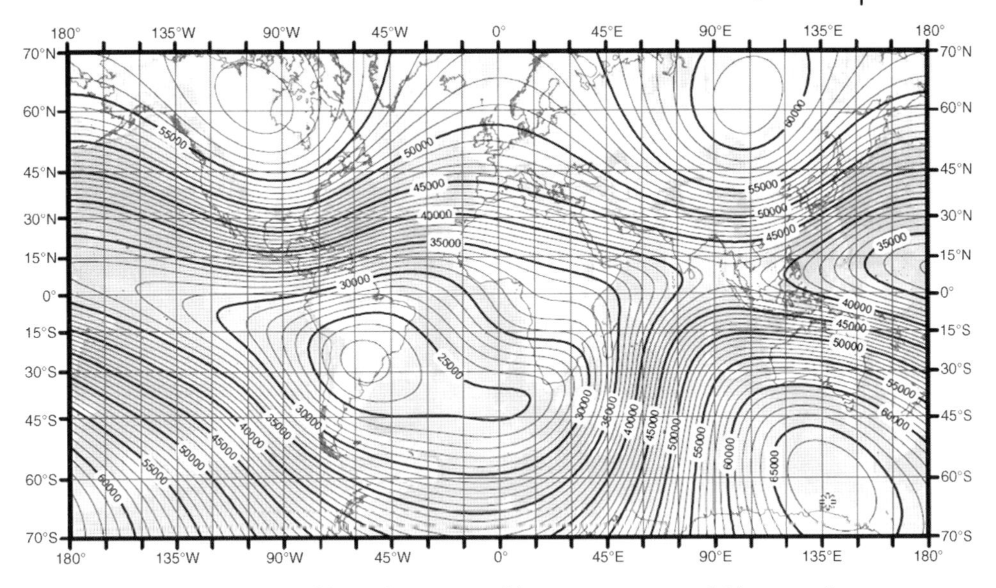

Figure 2.3 Mercator projection of the total intensity F of the main geomagnetic field computed using the 2010 World Magnetic Model. Contour interval is 1000 nT. From Maus *et al.* (2010). (For a color version of this figure, please see the color plate at the beginning of this book.)

A striking reminder of the temporal nature of the main internal field is presented in Figure 2.5 (Finlay *et al.*, 2010). This shows how the minimum in F at Earth's surface, and the location of this point, has changed and moved in the past century. Clearly, the SAA is no longer in the south Atlantic, but is moving westward over South America. This is a manifestation of secular variation of the geomagnetic field, directly related to processes in the dynamo source (Barton, 1989). Efforts to better understand these effects include analysis of maritime observations spanning the

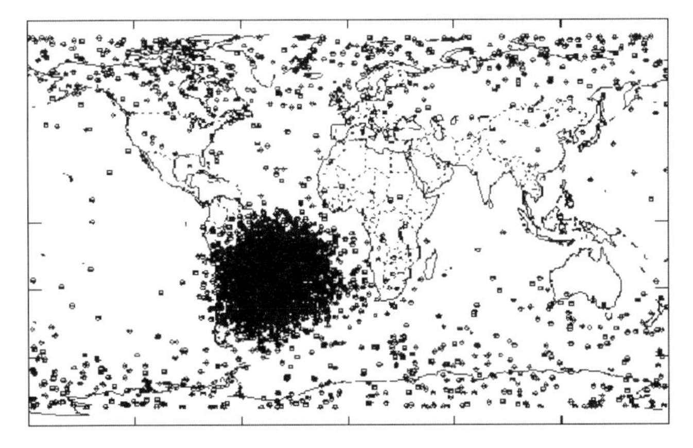

Figure 2.4 Map showing location of single-event upsets (SEUs) in NMOS DRAMs on UoSat-2 at 700 km altitude over 1 year. From Underwood *et al.* (2000) referenced in Dyer and Rodgers (1998).

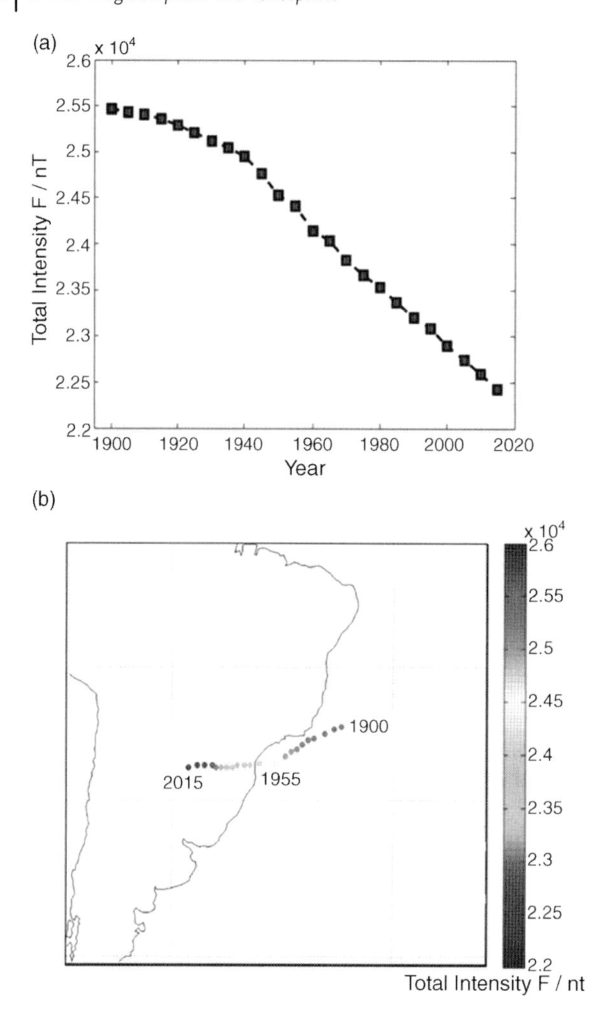

Figure 2.5 Variation in (a) strength and (b) location of the minimum in the total field intensity F in the South Atlantic Anomaly region during the past century. From Finlay *et al.* (2010). (For a color version of this figure, please see the color plate at the beginning of this book.)

years 1590–1990 to develop a model of the magnetic field at the core–mantle boundary (Jackson, Jonkers, and Walker, 2000; Jonkers, Jackson, and Murray, 2003).

This book focuses on remote sensing the magnetosphere from the ground, mainly by monitoring perturbations of geomagnetic field lines extending into space. For example, if the resonant frequency of a field line is determined using observations, then the density near the field line apex may be found using Equation 1.4. This requires a suitable model of the magnetic field in space.

At low latitudes, the geomagnetic field is relatively unaffected by the solar wind and other external sources, and a dipole configuration is usually assumed to be an adequate description of the field geometry (Allan and Knox, 1979a; Radoski,

1967, 1972). In essence, the propagation of ULF plasma waves in the magnetosphere is represented by an equation of the form(Singer, Southwood, and Kivelson, 1981):

$$\left[\partial^2(B_0 \times \xi)/\partial t^2\right] = V_A \times V_A \times (\nabla \times \nabla \times (B_0 \times \xi)), \tag{2.9}$$

where B_0 is the unperturbed magnetic field and ξ is the plasma or field displacement. Solving this for a nonuniform field, such as a dipole field, is not straightforward, and approximations or analytic expressions for certain conditions are often used to determine the density (Orr, 1973; Taylor and Walker, 1984). However, further complications arise at very low and high latitudes. At very low latitudes, geomagnetic field lines do not reach high altitudes and so crustal and other regional nondipolar anomalies may be important. It may then be appropriate to adjust the dipole configuration using the IGRF (Hattingh and Sutcliffe, 1987).

External factors become important when considering the mapping into space of mid- and high-latitude field lines. The dipole model is usually regarded as inaccurate for field lines extending beyond about $5R_E$ (Singer, Southwood, and Kivelson, 1981). In order to discuss models appropriate for these situations, we first need to consider the structure of the magnetosphere.

2.2 Structure of Earth's Magnetosphere

The magnetosphere is the volume of space mapped out by field lines extending from the ground and in which the motions of charged particles are dominated by geomagnetic effects. Since the solar wind is a supersonic conducting fluid, it is excluded from entering the magnetosphere directly, so the geomagnetic field is compressed on the sunward side. The location of the boundary, called the magnetopause, is determined by the balance of solar wind pressure and magnetic pressure. With some simplifying assumptions, this condition can be described by

$$Knm_p V_{sw}^2 \sin^2\theta = [2WB(r)]^2/2\mu_0, \tag{2.10}$$

where K describes the flow reduction due to diversion around the magnetopause and depends on the ratio of specific heat (the polytropic index) γ, m_p is the proton mass, n is the number density, θ is the angle between the magnetopause surface and the solar wind velocity vector, and $W \approx 1.22$ describes the compression of the dipole field.

As magnetic field strength varies with radial distance, $B(r) = B_0/r^3$, the position of the subsolar point of the magnetopause r_{mp} depends on the dynamic solar wind pressure $D_P = \rho V_{sw}^2$:

$$r_{mp} = k(D_p)^{-1/6}. \tag{2.11}$$

A detailed discussion on the fit to observations is given in Shue et al. (1997). This distance is typically around $10R_E$.

An approximately parabolic bow shock is formed typically 2–3 R_E in front of the magnetopause as the super-Alfvénic solar wind is decelerated to subsonic speed in the intervening magnetosheath. Here the shocked solar wind plasma, which is hotter and denser than in the solar wind (due to conversion of kinetic to thermal energy and compression), is deflected to flow tailward around the magnetopause. These regions are ideally studied using constellations of closely spaced spacecraft, such as the four-satellite Cluster mission (Paschmann *et al.*, 2005) and the five-satellite THEMIS constellation (Korotova *et al.*, 2011). Recent results show that the characteristic thickness of the bow shock, that is, the distance over which half the increase in electron temperature occurs, is of order 20 km (Schwartz *et al.*, 2011).

The magnetosheath is home to a rich variety of low-frequency plasma waves formed through a range of mechanisms (e.g., Constantinescu *et al.*, 2007; Schwartz, Burgess, and Moses, 1996). Of particular interest are the foreshock regions formed upstream of the bow shock by field-aligned backstreaming electrons and ions that have undergone specular reflection from the magnetopause. Geometric considerations lead to energy-dependent acceleration of particles (Fitzenreiter, 1995) resulting in peaked velocity distributions and hence the excitation of waves through beam instabilities. A cyclotron resonance interaction of backstreaming ions with the incoming solar wind produces ULF waves typically in the Pc3 range especially under radial IMF conditions (Blanco-Cano, Omidi, and Russell, 2009).

The shape of the dayside magnetopause resembles an oblate spheroid and is readily described using simple models (Mead and Beard, 1964; Olson, 1969). In contemporary applications, the global magnetosphere is described using a physics-based numerical model referenced to satellite data at the outer boundary and incorporated into a comprehensive space weather modeling framework (Toth *et al.*, 2005). A series of semiempirical models of the geomagnetic field, based on extensive satellite observations and including external magnetospheric sources, have been described by Tsyganenko (1990, 1995) and Tsyganenko and Sitnov (2007). Figure 2.6 illustrates the shape of a midlatitude magnetic field line, in 60 min snapshots in local time and under quiet solar wind conditions, determined using Tsyganenko's 2004 model. Clearly evident are the compression and stretching of the field on the sunward and night sides. In addition, the field lines also exhibit warping in the longitudinal direction, which means that field line oscillations cannot be described as simply toroidal (i.e., azimuthal) or poloidal (radial), since the polarization varies with local time (Degeling *et al.*, 2010). This is discussed further in Chapter 3.

The structure of the magnetosphere and its main current systems is represented schematically in Figure 2.7. Several distinct plasma regimes are present in the magnetosphere. A dense population of cold (electron volt, eV energy) particles, primarily from the underlying atmosphere, characterizes the inner magnetosphere (also called the plasmasphere). Warm to hot (tens of keV) electrons and ions occur in the ring current region. Trapped relativistic (MeV) particles are present in the Van Allen radiation belts, while the plasmatrough or outer magnetosphere includes low densities of particles that have convected from

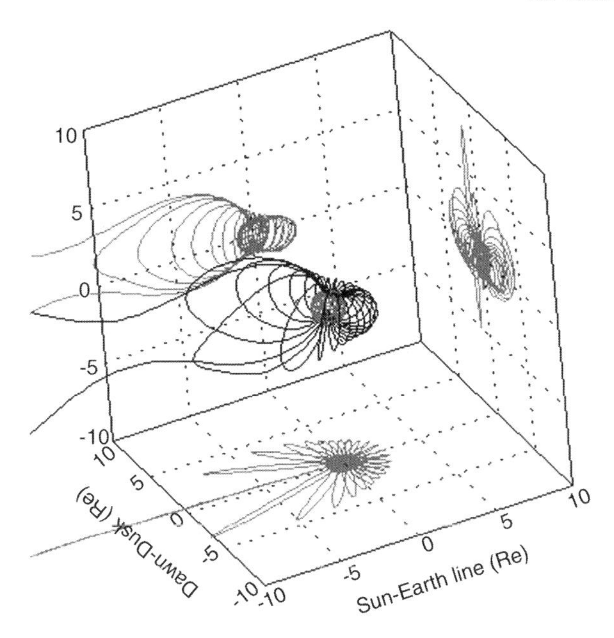

Figure 2.6 Geomagnetic field line traces computed using the 2004 Tsyganenko model for 57° geographic latitude and quiet magnetic conditions: dynamic pressure $D_P = 3$ nPa, IMF $B_y = 0$ nT, $B_z = 0$ nT, and magnetic disturbance index $D_{st} = 5$ nT.

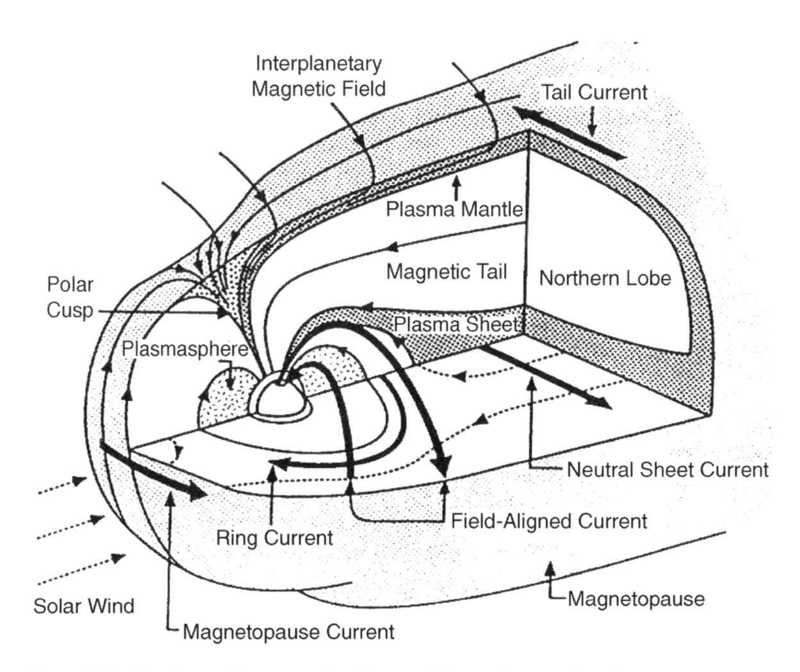

Figure 2.7 Structure of the magnetosphere and the main current systems.

the inner magnetosphere and also via reconnection from the solar wind, and locally accelerated populations.

Spectrograms of particle energies and fluxes obtained from low-altitude satellites near the polar cusps show features suggestive of morphologically different regions (Newell and Meng, 1988), associated with specific types of Pc1–2 ULF wave activity (Menk *et al.*, 1992). The cusp regions are located near magnetic noon and roughly denote the demarcation between closed field lines on the dayside and open field lines that are swept antisunward to form the magnetotail. The magnetic field strength is locally weak in the exterior cusp, affording magnetosheath plasma most direct access to low altitudes. The low-altitude cusp is thus characterized by a high flux of relatively low-energy ($\sim$100 eV–few keV) electrons and ions confined to a region around 1° wide in latitude (Newell and Meng, 1988).

Boundary layers lie immediately inside the magnetopause in the transition region between the magnetosheath and magnetosphere proper. The low-latitude boundary layer (LLBL) extends tailward from near the subsolar point, around the flanks of the magnetosphere, to 5–7R_E off the equatorial plane. It is populated by injected magnetosheath plasma structures that convect tailward at near the magnetosheath flow velocity. The LLBL is thus a site of momentum transfer from the magnetosheath, by mechanisms including waves, particle diffusion, wave-induced particle acceleration, and charged particle drift. The entire magnetopause boundary layer maps into a small region adjacent to the cusp at ionospheric altitudes.

In addition, a plasma sheet boundary layer (PSBL) maps from the equatorial tail region to the auroral ionospheres. The PSBL comprises energized solar wind plasma (arising from reconnection sites in the distant tail) superimposed upon a significant and variable ionospheric component that is often present in the form of ion streams. Typical particle densities are of order 0.1–1 cm^{-3} and at temperatures $\sim$1–5 keV.

Tailward of the exterior cusp is the plasma mantle, populated by de-energized magnetosheath plasma injected from the entry layer convecting antisunward across the polar cusp, and particles of ionospheric origin accelerated up from the low-altitude cusp. Particle densities here are typically in the range 0.01–1 cm^{-3} and at temperatures $\sim$100 eV. Particles in the plasma mantle exhibit latitude–energy dispersion since the less energetic ions and electrons are available within the convection region for a longer time.

The characteristic particle populations of some of the important regions of the magnetosphere represented in Figure 2.7 are summarized in Table 2.1. These regions are described in more detail later in this chapter.

Schematic diagrams such as Figure 2.7 convey the impression that the magnetosphere is a well-ordered stable system. This is untrue. The shape and location of the boundaries are highly variable as are the embedded particle populations and current systems. Various space–weather agencies provide real-time monitors of the magnetopause location, which under very active solar wind conditions may move inside geostationary orbit, while under extremely quiet conditions the bow shock has been observed 50–60R_E upstream (Le, Russell, and Petrinec, 2000).

Two aspects of the outer boundary are noteworthy. First, the viscous shear between the solar wind-like flow in the magnetosheath and the magnetospheric

Table 2.1 Particle characteristics in regions of the magnetosphere.

Quantity	Inner radiation belt	Plasmasphere	Ring current and outer radiation belt
L-shell	$1.2 < L < 2.5$	$1.2 < L < 5$	$3 < L < 6$
Particle density		$100\ cm^{-3}$	$\leq 1\ cm^{-3}$
Electron energy	50 keV–10 MeV	$\leq$10 eV	$\leq$20 keV–10 MeV
Ion energy	1–100 MeV	$\leq$10 eV	1 keV–30 MeV
Main constituents	H^+, e	H^+, e	H^+, He^+, e
Source region		Ionosphere	Solar wind
Loss region	Atmosphere	Ionosphere, magnetosphere	Atmosphere, magnetosphere

plasma stimulates the Kelvin–Helmholtz instability (KHI) at the low-latitude boundary layer (Lee, Albano, and Kan, 1981; Walker, 1981). Surface waves may be generated driving elliptical motion of the plasma and hence field line perturbations and elliptically polarized plasma waves. Under appropriate conditions, these may couple to kinetic pressure waves in the magnetosheath and fast mode magnetic pressure waves in the magnetosphere. If the wave angular frequency $\omega < k_y^2 V_A^2$, where k_y is the azimuthal wave number of the waves at the boundary and V_A is the Alfvén speed, then the waves decay exponentially in the radial direction either side of the boundary with amplitude varying as (Walker *et al.*, 1992)

$$\exp\left(-\sqrt{k_y^2 V_A^2 - \omega^2}\right). \tag{2.12}$$

KHI waves therefore do not generally penetrate far into the magnetosphere and are known as surface waves. The KH instability grows most rapidly at a period given by

$$T = \frac{2\pi}{0.6}\frac{d}{V_0} \approx 10d/V_0, \tag{2.13}$$

where d is the scale thickness of the boundary and V_0 is half the solar wind speed in the magnetosheath. The oscillations are therefore in the millihertz frequency range, leading to the excitation of Pc3–Pc5 plasma waves (Dunlop *et al.*, 1994; Lee and Olson, 1980; Walker, 1981). This is thought to be an important source of ULF waves in the morning and afternoon sectors of the outer magnetosphere.

The second feature of note is the possibility of overreflection from the magnetopause at high solar wind speeds. The shear flow between the plasma in the magnetosheath and that in the magnetosphere affects the reflection condition for waves in the magnetosphere arriving at the magnetopause, and when taking into account the boundary layer thickness may result in the formation of overreflection modes at the magnetopause (Walker, 2000). In essence, the reflected magnetospheric waves can be amplified via energy coupling from the magnetosheath plasma flow.

No discussion of magnetospheric dynamics is complete without reference to sources and effects of current systems. This is done in Section 2.3, allowing us to consider then the properties of the radiation belts and inner magnetosphere.

2.3 Magnetospheric Current Systems

Several major current systems flow in the magnetosphere, illustrated in Figure 2.7, and play a fundamental role in transporting plasma and magnetic flux throughout the magnetosphere–ionosphere system. These are briefly described.

2.3.1 Magnetopause Current

As described by Equation 2.10, the position of the magnetopause depends on the balance between solar wind and magnetic pressure. Simplistically, we may imagine that charged particles in the solar wind are reflected (through half gyro-orbits) by the Lorentz force arising from the encounter with tangential geomagnetic field lines in the outer magnetosphere. The different and opposite gyroradii for electrons and ions cause charge separation, hence an electric field and current $\boldsymbol{J}$ on the boundary. In effect, the conducting solar wind plasma encounters the relatively strong geomagnetic field, producing a current and therefore a $\boldsymbol{J} \times \boldsymbol{B}$ force that opposes the plasma pressure. The current flows from dawn to dusk across the dayside magnetopause, then oppositely across the magnetopause tailward of a neutral point near the cusps. This current, also called the Chapman–Ferraro current (Chapman and Ferraro, 1930; Olson, 1982), results in a global magnetopause current sheet with average density of order $10^{-7}\,\mathrm{A\,m^{-2}}$.

The magnetopause current produces a magnetic field that effectively cancels the geomagnetic field outside the magnetopause and doubles the dipole field inside the magnetopause. Consequently, during enhanced solar wind conditions when the magnetopause is compressed earthward, the ground-level field increases. This results in a short positive excursion in a disturbance index called Dst (see Section 2.3.3). The rotation of Earth under the magnetopause current system contributes to about 20% of the characteristic daily semiannual and annual variations observed in the magnetic field on the ground under quiet solar wind conditions (Olson, 1970).

2.3.2 Tail Current and Reconnection

The drift of charged particles in the distant tail of the magnetosphere produces a current directed from dawn to dusk flowing in the plasma sheet near the equatorial plane, where it is called the neutral sheet current, and closing across both the northern and southern sides of the magnetopause, like the Chapman–Ferraro current. The current system thus forms a θ-shaped configuration when viewed in the tailward direction from Earth.

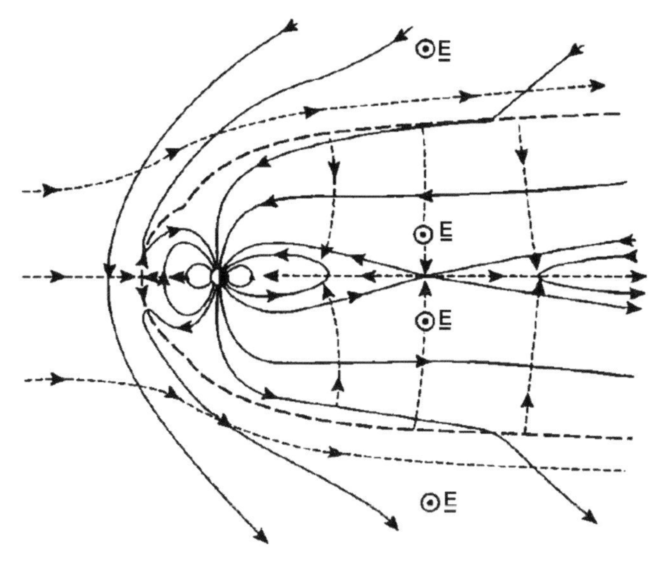

Figure 2.8 Schematic representation of the Dungey convection cycle and reconnection. Sun is to the left, solid lines denote magnetic field lines, and dashed lines represent plasma streamlines. From Cowley (1995).

The tail current system actually results from interaction between the IMF and the geomagnetic field lines, and between the solar wind plasma and plasma in the boundary layer. Momentum transfer from the solar wind drives convection of field lines from the dayside outer magnetosphere across the polar caps and into the nightside. In the original Chapman–Ferraro picture, the magnetosphere was a closed entity in space and the solar wind was excluded from the magnetosphere since the solar wind plasma is frozen-in to the IMF, and Earth's plasma to Earth's field. However, at high current densities, the frozen-in flux condition is relaxed and reconnection occurs between the interplanetary and geomagnetic field lines. This allows solar wind energy and plasma to couple to the magnetosphere by a dynamo process, driving a global convection pattern known as the Dungey cycle (Dungey, 1961). This is illustrated in Figure 2.8.

The cycle begins when reconnection at the subsolar magnetopause opens the previously closed magnetosphere to solar wind flux. The newly opened flux tubes and plasma are swept antisunward by the flowing solar wind and the embedded IMF to form the tail. The field lines become greatly stretched and eventually reconnect near the equatorial plane in the plasma sheet. This closes the open flux. The newly closed field lines relax earthward under magnetic tension and return flux to the dayside to complete the global convection pattern.

This cyclical flow is associated with a large-scale dawn-to-dusk electric field, as shown in Figure 2.8. From Faraday's law, the voltage across the system associated with this electric field is related to the total magnetic flux throughout and is a measure of the overall strength of the flow. Mapped to the ionosphere, this voltage is of order 100 kV and is connected with ionospheric flows of several hundred meters per second.

Dungey's picture gives the impression of a steady cycle with equal reconnection rates at the day- and nightsides. In fact, instantaneous reconnection rates on the day- and nightsides can be quite different. Reconnection at the subsolar magnetopause is most efficient when the IMF is directed southward, and the IMF orientation therefore modulates geomagnetic activity. The reconnection rate depends on the solar wind speed, the IMF B_z, and the effective length of the reconnected field line. The geomagnetic field is open (via reconnection) for about 6% of the time and the open flux is of order 0.5 GWb compared to the total flux in the magnetosphere system that is of order 8 GWb (Milan, Provan, and Hubert, 2007). About 2.7 GWb open flux per day enters the magnetosphere through dayside reconnection. The magnetic flux flowing through the magnetosphere is only about 10–20% of that arriving in the solar wind at the dayside magnetopause. However, reconnection is the dominant mechanism for momentum and flux transfer from the solar wind to the magnetosphere, and in the tail results in auroral and other activity often classed as substorms.

2.3.3 Ring Current

Energetic ions and electrons trapped in the midpart of the magnetosphere drift azimuthally in opposite directions around Earth in a doughnut-shaped region to form a ring current with density of order $10^{-8}\,\mathrm{A\,m^{-2}}$. Ring current particles typically have energy in the range 10–200 keV and are injected into this region during storms and substorms, also arriving by convection from the nightside plasma sheet. However, radiation belt particles with MeV energies also contribute to the ring current. Under quiet conditions, most ring current ions are protons. However, under active conditions, O^+ ions from the ionosphere form a significant component. The mass density and ion composition can be determined by remote sensing using ground-based magnetometers, as will be seen in Chapters 6 and 7.

Consistent with Lenz's law, the magnetic field produced by the ring current opposes the main geomagnetic field, resulting in a decrease on the ground. This is characterized by the D_{st} index derived from a network of low-latitude magnetometers. The Dessler-Parker-Sckopke (DPS) theorem quantifies this effect (Carovillano and Siscoe, 1973; Sckopke, 1966), although smaller contributions to D_{st} also arise from magnetopause and magnetotail currents, while ionospheric currents provide a negligible contribution (Vasyliunas, 2006). An enhanced ring current with total energy of order 10^{15} J is associated with a moderate magnetic storm and results in a reduction of the ground-level magnetic field of order 100 nT for a period of a few hours (the main phase), after which the ring current begins to decay through charge exchange reactions with neutral hydrogen atoms, Coulomb collisions with low-energy particles, or precipitation of ring current particles into the atmosphere.

2.3.4 Field-Aligned Currents

The magnetosphere and ionosphere are coupled through the motion of particles along field lines producing field-aligned currents. The existence of these currents was first proposed by the Norwegian scientist Kristian Birkeland in 1908, and therefore they are often called Birkeland currents. In a series of remarkable experiments, Birkeland constructed a terrella – a small magnetized sphere representing a model of Earth in space – and created artificial auroras around its magnetic poles using electron beams. He thus demonstrated that auroras are caused by electrically charged particles traveling earthward along magnetic field lines and energizing gas molecules in the polar atmosphere. He also reasoned that these energetic electrons affect the geomagnetic field, causing polar magnetic storms. Norway honors Birkeland on its 200 kroner banknote.

The existence of Birkeland currents was proven in 1976 by observations from the Triad spacecraft carrying magnetometers over the polar ionosphere (Iijima and Potemra, 1976). The currents are mostly carried by electrons and broadly form two rings, one around 70° magnetic latitude consisting of current into the ionosphere over the evening and nightside and out of the ionosphere on the morning and dayside (called the region 1 current), and a lower latitude region 2 system of oppositely directed current. These currents are a consequence of the global circulation pattern, since region 1 currents map to the interface between high-latitude field lines convecting tailward and those returning to the dayside.

These currents have important consequences apart from characterizing the location of auroras. Spacecraft in low Earth orbit use magnetometers for attitude control, by determining their orientation with respect to the main geomagnetic field. However, as these spacecraft pass over the polar regions, the strong, localized magnetic fields produced by the Birkeland currents may be sufficient to affect the spacecraft's attitude control system. Magnetometer data from constellations of communication satellites can be used to map the precise location and magnitude of the field-aligned current systems (Waters, Anderson, and Liou, 2001).

2.3.5
Ionospheric Currents

The upward and downward field-aligned currents are closed by Pedersen currents in the ionosphere. This forms a solenoidal current geometry, so that the magnetic field on the ground is mostly due to Hall currents in the ionosphere (Fukushima, 1969). At latitudes between the region 1 and 2 current systems, intense Hall currents result in electrojets that flow toward midnight around the auroral ovals and are associated with intense auroral displays. The westward auroral electrojet (around the dawn side) produces magnetic variations that decrease the surface-level field.

Magnetospheric convection and plasma flow over the polar regions result in a convection electric field that maps to the polar caps and causes a dawn-to-dusk polar cap potential of order 50–100 kV. Rotation of Earth under this field results in a diurnal magnetic field variation.

Magnetic observatories also show systematic seasonal and diurnal variations due to motions of the upper atmosphere driven by the variation in solar radiation with local time and latitude and the gravitational pull of the Sun and the Moon. These result in winds and tidal forces that drive motions of charged particles in the ionosphere, and the resultant electric currents cause a quiet day variation in the ground-level magnetic field. These ionospheric currents also give rise to currents within the conducting layers of Earth. The precise magnitude and direction of these currents depend on the electrical conductivity profile. The quiet-time diurnal ionospheric currents thus form a convenient source for mapping the conductivity profile of Earth.

2.4 The Radiation Belts

The first US spacecraft, Explorer 1, in 1958 detected intense charged particle radiation trapped by the geomagnetic field in what are now known as the Van Allen radiation belts. The two ("outer" and "inner") belts are highly variable in shape and density, extending at times beyond geostationary orbit ($6.6 R_E$ altitude; "GEO") and down to low Earth orbit (300–1000 km; "LEO") near the South Atlantic anomaly. For example, during the October–November 2003 Halloween storm, a new inner belt formed with a decay time for 2–6 MeV electrons at $L = 1.5$ of ~180 days (Baker *et al.*, 2007), although the inner belt usually comprises energetic protons resulting from the interaction of cosmic rays with the atmosphere.

The radiation belts have been well mapped using the low-altitude polar orbiting SAMPEX spacecraft (Li *et al.*, 2001). Data also come from the GPS spacecraft that orbit at 20 200 km altitude in the heart of the radiation belts, although the high radiation doses in these regions require significant shielding to protect spacecraft and humans (Hastings and Garrett, 1996). Figure 2.9 shows the yearly average of 2–6 MeV electron flux measured by SAMPEX during 1993–2001 as a function

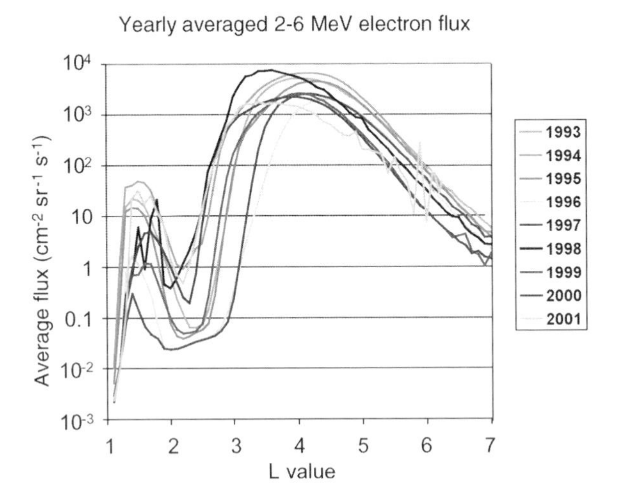

Figure 2.9 Yearly averaged 2–6 MeV electron flux measured at low altitudes by the SAMPEX spacecraft during 1993–2001, showing location and intensity of the radiation belts. Data courtesy of Shri Kanekal and SAMPEX Data Center staff. (For a color version of this figure, please see the color plate at the beginning of this book.)

of L. The shape and position of the inner and outer radiation belts are obvious. The outer belt was weakest during years of low geomagnetic activity (1996 and 1997) and most intense during years of high activity in the descending and ascending phases of the sunspot cycle (1994 and 1998–2001). It is to be kept in mind that these measurements are actually of precipitation from the radiation belts to 550 km altitude and that the inner belt is usually characterized by proton rather than electron fluxes.

It is well known that large-scale high-speed solar wind streams often trigger magnetic storms that produce rapid changes in size and intensity of the radiation belts, but there is insufficient understanding of transport, acceleration, and loss processes to predict the radiation belt responses. However, the occurrence of such streams and sudden large increases in the MeV electron flux correlate with the appearance of large-amplitude millihertz-frequency (Pc5) plasma waves throughout the magnetosphere that may energize trapped particles (Elkington, 2006; Kim *et al.*, 2006). This is seen in Figure 2.10 that compares solar wind speed, the power of millihertz-frequency (ULF) Pc5 waves recorded on the ground, relativistic electron flux at geosynchronous orbit, and the D_{st} parameter, during January–June 1995 (Mathie and Mann, 2000a). Vertical lines identify magnetic storm onsets as indicated by the sudden increase in the ring current (drop in D_{st}). It is clear that high radiation belt electron fluxes are connected with enhanced Pc5 activity during storms triggered by high solar wind speeds.

Particles may be accelerated to radiation belt energies through stochastic diffusion or resonant interaction with the radial electric field of plasma waves (Thorne, 2010). Possible sources of such waves are summarized in Chapter 4 but include fluctuations in the solar wind pressure that transmit compressional mode ULF wave

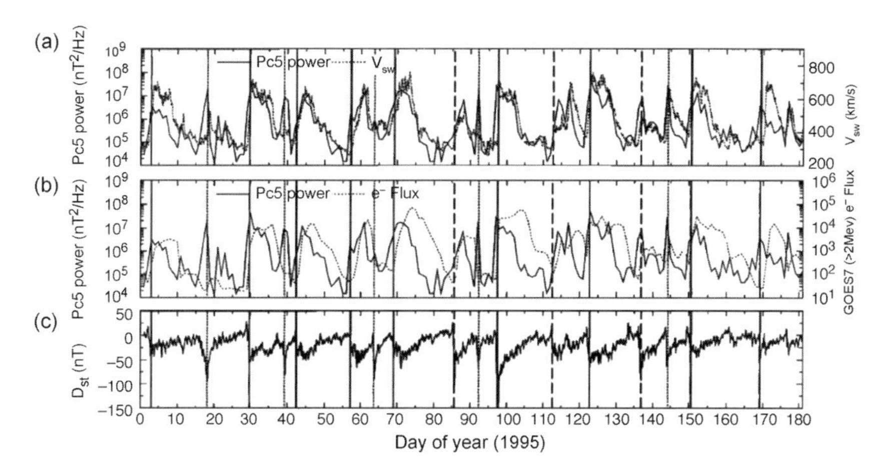

Figure 2.10 Comparison between solar wind speed, daily averaged >2 MeV electron flux measured at geostationary orbit measured by the GOES 7 satellite, daily averaged power in the 1–10 mHz (Pc5) range from an auroral zone ground station, and geomagnetic activity represented by the D_{st} index, during January–June 1995. Vertical lines denote magnetic storm onsets, with solid lines depicting storms with significant electron flux increases. From Mathie and Mann (2000a).

energy into the magnetosphere, cyclotron resonance with backstreaming ions in the upstream solar wind, shear-flow instabilities at the magnetopause (such as the Kelvin–Helmholtz instability and overreflection), magnetospheric cavity or wave-guide modes, and internally generated drift and drift-bounce resonance waves. It has been shown that under active conditions, drift-resonance interaction with Pc3–Pc5 waves may cause earthward radial diffusion and hence acceleration of electrons to megaelectron volt energies deep into the magnetosphere within a few hours (Loto'aniu *et al.*, 2006; Ukhorskiy *et al.*, 2005). Recent multisatellite and ground observations link the energization (probably via drift resonance) of relativistic radiation belt electrons with compressional poloidal mode Pc5 waves driven by quasi-periodic fluctuations in the solar wind dynamic pressure (Tan *et al.*, 2011). These wave modes are able to transfer energy effectively across magnetic field lines.

Loss mechanisms for radiation belt particles include outward radiation diffusion, loss to the magnetopause, collisions with neutral particles, and wave-mediated relativistic electron precipitation (REP) into the atmosphere (O'Brien, Looper, and Blake, 2004). REP events comprising <1 s microbursts of >1 MeV electrons have been observed from the SAMPEX satellite at $L = 4$–6 during the local morning, but they also occur on timescales of minutes and hours. The latter may be associated with $\sim$1 Hz EMIC waves (Clilverd *et al.*, 2010). Some consequences of REP are outlined in Section 2.8.

2.5 The Inner Magnetosphere

The inner magnetosphere, also called the plasmasphere, is populated by cold ($T_e \leq 10$ eV) dense plasma originating from, and in dynamic equilibrium with, the underlying ionosphere (see Section 2.6). Properties of the plasmasphere are determined by the pressure balance and hence particle flow along nearly corotating closed flux tubes linking to the ionosphere and by erosion of the outer boundary, the plasmapause, under the influence of variable convection processes in the outer magnetosphere.

Discovered in 1963 with VLF whistler and spacecraft particle measurements (Lemaire and Gringauz, 1998), the plasmapause is usually described as a sudden order-of-magnitude change in electron density near the equatorial plane. The plasmapause is generally regarded as the separatrix between corotating plasma drift paths on closed field lines confined to the inner magnetosphere, and convective plasma motion on drift paths that allow escape into interplanetary space (Nishida, 1966). In reality it is a highly variable three-dimensional feature whose position is controlled by the balance between factors including the corotation electric field, the magnetospheric convection electric field, the ionospheric dynamo electric field, and refilling from the ionosphere. The corotation field arises from rotation with Earth of the neutral atmosphere and hence the lower ionosphere. This produces an electric field $\boldsymbol{E}_c = -\boldsymbol{v} \times \boldsymbol{B}$ in a nonrotating reference frame, where $\boldsymbol{v}$ is the corotation velocity. The corotation field usually dominates the convection electric field and the dynamo field, which is produced by ionospheric winds moving conducting plasma through the geomagnetic field.

Several empirical models have been developed to describe the approximate L-value of the inner boundary of the plasmapause. An early one gives (Orr and Webb, 1975)

$$L_{pp} = 6.52 - 1.44Kp + 0.18Kp^2 \tag{2.14}$$

for times centered on 0200 LT ± 2 h, where K_p represents the average nighttime level of magnetic activity. A more sophisticated model (O'Brien and Moldwin, 2003) provides

$$L_{pp} = a_1\left[1 + a_{MLT}\cos(\phi - a_\phi)\right]Q + b_1\left[1 + b_{MLT}\cos(\phi - b_\phi)\right], \tag{2.15}$$

where Q relates to a magnetic activity index, a_{MLT} and b_{MLT} provide a local time variation, a_ϕ and b_ϕ account for the bulged shape of the plasmapause, and $\phi = 2\pi(MLT/24)$.

Empirical models of the plasmasphere density have been developed using spacecraft observations. One such description, which has been compared with earlier efforts, is given by (Sheeley *et al.*, 2001)

$$N_e = 1390(3/L)^{4.83} \pm 440(3/L)^{3.60}, \quad 3 \leq L \leq 7. \tag{2.16}$$

A review of empirical models of plasmaspheric density developed using *in situ* and ground-based observations appears in Reinisch et al. (2009). Not accounted for in such models is an annual and longitudinal variation in plasmaspheric electron and ion (H^+, He^+, and O^+) densities (Menk *et al.*, 2012). Furthermore, all these models give the mistaken impression that the plasmasphere is well ordered and the plasmapause position well defined. Both are highly variable. Sometimes complex multiple plasmapause-like features are present, while after extended quiet periods the plasmasphere is saturated (i.e., flux tubes completely filled from the underlying ionosphere) and no plasmapause is evident. Under these conditions, density should decrease with radial distance as L^{-4}, since flux tubes increase in volume as a function of L^4. The process of refilling was most recently summarized, with measurements of refilling rates and upward fluxes, in Obana, Menk, and Yoshikawa (2010) and is discussed in more detail in Section 7.4.

Deep within the plasmasphere, plasma motion is generally controlled by flux transfer through ionosphere–protonosphere coupling and radial $E \times B$ drift (see Section 3.2) of flux tubes driven by neutral winds in the ionosphere. However, although the plasmapause position is controlled by solar wind-induced convection, it does not simply move radially with changing magnetic activity. There are three issues here. First, flux tubes drift radially under the influence of the magnetospheric electric field that may penetrate to low altitudes under magnetically disturbed conditions. Within the plasmasphere, these effects are superimposed upon the diurnal density variations arising from ionization interchange with the underlying ionosphere. Also, the plasmaspheric refilling rate, and hence the plasmapause shape, may differ for different ion species, and the O^+ concentration can be enhanced by an order of magnitude or more near the storm-time plasmapause. Figure 2.11 shows, for example, the radial variation in mass-loaded ion density for two combinations of ion species compared with the density profile if only H^+

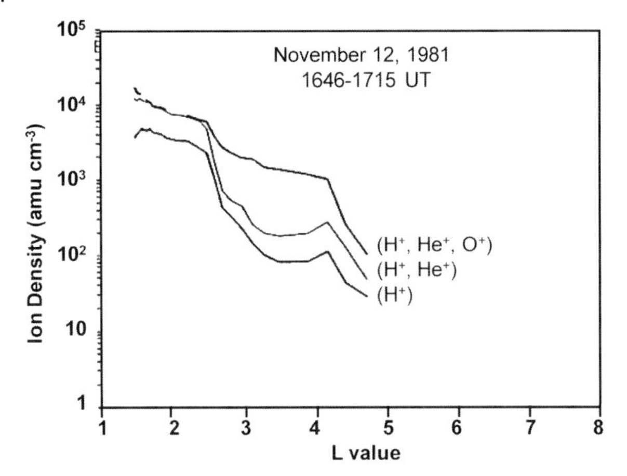

Figure 2.11 Radial variation in mass-loaded ion density for two combinations of ion species compared with the H^+ density, inbound pass of the DE-1 spacecraft on November 12, 1981. From Fraser *et al.* (2005).

were present, for an inbound pass of the DE-1 spacecraft across the plasmapause (Fraser *et al.*, 2005). The mass-loaded profiles, especially when O^+ is included, are quite different from the profile based only on H^+ measurements.

Second, although the dayside plasmasphere is somewhat shielded from the perturbing electric fields, the dayside plasmapause responds promptly to over- or under-shielding associated with sudden changes in the convection electric field and can thus rapidly develop localized dents, shoulders, and other features (Goldstein *et al.*, 2002).

Third, a sudden increase in the dawn-to-dusk convection electric field causes an outward plasmapause bulge usually located around 18 LT to rotate rapidly sunward toward noon, forming a plume that allows convection and drainage of plasmaspheric plasma outward into the outer magnetosphere (Grebowsky, 1970).

Observations from the EUV imager experiment on board the IMAGE spacecraft reveal a rich variety of irregularity features at the plasmapause (Sandel *et al.*, 2003), including noncorotating plumes and radial "fingers" of depleted or enhanced density, and density variations on all scales.

The density change and hence jump in Alfvén speed at the plasmapause form a boundary for plasma waves propagating in the magnetosphere. This has several important consequences: (i) MHD models predict separate cavity mode resonances in the inner and outer magnetospheres, with characteristic eigenperiods, amplitude, and phase structure. The spectrum of ULF waves should thus differ inside and outside the plasmasphere. (ii) The density gradient produces a step in the radial field line eigenfrequency profile, and hence a rapid change in phase and polarization of ULF waves detected on the ground (Lanzerotti, Fukunishi, and Chen, 1974). (iii) Compressional (fast) mode MHD waves propagating earthward in the equatorial plane, on encountering the plasmapause, can couple energy to the field-guided Alfvén mode, resulting in characteristic dispersion properties across the ground. It is not clear if the physical mechanism is one of refraction or diffraction. (iv) The amplitude of earthward propagating ULF waves increases initially at the plasmpause due to increasing density, then decreases as the waves propagate inward due to continual partial reflection of wave energy (Allan and Knox, 1979a). (v) Surface wave modes may be excited at a steep plasmapause, but their existence is yet to be proven. (vi) Spatially localized guided poloidal mode waves may be produced near the plasmapause (Klimushkin, Mager, and Glassmeier, 2004).

The plasmapause thus exerts a profound influence on propagating ULF waves including those involved in accelerating trapped particles. However, since the plasmapause is highly variable, its behavior is difficult to monitor with spacecraft observations. In Chapter 7 we present examples of remote sensing this region using ground-based magnetometer observations.

2.6 Formation and Properties of the Ionosphere

The ionosphere is a partially ionized region that forms the lower boundary to the magnetosphere and the upper boundary to the neutral atmosphere. In space physics studies, the ionosphere is sometimes ignored or approximated as a thin sheet with high conductivity. However, the neutral atmosphere does not support electric current, so magnetospheric ULF waves convert in the ionosphere from plasma waves to electromagnetic waves in the neutral atmosphere, with significant changes to wave properties observed on the ground. Furthermore, wave fields drive motions of the ionospheric plasma that may affect the propagation of electromagnetic signals through or reflected from the various ionospheric layers. This impacts upon radio astronomy, satellite communication and navigation, and over-the-horizon radar surveillance applications. These topics are discussed further in Chapter 8.

It is instructive to consider how temperature varies with altitude, shown in Figure 2.12, using predictions from the IRI-2007 (International Reference Ionosphere) model (Bilitza and Reinisch, 2008), in this case for 12 LT at -33.0° latitude, 152.0° longitude on January 1, 2001 (i.e., midday at Newcastle, Australia, solar maximum). Neutral particle temperature T_n increases sharply at $\sim$100 km altitude from troposphere and stratosphere values of 150–300 K (not shown here – the model commences at 50 km) to $\sim$1200 K at 300 km due to the absorption of incident solar UV, EUV, and X-ray radiation, which causes photoionization and the production of free electrons. Electron temperatures increase with altitude at a faster rate than for neutrals, since the ambient electrons are more efficient in removing excess kinetic energy from the continually released energetic photoelectrons. The resulting hot electron gas is cooled mainly by collisions with neutral particles at lower altitudes and with ions at higher altitudes. Electron and ion temperatures are roughly equal above about 1000 km and are higher at solar maximum than minimum. Not surprisingly, large diurnal effects occur at lower altitudes.

At low altitudes, atmospheric gases are mixed by turbulence and their relative proportions are almost constant. Beyond the turbopause at about 100 km, the gas constituents separate by mass and are regarded as being in diffusive equilibrium. In

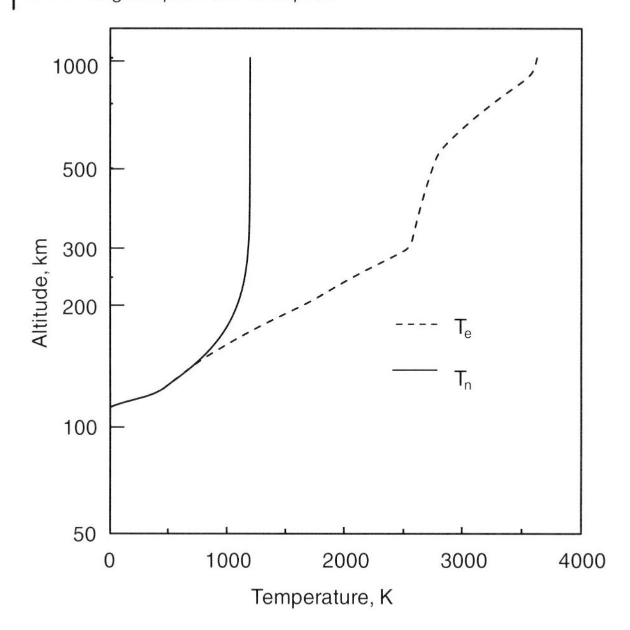

Figure 2.12 Altitude variation of electron and neutral particle temperature (dashed and solid lines, respectively) through the ionosphere for midday, solar maximum at 33.0° latitude, 152.0° longitude on January 1, 2001.

such a situation, the density of a given volume of gas with thermal pressure P at altitude z depends on the balance between the pressure gradient across the gas and the gravitational force:

$$\frac{\partial P}{\partial z'} = -Nmg = \frac{\partial}{\partial z'}(NkT), \tag{2.17}$$

where k is the Boltzmann constant, N is the mass density, T is the temperature and

$$z' = \int_{z_0}^{z} (g/g_0)\mathrm{d}z \tag{2.18}$$

accounts for the height variation of gravity g. It is useful to define a density scale height H_N representing the logarithmic decrement of density with altitude (Bauer, 1969):

$$\frac{1}{N}\frac{\partial N}{\partial z'} = \frac{\partial(\ln N)}{\partial z'} = -\frac{1}{H_\mathrm{N}}. \tag{2.19}$$

It is sometimes assumed that pressure scale height $H = kT/mg$ is the same as the density scale height H_N, but this only applies to an isothermal situation. Nevertheless, the dependence of scale height on particle density means that the lightest molecules and atoms have largest scale heights.

The formation and properties of the ionosphere have been described in detail in various texts (e.g., Kelley, 2009; Luhmann, 1995; Ratcliffe, 1972; Rees, 1989; Rishbeth, 1973; Schunk and Nagy, 2009). The Sun may be regarded as a blackbody

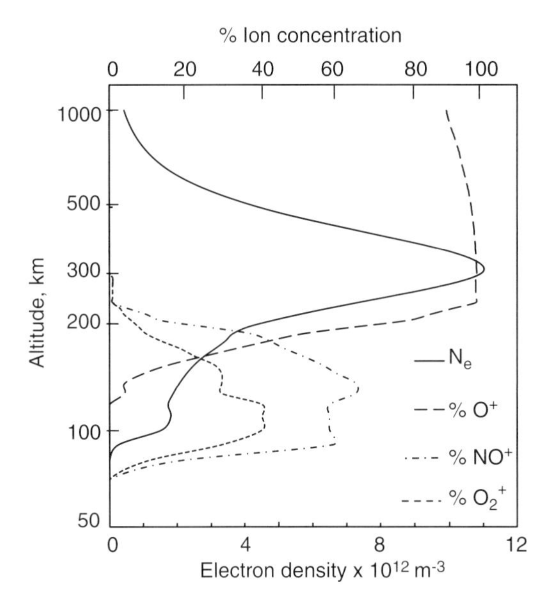

Figure 2.13 Altitude variation of electron density (solid line) and ionic composition under the same conditions as for Figure 2.12.

radiator with surface temperature $T_E \approx 6000$ K, so $kT_E \approx 0.5$ eV, although there is considerable energy flux to 20 eV. The major neutral molecular constituents in the ionosphere are N_2 and O_2 and their abundance is controlled by geochemical and biological processes. Their ionization potentials are 15.58 and 12.08 eV, respectively. The main atomic species is O, produced from dissociation of O_2 by solar UV photons with threshold energy 5.12 eV. The ionic composition is shown by the dashed and dotted lines in Figure 2.13, generated using the IRI model under the same conditions as before. NO^+ and $O_2{}^+$ dominate in the lower ionosphere, but O^+ is the main constituent above 300 km. Not shown here are H^+ that becomes dominant above about 3000 km altitude, $N_2{}^+$ that rapidly recombines to form NO^+, and He^+ that is never a dominant species in the ionosphere.

Also shown in Figure 2.13 is the electron density profile. The parabolic region with peak density ~350 km altitude is called the F region, while density ledges near 150 and 110 km correspond to the F1 and E regions, respectively. In fact, the daytime F region peak is identified as the F2 region, while the F1 region disappears at night. Not evident in Figure 2.13 is the D region, which has relatively low electron density and occurs around 65–80 km.

The distribution of particle density N at any level in the ionosphere depends on production, loss, and transport processes and is described by the continuity equation:

$$\frac{\partial N}{\partial t} = q - l(N) - \nabla \bullet (N\boldsymbol{v}), \tag{2.20}$$

where q is the rate of production, $l(N)$ is the loss rate of charged particles due to chemical process, and $\boldsymbol{v}$ is the drift velocity leading to loss by advection. The transport term is negligible below about 200 km altitude. Under photochemical equilibrium, ionization production and loss processes are balanced, that is, $q = l(N)$.

The production of photoionization is described by

$$q_{pj} = N_j \sigma_{ij} Q_\infty \exp(\tau), \tag{2.21}$$

where σ_{ij} is the ionization cross section of the jth constituent $[m^2]$, N_j is the number density of that constituent, and Q_∞ is the number of photons per second incident on the atmosphere. τ is called the optical depth and describes the attenuation of solar irradiance with distance through the atmosphere:

$$\tau(z) = \sigma \sec\chi N_\mathrm{T}(z), \tag{2.22}$$

where χ is the solar zenith angle and $N_\mathrm{T}(z)$ is the total number of particles from ∞ to altitude z:

$$N_\mathrm{T}(z) = \int_\infty^z n(z')\mathrm{d}z'. \tag{2.23}$$

The production of ionization is greatest at the altitude where $\tau = 1$ and when the Sun is vertically overhead, that is, $\chi = 0$. Photoionization production thus depends on the respective ionization potentials for atoms and molecules, their relative concentrations, and the flux of radiation at different wavelengths. In combination with altitude-dependent electron loss terms and transport, this leads to the formation of the various ionospheric layers.

The E region arises from the absorption of EUV radiation, including the solar H Lyman β (102.6 nm, $\tau = 1$ at 105 km) and C III (97.7 nm, 120 km) lines and soft X-rays (1–20 nm). Metallic ions from meteors and auroral particle fluxes are important in maintaining the nighttime E region. The most heavily absorbed part of the spectrum, and hence peak electron and ion production, occurs in the F1 region, including the He II Lyman α (30.4 nm, 130 km), He I (58.4 nm, 164 km), and H Lyman continuum (91.1–84.0 nm, 105–120 km). In Figure 2.13, this merges with the F2 region, which arises not so much from absorption of particular spectral lines but from the height variation of two-step recombination processes and vertical transport, while the topside ionosphere results from diffusion from below. Energetic particle precipitation also causes ionization and is responsible for maintenance of the ionosphere during polar winter.

Electron loss is mostly through recombination with ions, described by

$$l(N) = \alpha N(e)N(A^+) = \alpha N^2, \tag{2.24}$$

if $N(e)$ is the electron number density per unit volume, $N(A^+)$ is the number of positive ions per unit volume, and α is the recombination coefficient. Important recombination reactions in the E region are $NO^+ + e \rightarrow N + O$, $O_2{}^+ + e \rightarrow O + O$, and $N_2{}^+ + e \rightarrow N + N$ (all dissociative recombination). Radiative recombination, $O^+ + e \rightarrow O^* + h\nu$, results in airglow but has a slow reaction time. In the F1 region, a charge exchange reaction produces O_2^+ or N_2^+ molecular ions, which then undergo dissociative recombination: $O^+ + O_2 + e \rightarrow O + O + O$ and $O^+ + N_2 + e \rightarrow O + N + N$.

A mathematical description of the absorption of solar radiation and consequent formation of the ionosphere was first provided by Chapman (1931) and is not repeated here. The theory allowed prediction of the normalized ionization production rate with height, time of day, season, and latitude and solar zenith angle. The production rate is given by

$$q = q_0 \exp[1 - y - \exp(-y)\sec\chi], \tag{2.25}$$

where q_0 is the peak production rate for an overhead Sun at height h_0 and

$$y = \frac{h - h_0}{H}. \tag{2.26}$$

Under photochemical equilibrium, ionization production and loss processes are balanced, that is, $\partial N/\partial t = 0$, and from Equation 2.20, $q = \alpha N^2$. Putting $N_0 = [q_0/\alpha]^{1/2}$ and using Equation (2.25) gives

$$N = N_0 \exp\left[\frac{1}{2}[1 - y - \exp(-y)\sec\chi]\right]. \tag{2.27}$$

For several hours near noon each day, sec $\chi \approx 1$. Then, $h_0 = h$ and so

$$N \approx N_0 \exp\left[\frac{-y^2}{4}\right] \approx N_0\left[1 - \frac{y^2}{4}\right]. \tag{2.28}$$

The electron density profile is therefore parabolic in shape, called a Chapman profile.

The presence of free charges in the ionosphere results in a current density and hence conductivity, which is anisotropic due to the effects of the magnetic field and collisions. The various components are: (i) direct conductivity $\sigma_0 = \sigma_D$ associated with the current that is parallel to the geomagnetic field; (ii) Pedersen conductivity $\sigma_1 = \sigma_P$ associated with the current that is parallel to an imposed electric field, which is itself perpendicular to the magnetic field; (iii) Hall conductivity $\sigma_2 = \sigma_H$ associated with the current that is perpendicular to an imposed electric field, which is itself perpendicular to the geomagnetic field.

Expressing the electric field as a vector with x-, y-, and z-components, assuming a horizontal ionosphere and vertical magnetic field, and using Ohm's law (Blelly and Alcaydé, 2007) gives

$$J = \begin{pmatrix} \sigma_P & \sigma_H & 0 \\ -\sigma_H & \sigma_P & 0 \\ 0 & 0 & \sigma_D \end{pmatrix} \begin{pmatrix} E_x \\ E_y \\ E_z \end{pmatrix}, \tag{2.29}$$

where

$$\sigma_P = \left[1 - \frac{1}{m_e \nu_{ei}} \frac{\alpha}{(\alpha^2 + \beta^2)}\right] \frac{e^2 N_e}{m_e \nu_{ei}}, \tag{2.30}$$

$$\sigma_H = \frac{\beta}{(m_e \nu_{ei})(\alpha^2 + \beta^2)} \frac{e^2 N_e}{m_e \nu_{ei}}, \tag{2.31}$$

$$\sigma_D = \left[\frac{1/m_e\nu_{en} + 1/m_i\nu_{in}}{1/m_e\nu_{ei} + 1/m_e\nu_{en} + 1/m_i\nu_{in}}\right]\frac{e^2N_e}{m_e\nu_{ei}}, \tag{2.32}$$

$$\alpha = \frac{1}{m_e\nu_{ei}} + \frac{\nu_{en}}{m_e\left(\nu_{en}^2 + \omega_e^2\right)} + \frac{\nu_{in}}{m_i\left(\nu_{in}^2 + \omega_i^2\right)}, \tag{2.33}$$

$$\beta = \frac{\omega_e}{m_e\left(\nu_{en}^2 + \omega_e^2\right)} - \frac{\omega_i}{m_i\left(\nu_{in}^2 + \omega_i^2\right)}, \tag{2.34}$$

$$\omega_i = eB/m_i \quad \text{and} \quad \omega_e = eB/m_e, \tag{2.35}$$

where ν_{in}, ν_{ei}, and ν_{en} are collision frequencies for the ion–neutral, electron–ion, and electron–neutral species, respectively. The ionosphere conductivity thus depends on particle masses, magnetic field strength, ionization and its temporal and spatial variations, and collision frequencies that can in turn be determined from the relevant number densities together with the temperature, mean molecular mass, and density of the neutral components. This treatment is often simplified to a two-fluid case, and further by ignoring collision terms (Song, Gombosi, and Ridley, 2001).

If the magnetic field is oblique, then the conductivity depends on the dip angle I (Sciffer and Waters, 2002):

$$\sigma = \begin{pmatrix} \sum_D \cos^2 I + \sum_P \sin^2 I & \sum_H \sin I & (\sum_D - \sum_P)\sin I \cos I \\ -\sum_H \sin I & \sum_P & \sum_H \cos I \\ (\sum_D - \sum_P)\sin I \cos I & -\sum_H \cos I & \sum_D \sin^2 I + \sum_P \cos^2 I \end{pmatrix}. \tag{2.36}$$

If the ionosphere is approximated as a thin conducting sheet, then height integrated-conductivities are specified:

$$\sigma_P = \int_0^z \sigma_P dz \quad \text{and} \quad \sigma_H = \int_0^z \sigma_H dz. \tag{2.37}$$

Figure 2.14 shows height profiles of the direct, Pedersen and Hall conductivities and phases for noon at solar minimum and maximum and dip angle $I = 70°$ (Sciffer, Waters, and Menk, 2005). Above 200 km, the direct conductivity, along $\mathbf{B}$, is about 10^7 times larger than the Pedersen and Hall conductivities. However, the Pedersen conductivity peaks in a relatively narrow region near 140 km, and the Hall conductivity peaks near 120 km. This gives rise to anisotropic ionosphere conductivity, resulting in conversion of the ULF wave modes (Hughes, 1974), discussed further in Chapter 8. At night, the conductivities are an order of magnitude smaller, but the qualitative results are unchanged.

2.7 Geomagnetic Disturbances

Secular variations of the geomagnetic field arise from the internal core field. However, shorter period variations (of order a day or less) are due to external contributions to the geomagnetic field associated with the solar–terrestrial interaction, including magnetic storms and ULF pulsations.

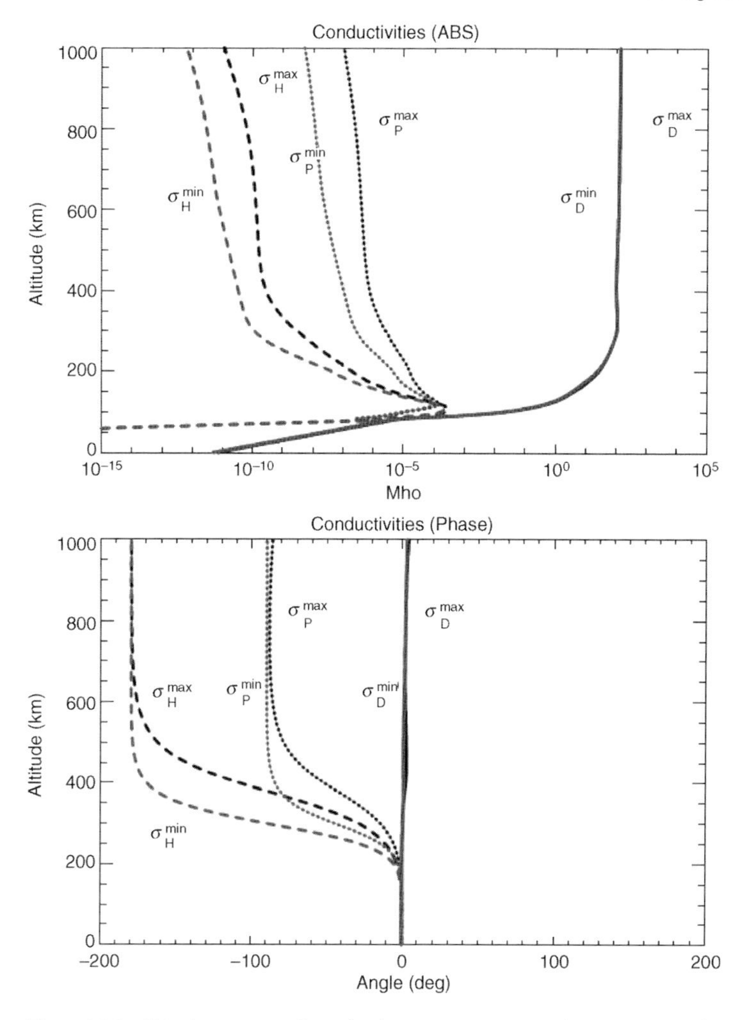

Figure 2.14 Altitude variation of amplitude and phase of direct (solid lines), Pedersen (dotted) and Hall (dashed) conductivities through the ionosphere for midday, solar minimum and maximum conditions at dip angle $I = 70°$. From Sciffer, Waters, and Menk (2005).

During the International Polar Year, 1932, a system was developed to characterize the level of disturbance of the geomagnetic field due to these external effects. This is based on indices describing the level of disturbance each 3 h, called K indices. The K indices from typically 12 midlatitude observatories are averaged to produce the three hourly planetary magnetic activity index, K_p. Today this is the most widely used measure of geomagnetic disturbance. The K and K_p indices are quasi-logarithmic, ranging from 0_0 to 9_+, and an equivalent "linearized" 3 h version of K is the a index, with a daily average value A and global daily average value A_p. The a_a index has been recorded since 1868 and is based on the three hourly average of K indices from two near-antipodal subauroral observatories in England and Australia. Detailed historical analysis of this index shows that the average duration of a geomagnetic storm (above a threshold of 40 nT) is 1.4 days, the magnitude of storms has not changed over the lifetime of this index, but the number of storms per solar cycle has almost doubled since 1915, most likely due to an increase in solar activity (Clilverd *et al.*, 2002, 2005; Lockwood, Stamper, and Wild, 1999). This is a surprising but clearly evident trend. Figure 2.15 shows the annually averaged a_a index and sunspot count since 1868. An increasing trend in both is evident. This is highlighted by the simple linear and cubic fits that are overplotted on both axes. The latter points to an 80–90

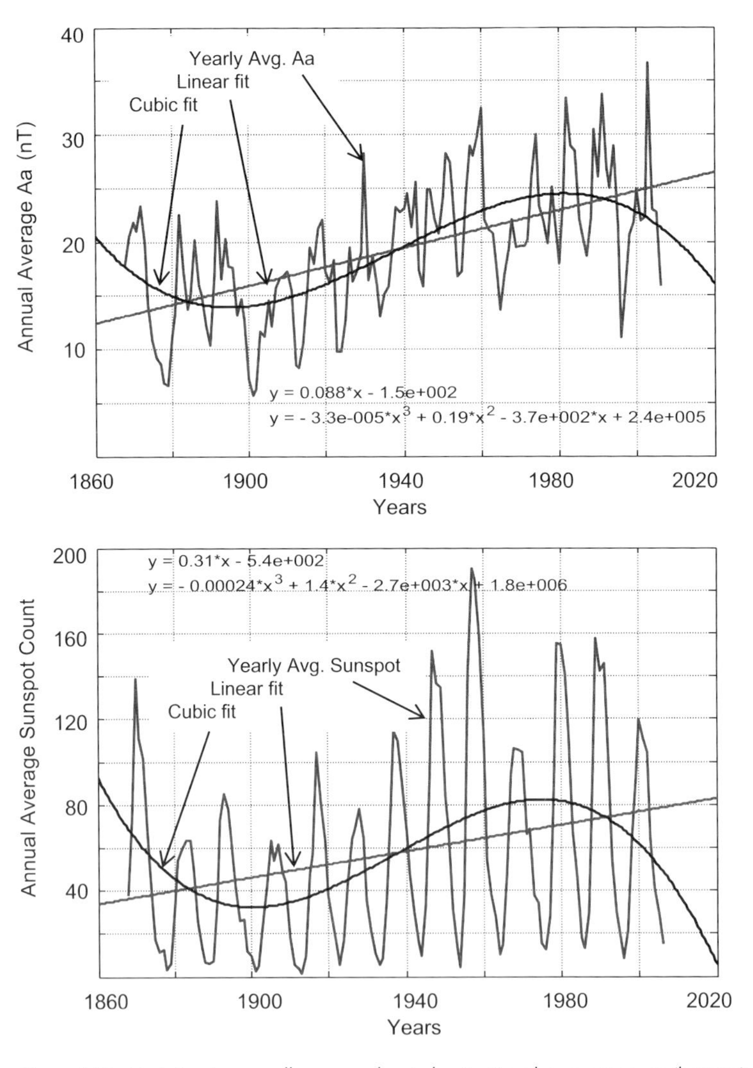

Figure 2.15 Variation in annually averaged a_a index (top) and sunspot count (bottom) since 1868. Simple linear and cubic fits are superimposed.

year variation in geomagnetic and solar activity, often called the Gleissberg cycle (Demetrescu and Dobrica, 2008; Peristykh and Damon, 2003).

There have been a number of efforts to develop a coupling function and hence an empirical model describing the solar wind drivers of geomagnetic activity. One well-known parameter is (Perrault and Akasofu, 1978)

$$\varepsilon = V_{SW} B^2 \sin^4(\theta_c/2), \tag{2.38}$$

where V_{SW} is the solar wind velocity and hence the rate at which field lines are convected to the magnetopause, and $\theta_c = \tan^{-1}(B_y/B_z)$ is the IMF clock angle. Recent work (Newell *et al.*, 2007) suggests that the rate at which magnetic flux is opened at the magnetopause, $d\Phi_{MP}/dt$, correlates best with a variety of magnetospheric activity indices (except D_{st}), via

$$\frac{d\Phi_{MP}}{dt} = V_{SW}^{4/3} B_T^{2/3} \sin^{8/3}(\theta_c/2), \tag{2.39}$$

where B_T is the magnitude of the IMF. This relationship quantifies the importance of reconnection, discussed in Section 2.3.

Coronal mass ejections (CMEs) occur most frequently near the peak of the sunspot cycle and can trigger intense geomagnetic storms. However, during the declining and minimum solar phases, coronal holes and accompanying high-speed solar wind streams and CIRs (Gosling and Pizzo, 1999) become the dominant cause of geomagnetic activity on Earth. These storms tend to be less intense but recur over multiple solar rotations because the features rotate with the Sun. This results in periodicities of 27, 13.5, 9, and 7 days in a range of parameters, including ionospheric total electron content and the ratio of thermospheric oxygen to nitrogen (O/N_2) (Mannucci *et al.*, 2012).

Substorms are the most frequent major disturbance in Earth's magnetosphere. Typically one to four substorms occur daily, each in the local evening/night sector and lasting 1–5 h. They are characterized by the onset of active auroral displays and intense irregular activity, including Pi1 and Pi2 pulsations, recorded by magnetometers at high latitudes. It has long been known that substorms are associated with southward IMF B_z and hence magnetopause reconnection. However, during years of high-speed solar wind streams, substorms occur more frequently, are on average 30% more intense, and transfer twice as much magnetic energy to the auroral ionosphere ($\sim 6 \times 10^{14}$ J) than during years of few or no high-speed streams (Tanskanen *et al.*, 2005).

Geomagnetic storms are larger events whose temporal development can be monitored using the D_{st} index. The causes of the initial and main phase signatures were outlined in Section 2.3. Storms are triggered by a variety of solar disturbances, including high-speed streams, CMEs, and other transients, and may deposit 10^{12}–10^{13} W of energy into the magnetosphere (Baker, Turner, and Pulkinnen, 2001; Koskinen and Tanskanen, 2002). Storm activity is modulated by the IMF and a well-established semiannual variation (equinoctial maximum) in geomagnetic activity is due to the annual variation in the southward component of the IMF, which is ordered in solar equatorial coordinates, relative to the solar–magnetospheric reference frame that orders the interaction of the solar wind with Earth (Russell and McPherron, 1973).

During geomagnetic storms, energy is derived from the solar wind flow and dissipated via ring current injection and decay, ionospheric Joule heating, particle precipitation into the atmosphere, and other related processes. The significance of the various energy dissipation processes depends on the power input from the solar wind. For example, during moderate CME-driven storms, Joule heating and auroral particle precipitation account for most of the energy dissipation (Baker, Turner, and Pulkinnen, 2001). Altogether up to 10^{17} J may be dissipated into the magnetosphere and ionosphere during major storms. Global resonances of the magnetosphere by Pc5 plasma waves may account for 30% or more of the energy deposited in the ionosphere, via Joule heating, during a substorm cycle (Rae *et al.*, 2007b).

2.8 Space Weather Effects

The term space weather describes the collective effect of the variable wave fields and particle populations on the magnetosphere and impacts on technological systems including elements of critical infrastructure (Daglis, 2004; Koskinen *et al.*, 2001). The consequent business risks and potential societal and economic impacts are well documented (Baker *et al.*, 2008; Hapgood, 2010; Turner, 2012). Research aimed at understanding space weather processes is an international priority area, addressed by several spacecraft missions designed to provide information on plasma wave and particle properties in the outer magnetosphere. The missions and responsible agencies include Cluster (four-satellite constellation by ESA), Double Star (two satellites, joint ESA/China), THEMIS (five-satellite constellation, NASA), GOES (series of US satellites providing continuous operation at a variety of longitudes at geostationary orbit), RBSP (two-satellite mission through the radiation belts, NASA), ERG (Japan), and Resonance (Russia).

About 600 satellites operate in orbit around Earth, worth $\sim$US\$2 $\times$ 10^{11} to deploy. The 250 commercial communication satellites in geosynchronous orbit provide an annual revenue stream of order \$2.5 $\times$ 10^{10} or $>$\$2.5 $\times$ 10^{11} over the life of the satellites. This figure underestimates the actual economic value of these operations, given the interdependence of networked services. Space weather processes may cause operational anomalies with all these satellites (Bedingfield, Leach, and Alexander, 1996), as well as single-event upsets in portable electronics, distortion of HF radio systems, and compromised operation of precision timing and navigation services such as GPS (Coster and Komjathy, 2008).

Since the skin depth in Earth of long-period geomagnetic variations may be many kilometers, these variations can induce an EMF in the conducting surface of Earth, causing geomagnetically induced currents (GICs) in long conductors, including power networks, pipelines, telecommunication cables, and railway systems (Pirjola, 2002). GICs in long gas pipelines spanning Australia are clearly linked with Pc3–Pc5 magnetic pulsations (Marshall, Waters, and Sciffer, 2010). Such GICs may damage large transformers in power distribution networks at high and lower latitudes (Marshall *et al.*, 2011, 2012).

Single-event upsets, latchup, or burnout are a growing concern in satellite and avionics applications where high computational density is required with minimum mass (constraining shielding thickness). In addition, the energetic radiation associated with solar particle events and the consequent interactions with atmospheric constituents provide a potentially significant radiation dose to aircrew and passengers at aircraft altitudes (Getley, 2004).

Spacecraft interact with and are affected by their environment, and spacecraft design must take these factors into account. Detailed discussions of spacecraft–environment interactions have been provided by Hastings and Garrett (1996) and Tribble (2003). Virtually all satellites experience SEUs that are caused at LEO by energetic particles from the inner radiation belt and at GEO by dielectric charging of order 10^6–$10^7\,\mathrm{V\,m^{-1}}$ due to bombardment by energetic electrons. An example of this was shown in Figure 2.4. At geostationary orbit, SEU rates range from 10^{-10} to 10^{-4} errors per bit-day (Tribble, 2003). During the Halloween 2003 magnetic storm, the radiation belts were drained and then reformed much closer to Earth, causing space weather effects on 59% of NASA's Earth and space science missions (Barbiera and Mahmot, 2004). Over half the anomalies experienced by operators of commercial satellites in 2003 also occurred during the October magnetic storms (Baker *et al.*, 2008). Severe and extreme geomagnetic storms that can affect spacecraft operations and power systems on the ground occur about 100 times per solar cycle (Turner, 2012).

Relativistic electron precipitation may also drive chemical changes in the polar atmosphere, possibly over long timescales and large geographical areas. Such REP events cause changes in the fast recombination region of the atmosphere affecting electron density, atmospheric NO_x concentrations at polar latitudes (Newnham *et al.*, 2011), ozone concentrations (Rodger *et al.*, 2010; Veronnen *et al.*, 2005), OH concentrations (Verronen *et al.*, 2011), and polar surface air temperature (Seppälä *et al.*, 2009), in addition to variations in the amplitude and phase of propagating VLF signals (Clilverd *et al.*, 2009).

It is well established that REP into the atmosphere is associated with magnetic storms (Clilverd *et al.*, 2010; Meredith *et al.*, 2011). Present climate models do not include REP effects as they are considered to be of minor importance based on relative energy considerations. Pc5 waves and electromagnetic ion cyclotron waves in the outer magnetosphere are likely involved in the acceleration of electrons to relativistic energies in the radiation belts (Kim *et al.*, 2006; Rostoker, Skone, and Baker, 1998), and also in the destabilization and precipitation of these electrons to the atmosphere. This leaves open the question of forcing of the atmosphere and climate from above under the possible influence of wave–particle interactions.

3
ULF Plasma Waves in the Magnetosphere

3.1
Basic Properties of a Plasma

Over 99% of matter in the visible universe is in the plasma state, from stellar interiors and atmospheres to gaseous nebulae and interstellar gas. The plasma universe starts immediately beyond Earth's neutral atmosphere; however, plasmas are also encountered in discharge tubes and lightning. A plasma[1] is a gas of mobile positively and negatively charged particles that are largely free to move collectively in response to electric, magnetic, and other forces. These motions in turn produce electric currents and magnetic fields that affect the plasma particles, resulting in propagating waves and instabilities. Here, we consider electrically neutral plasmas described using equations of electromagnetism and fluid mechanics. Coulomb interactions take place, but not electron attachment, dissociation, recombination, excitation, or de-excitation.

Since motions in plasmas are dominated by collective rather than individual particle effects, plasma scale sizes are large compared to the dimensions over which individual particle effects dominate, described by the electron Debye length λ_{De} (Langmuir, 1928):

$$\lambda_{\mathrm{De}} = \sqrt{\frac{\varepsilon_0 k_{\mathrm{B}} T_{\mathrm{e}}}{N_{\mathrm{e}} e^2}} \approx 69\sqrt{\frac{T_{\mathrm{e}}(\mathrm{K})}{N_{\mathrm{e}}(\mathrm{m}^{-3})}}\ \mathrm{m} \approx 743\sqrt{\frac{T_{\mathrm{e}}(\mathrm{eV})}{N_{\mathrm{e}}(\mathrm{cm}^{-3})}}\ \mathrm{cm} \tag{3.1}$$

where ε_0 is the permittivity of free space, k_{B} is the Boltzmann constant, T_{e} is the electron temperature, N_{e} is the electron number density, e is the electron charge, and ions are immobile compared to electrons. In practical terms, the Debye length is the characteristic distance in a plasma over which charged particles screen out an electric field by 1/e. Spacecraft always interact with the surrounding plasma, acquiring an electric potential and hence a Debye sheath that may affect the operation of payloads, including counters, probes, booms, and antennas. This

1) The term "plasma" was first used by Langmuir in 1928.

Magnetoseismology: Ground-based remote sensing of Earth's magnetosphere, First Edition. Frederick W. Menk and Colin L. Waters.

Table 3.1 Typical properties of some space plasmas.

Plasma type	N_e (cm^{-3})	B (nT)	T_e (eV)	λ_D (m)	f_{pe} (Hz)
Interstellar gas	0.1	0.1	0.1	10	10^3
Solar wind	10	1	10	10	3×10^4
Magnetosphere ($6R_E$)	5	200	10^3	10^2	2×10^4
Ionosphere (300–400 km)	10^6	10^4	0.1	10^{-3}	10^7
Solar corona	10^6	10^5	10^2	0.1	10^7
Solar core	10^{26}	10^{11}	10^3	10^{-11}	

includes, for example, the electric field instrument on the Cluster satellites, since at $8.6R_E$ above the polar cap λ_{De} is 24 m (Engwall, 2004). Table 3.1 lists typical properties for some plasmas, including Debye length.

Collective effects can only be exhibited if sufficient particles exist to shield discrete particle effects. For a spherical particle distribution, the characteristic number density, called the plasma parameter, is

$$N_D = N_e \frac{4}{3}\pi\lambda_{De}^3 \approx 1.7 \times 10^9 \left(T_e^{3/2}/N_e^{1/2}\right) \tag{3.2}$$

where T is in eV and N_e is in cm^{-3}.

For a gas to exhibit collective (i.e., plasma) behavior, its dimensions $L \gg \lambda_{De}$ and $N_D \gg 1$.

When an electron in a "cold" plasma is displaced from equilibrium, an electric field and hence force are established that accelerate the electron back, causing overshoot and simple harmonic oscillation. This occurs at the plasma frequency:

$$\omega_{pe} = 2\pi f_{pe} = \sqrt{\frac{N_e e^2}{m_e \varepsilon_0}}. \tag{3.3}$$

Substituting values for the constants (where N_e is in m^{-3}) gives

$$f_{pe} = 8.98\sqrt{N_e}\ \text{Hz}, \tag{3.4}$$

which determines the fundamental timescale of the plasma. There is a different frequency for each species in a multicomponent plasma, where the resultant plasma frequency is the sum of the component frequencies, although the electron frequency term dominates. Measuring the plasma frequency, especially using remote sounders, is a useful way to determine the plasma density. For example, a radio signal of angular frequency ω propagating in a plasma is reflected when $\omega = \omega_{pe}$. This is the principle behind the operation of ground-based ionosondes and the Radio Plasma Imager (RPI) experiment on the IMAGE spacecraft (Reinisch *et al.*, 2000).

3.2 Particle Motions

3.2.1 Motions of Isolated Charged Particles

In order to understand collective particle behavior, we consider first the forces acting on individual charged particles. In a uniform magnetic field but with no imposed electric field or collisions, isolated charged particles with velocity components v_{par} and v_{perp} parallel and perpendicular to $\boldsymbol{B}$ will perform circular motion around the field lines at the cyclotron frequency:

$$\omega_{\text{c}} = \frac{v_{\text{perp}}}{r_{\text{L}}} = \frac{qB}{m}, \tag{3.5}$$

where r_{L} is the radius of orbit, called the Larmor radius, and q is the particle's charge. Taking both components into account, the particle describes a helical trajectory with the guiding center moving at v_{par} along the field line. This is illustrated in Figure 3.1. The direction of the particle orbit is such that the magnetic field generated by it opposes the external field. The plasma particles thus tend to reduce the main field, and the plasma is diamagnetic.

Next we include uniform and orthogonal magnetic and electric fields. A charged particle experiences forces described by the Lorentz force equation:

$$m\frac{\text{d}\boldsymbol{v}}{\text{d}t} = m\frac{\text{d}^2\boldsymbol{r}}{\text{d}t^2} = q(\boldsymbol{E} + \boldsymbol{v} \times \boldsymbol{B}), \tag{3.6}$$

where $\boldsymbol{E}$ is in units of V m^{-1} and $\boldsymbol{B}$ in Tesla. During each gyro-orbit, a particle will be accelerated perpendicular to $\boldsymbol{B}$ due to the electric field, so that the orbit is not closed and the particle drifts orthogonal to both $\boldsymbol{B}$ and $\boldsymbol{E}$. This is known as $E \times B$ drift and the velocity of the guiding center, the drift velocity, is given by

$$\boldsymbol{v}_{\text{E}} = \frac{\boldsymbol{E} \times \boldsymbol{B}}{B^2} \text{m s}^{-1}. \tag{3.7}$$

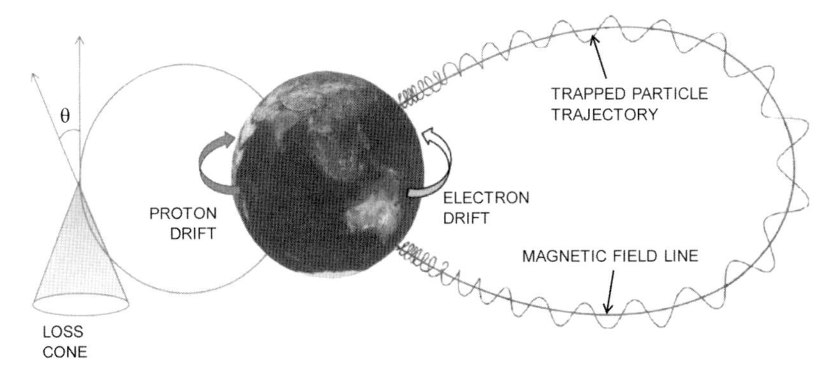

Figure 3.1 Charged particle motions in the magnetosphere, showing gyration around a field line, bouncing between mirror points, and azimuthal drift along L shells. Pitch angle is θ. (For a color version of this figure, please see the color plate at the beginning of this book.)

Electrons and ions gyrate in opposite directions, but are also accelerated by the $\boldsymbol{E}$ field in opposite directions. Both particles therefore $E \times B$ drift in the same direction and no current flows. The effect of this drift is to move charged particles out of the magnetic and electric field regions.

For a general force $\boldsymbol{F}$, Equation 3.6 is modified by substituting $\boldsymbol{F} = q\ \boldsymbol{E}$. For example, in a gravitational field, $\boldsymbol{F} = m\ \boldsymbol{g}$, although in reality this force is negligible.

In a time-varying electric field directed perpendicular to a static magnetic field, particles execute gyromotion and $E \times B$ drift but also experience a force due to $E(t)$, resulting in polarization drift that is orthogonal to the other two drifts and 90° out of phase with $E(t)$. The drift velocity is

$$\boldsymbol{v}_{\mathrm{P}} = \frac{m}{qB^2}\frac{\mathrm{d}\boldsymbol{E}}{\mathrm{d}t} = \pm\frac{1}{\omega_{\mathrm{c}} B}\frac{\mathrm{d}E}{\mathrm{d}t}, \tag{3.8}$$

where the $\pm$ sign denotes opposite drift for ions and electrons. This results in a polarization current:

$$J_{\mathrm{p}} = ne\left(v_{\mathrm{ip}} - v_{\mathrm{ep}}\right) = \frac{ne}{eB^2}\frac{\mathrm{d}E}{\mathrm{d}t}\left(m_{\mathrm{e}} + m_{\mathrm{i}}\right) = \frac{\rho}{B^2}\frac{\mathrm{d}E}{\mathrm{d}t}, \tag{3.9}$$

where ρ is the mass density of the population.

In the magnetosphere, the magnetic field varies with distance, so the motion of a particle's guiding center becomes more complex. We consider two cases, the first being the effect of the spatial gradient in $\boldsymbol{B}$. From Equation 3.5, the radius of gyration depends on the magnitude of B and thus, in a magnetic gradient, the gyration orbits are not closed and the guiding center will drift perpendicular to both $\boldsymbol{B}$ and the gradient in B, in the direction of constant B. This is called grad-B drift and is in opposite directions for ions and electrons. In the magnetosphere, this drift results in charge separation and formation of the ring current, described in Section 2.3. An expression for grad-B drift can be obtained by representing the magnetic field as a static background component $\boldsymbol{B}_0$ and a spatially varying perturbation term, using the Lorentz equation and a Taylor expansion (Chen, 1984).

Geomagnetic field lines are also curved. The guiding center motion of particles along the field lines therefore produces a centrifugal force perpendicular to $\boldsymbol{B}$ and in the plane of curvature. The resultant curvature drift is in the same direction as the grad-B drift.

It can be shown that averaged over several gyroperiods and ignoring relativistic effects and plasma pressure, an expression for the combined grad-B and curvature drifts of the guiding center of a charged particle is

$$\boldsymbol{v}_{\mathrm{R}} + \boldsymbol{v}_{\nabla\mathrm{B}} = \frac{m}{q}\frac{\boldsymbol{R}_{\mathrm{B}} \times \boldsymbol{B}_0}{R_{\mathrm{B}}^2 B^2}\left(v_{\mathrm{par}}^2 + \frac{1}{2}v_{\mathrm{perp}}^2\right), \tag{3.10}$$

where R_{B} is the radius of curvature. Consider, for example, an isotropic population of 1 eV protons and 30 keV electrons at $5\,R_{\mathrm{E}}$ altitude in the equatorial plane where

$B = 3 \times 10^4$ nT. Due to grad-B and curvature effects, the ions drift westward at the velocity of 0.39 m s^{-1} and the electrons eastward at the velocity of 1.17×10^4 m s^{-1}, the latter taking 4.8 h to drift around Earth. For a number density of 10 el cm^{-3}, this results in a ring current density of order 2×10^{-8} A m^{-2}.

Earth's radiation belts are formed by particles that are confined to or trapped by the geomagnetic field lines. These particles exhibit three types of periodic motions: gyromotion around the field lines, bounce motion between conjugate points, and drift motion around Earth. The fundamental physics of these processes is described in detail in a number of texts (e.g. Alfvén and Fälthammar, 1963; Schulz and Lanzerotti, 1974; Walt, 1994) and briefly outlined in the following sections.

3.2.2
First Adiabatic Invariant

Consider a gyrating charged particle moving with nonrelativistic velocity v_{par} parallel to a field line. The gyromotion results in a magnetic moment equal to the product of the current around the particle orbit and the area of the orbit:

$$\mu = \frac{q\omega_{\mathrm{c}}}{2\pi}\frac{\pi v_{\mathrm{perp}}^2}{\omega_{\mathrm{c}}^2} = \frac{1}{2}\frac{v_{\mathrm{perp}}^2 q}{\omega_{\mathrm{c}}} = \frac{1}{2}\frac{m v_{\mathrm{perp}}^2}{B} = \frac{W_{\mathrm{perp}}}{B}, \tag{3.11}$$

where W_{perp} is the component of the particle's kinetic energy perpendicular to the magnetic field. It can be shown that μ is an approximate constant of motion, $\mathrm{d}\mu/\mathrm{d}t = 0$, for sufficiently small spatial variations in B and over time periods that are larger than the particle gyroperiod. The quantity μ is called the first adiabatic invariant.

This is an important result. Consider a gyrating particle with guiding center moving earthward along a field line. The total energy of the particle is conserved:

$$W = W_{\mathrm{perp}} + W_{\mathrm{par}} = \text{constant}. \tag{3.12}$$

From Equation 3.11, since μ is constant, as B increases so does W_{perp}, and from Equation 3.12 W_{par} must correspondingly decrease. This means that as a particle moves earthward, its gyromotion must increase and the motion parallel to the field line decreases, until eventually W_{par} is zero, and the particle is orbiting in a fixed position along B. At this point, the parallel velocity reverses and the particle is reflected back up the field line. This location is called the mirror point. The gyrating charged particle thus bounces between mirror points in conjugate hemispheres, as illustrated in Figure 3.1.

The angle of the particle trajectory to the magnetic field is called the pitch angle θ, represented in Figure 3.1. Then, $v_{\mathrm{perp}} = v \sin\theta$ and using Equation 3.11,

$$\mu = \frac{1/2 m v^2 \sin^2\theta}{B_0} = \frac{1/2 m v^2}{B_{\mathrm{m}}} \quad \text{when } \theta = 90^\circ, \tag{3.13}$$

where B_0 is the magnetic field at an initial location (e.g., at the equatorial plane where $\theta = \theta_{\mathrm{eq}}$) and B_{m} is the magnetic field at the mirror point, where $\theta = 90^\circ$. Since

μ is invariant, as a particle gyrates earthward along a field line, the pitch angle increases as B increases.

Particle mirroring can be conceptualized as follows. A charged particle executing gyromotion about a field line establishes a magnetic dipole, which by Lenz's law opposes the background $\mathbf{B}$ field. In a nonuniform field, this provides a force parallel to the field line and opposite to the direction of motion of the guiding center.

Particles with large equatorial pitch angles have small parallel velocities and mirror at relatively high altitudes. If the pitch angle at B_0 is too small, then the particle interacts with the neutral atmosphere and does not mirror. From Equation 3.13, the smallest pitch angle of a mirroring particle θ_m is

$$\sin^2 \theta_m = \frac{B_0}{B_m}. \tag{3.14}$$

This equation defines the boundary of a region in velocity space called the loss cone, so that particles within this region are not confined to bounce along the field line. For example, a particle experiencing collision or interacting with a wave field may undergo redistribution of its pitch angle to the extent where it is scattered into the loss cone, meaning that the particle's parallel velocity is increased sufficiently that it no longer mirrors. This is more likely to happen for electrons because of their higher collision frequency, and accounts for particle loss from the magnetosphere into the atmosphere.

Knowledge of the pitch angle allows the bounce time between conjugate mirror points to be determined. Various approximate expressions exist for computing this (Hamlin *et al.*, 1961; Walt, 1994). In a dipole field, one useful approximation is

$$T_b \approx 0.117 \left(\frac{R_0}{R_E}\right) \frac{1}{\beta} \left[1 - 0.4635 \left(\sin \theta_{eq}\right)^{3/4}\right] \text{s}, \tag{3.15}$$

where R_0 is the distance from center of the dipole to the equatorial crossing of the field line and $\beta = v/c$.

For 1 MeV electrons and protons crossing the equatorial plane at 2000 km altitude, typical bounce times are 0.1 and 2 s, respectively. The same particles crossing the equatorial plane at $4R_E$ altitude, in the heart of the radiation belt, have bounce times of 0.3 and 5 s, respectively. Relativistic effects are important for these particle energies.

3.2.3
Second Adiabatic Invariant

A second adiabatic invariant J is obtained by defining the action integral

$$J_2 = \oint p \cdot \text{d}s, \tag{3.16}$$

where p is the momentum of the system, s is a position coordinate, and the action J_2 is invariant for slow changes in the system compared to the period of motion.

For a charged particle moving along the field-aligned x-direction between mirror points a and b,

$$J_2 = \int_a^b m v_x \mathrm{d}x = \frac{\pi m v_{\mathrm{perp}}^2}{\omega_\mathrm{c}} = \frac{2m\pi}{q}\left(\frac{W_{\mathrm{perp}}}{B}\right) = \frac{2m\pi}{q}\mu = K\mu, \tag{3.17}$$

where K is a constant and μ is the first adiabatic invariant. Thus, J_2 is a constant integrated over this path for slow temporal changes compared to the bounce time. This defines the drift path and surfaces mapped out by the particle. In a perfectly axisymmetric magnetic field, a drifting particle would circle Earth under gradient and curvature drift and return to the initial field line, tracing out an L-shell surface. However, the geomagnetic field is distorted due to the influence of the solar wind pressure and is therefore not axisymmetric, and conservation of the adiabatic invariant means the bouncing and drifting particle may trace out different L shells during the longitudinal drift in order to return to the original field line. This is called L-shell splitting and may be responsible for the formation of the region 2 current system (Schield, Freeman, and Dessler, 1969).

The longitudinal drift period can also be readily computed for a dipole field. A useful approximation is (Walt, 1994)

$$T_\mathrm{d} \approx \frac{2\pi q B_0 R_\mathrm{E}^3}{m v^2}\frac{1}{R_0}\left[1 - 0.3333\left(\sin\theta_{\mathrm{eq}}\right)^{0.62}\right]. \tag{3.18}$$

Typical drift periods around the magnetosphere for equatorially trapped ($\theta = 90°$) 1 MeV electrons and protons at 2000 km altitude are 3×10^3 and 2×10^3 s, respectively. The same equatorially trapped particles at $4R_\mathrm{E}$ altitude have drift times of 8×10^2 and 6×10^2 s, respectively. Note that the drift period decreases as R_0 increases.

3.2.4
Third Adiabatic Invariant

Conservation of the first and second adiabatic invariants causes particles to mirror in field-aligned trajectories and to return to the original field line. A particle undergoing gradient and curvature drift eventually drifts entirely around Earth. The third adiabatic invariant J_3 defines the drift path followed, requiring that the magnetic flux enclosed by the longitudinal drift path remains constant during slow changes in the magnetic field (relative to the circulation time). Thus, slow compressions or expansions of the geomagnetic field cause trapped particles to move outward or earthward during their longitudinal drift in order to conserve the magnetic flux. However, rapid changes in the geomagnetic field violate this invariant, resulting, for example, in loss of particles of given energies from the magnetosphere on specific drift paths. This may allow a “bump-on-tail” energy distribution and drift-bounce wave–particle instability to form (Ozeke and Mann, 2001).

The invariant is written as a flux invariant:

$$J_3 = q\oint B \cdot \mathrm{d}\boldsymbol{S} = q\Phi, \tag{3.19}$$

where $\mathrm{d}\boldsymbol{S}$ is an element of the surface enclosed by the equatorial drift path and Φ is the enclosed magnetic flux.

3.3
Low-Frequency Magnetized Plasma Waves

Magnetic field strength and plasma density change with location, so the spatial variation of Alfvén speed in the magnetosphere is not uniform. This feature combined with the various plasma boundaries (ionosphere, plasmapause, and magnetopause) yields a rich variety of resonance and wave coupling physics, even for wave frequencies much less than the ion gyrofrequencies. However, an understanding of these processes begins with a simplified description that assumes a plasma with no boundaries immersed in a uniform background magnetic field $\boldsymbol{B}$ and plane wave fronts.

A comprehensive mathematical treatment of ultralow-frequency, magnetized plasma waves with application to near-Earth space physics was given by Walker (2005). The properties of magnetized plasma wave modes were originally described by Alfvén (1948) and all three low-frequency modes are collectively known as Alfvén waves. Since the slow mode requires kinetic pressure, the cold plasma approximation considers only the propagation and interaction between the fast and shear modes. Most texts combine the equations from fluid mechanics and electromagnetism as the starting point and develop a vector wave equation that is then illustrated by deriving the wave mode solutions in various coordinate systems.

In the following sections, a similar development is presented that combines the fluid and electromagnetic equations in order to develop the shear Alfvén mode wave equation in terms of plasma displacement. Estimating the field line resonance frequencies for a dipole magnetic field is a common task in ULF wave research and is crucial for remote sensing applications. A short IDL code that solves the plasma displacement, toroidal mode differential equation is available (see Appendix 1). This mode is then used to illustrate how the same results may be obtained from the equations of electromagnetism alone.

In the magnetosphere, the shear mode resonances derive their excitation energy by coupling to the fast mode, as discussed by Tamao (1965). There is an additional wave mode coupling process through the ionosphere Hall current, an important topic discussed in Chapter 8. Here, we provide a treatment that begins with a simplified geometry by solving the coupled equations in one dimension while retaining the essential features of cold, magnetized plasma wave mode coupling. The aim is to provide introductory details of these ideas at the undergraduate level. We then consider the properties of cold magnetized plasma wave numerical solutions in more complex magnetic field geometries.

3.3.1 Equations of Linear MHD

Derivations of the possible low-frequency wave modes that propagate in a cold, magnetized plasma usually begin by merely stating the MHD equations. This can overlook the various assumptions in the subsequent analysis. A well-known relation is Ohm's law, often expressed as $V = IR$. However, "force" is a more tangible quantity, so we begin from here. The force acting on a fluid of density ρ immersed in a magnetic field $\boldsymbol{B}$ involves both electrical and kinetic effects, which may be described by the momentum equation:

$$\rho \frac{\mathrm{d}\boldsymbol{v}}{\mathrm{d}t} = \boldsymbol{J} \times \boldsymbol{B} - \nabla P, \tag{3.20}$$

where $\boldsymbol{v}$ is the fluid bulk velocity and P is the pressure (kinetic). The generalized Ohm's law is obtained by multiplying Equation 3.20 by the charge-to-mass ratio q/m, adding the ion and electron forms, deleting quadratic terms of the small quantities, and assuming a neutral plasma, to give

$$\frac{m_{\mathrm{i}} m_{\mathrm{e}}}{\rho e^2} \frac{\partial \boldsymbol{J}}{\partial t} = \frac{m_{\mathrm{i}}}{2\rho e} \nabla P + \boldsymbol{E} + \boldsymbol{v} \times \boldsymbol{B} - \frac{m_{\mathrm{i}}}{\rho e} \boldsymbol{J} \times \boldsymbol{B} - \frac{\boldsymbol{J}}{\sigma}. \tag{3.21}$$

The Lorentz force terms are also evident in Equation 3.21. The assumptions for *ideal MHD* are: (i) low-frequency perturbations (i.e., $\partial J/\partial t \approx 0$); (ii) the magnetic pressure is much larger than the kinetic pressure (i.e., $\nabla P \approx 0$); (iii) for small currents, the Hall term $\boldsymbol{J} \times \boldsymbol{B}$ is smaller than $\boldsymbol{v} \times \boldsymbol{B}$; and (iv) collisions between particles are neglected, so the conductivity is large ($\sigma \to \infty$). Therefore, for linear ideal MHD,

$$\boldsymbol{E} = -\boldsymbol{v} \times \boldsymbol{B}. \tag{3.22}$$

Faraday's law for low frequencies becomes

$$\frac{\partial \boldsymbol{B}}{\partial t} = \nabla \times (\boldsymbol{v} \times \boldsymbol{B}). \tag{3.23}$$

The low-frequency form of Ampère's law is

$$\nabla \times \boldsymbol{B} = \mu \boldsymbol{J}. \tag{3.24}$$

We can now distinguish between the steady-state and first-order perturbation (time-varying) velocity, magnetic field, current, and plasma density:

$$\boldsymbol{v} = \boldsymbol{v}_0 + \boldsymbol{v}_1, \quad \boldsymbol{B} = \boldsymbol{B}_0 + \boldsymbol{b}, \quad \boldsymbol{J} = \boldsymbol{J}_0 + \boldsymbol{j} \quad \rho = \rho_0 + \rho_1. \tag{3.25}$$

For example, $\partial \boldsymbol{B}_0/\partial t = 0$. If the plasma is initially at rest, ($\boldsymbol{v}_0 = 0$), then there are no Doppler effects and the operator $\mathrm{d}/\mathrm{d}t = \partial/\partial t$.

Keeping only linear terms, the combined momentum, Ampère and Faraday relations, become

$$\mu\rho \frac{\mathrm{d}\boldsymbol{v}}{\mathrm{d}t} = (\nabla \times \boldsymbol{b}) \times \boldsymbol{B}_0 + (\nabla \times \boldsymbol{B}_0) \times \boldsymbol{b}, \tag{3.26}$$

$$\nabla \times \boldsymbol{b} = \mu \boldsymbol{j}, \tag{3.27}$$

$$\frac{\partial \boldsymbol{b}}{\partial t} = \nabla \times (\boldsymbol{v}_1 \times \boldsymbol{B}_0). \tag{3.28}$$

The plasma displacement vector ξ is related to the magnetic field by integrating Faraday's law:

$$\boldsymbol{b} = \nabla \times (\xi \times \boldsymbol{B}_0). \tag{3.29}$$

3.3.2
The Wave Equation

Taking the curl of both sides of Equation 3.29 gives

$$\nabla \times \boldsymbol{b} = \nabla \times \nabla \times (\xi \times \boldsymbol{B}_0). \tag{3.30}$$

Putting $\nabla \times \boldsymbol{B}_0 = 0$ and substituting Equation 3.30 into Equation 3.26 yield

$$\frac{\partial^2 \xi}{\partial t^2} = \frac{1}{\mu \rho} \boldsymbol{B}_0 \times [\nabla \times \nabla \times (\boldsymbol{B}_0 \times \xi)]. \tag{3.31}$$

From Equation 3.22, the electric field is

$$\boldsymbol{E} = \boldsymbol{B}_0 \times \frac{\partial \xi}{\partial t}. \tag{3.32}$$

Using the Alfvén speed (Equation 1.5) and crossing both sides of Equation 3.31 by $\boldsymbol{B}_0$ give

$$\frac{\partial^2 \boldsymbol{E}}{\partial t^2} = V_A \times [V_A \times [\nabla \times (\nabla \times \boldsymbol{E})]]. \tag{3.33}$$

This is the linear MHD wave equation used by a number of authors (Cummings, O'Sullivan, and Coleman, 1969; Radoski, 1974; Westphal and Jacobs, 1962). In order to proceed, the choice of coordinate system and the spatial variation of the Alfvén speed must be specified. The next approximation for the magnetosphere, after Cartesian coordinates and a uniform $\boldsymbol{B}_0$, is dipole coordinates and the associated dipole field described by Equations 2.2–2.4.

3.4
The Shear Alfvén Mode in a Dipole Magnetic Field

3.4.1
Toroidal Oscillation of Field Lines

An equation for separate plasma oscillation modes in any magnetic field geometry was given by Singer, Southwood, and Kivelson (1981). Consider two field lines separated by a perpendicular distance δ_α. At another location along the field, the separation δ_α scales by $h_\alpha \delta_\alpha$. Plasma displacement $\boldsymbol{\xi}$ in the $\boldsymbol{\alpha}$-direction causes a magnetic perturbation $\boldsymbol{b}$ given by Equation 3.29. This expression is combined with the momentum equation and Ampère's law to give the wave equation for transverse oscillations:

$$\frac{\partial^2}{\partial s^2}\left(\frac{\xi}{h_\alpha}\right)+\frac{\partial}{\partial s}\left(\frac{\xi}{h_\alpha}\right)\frac{\partial}{\partial s}\left[\ln\left(h_\alpha^2 B_0\right)\right]+\frac{\omega^2}{V_A^2}\left(\frac{\xi}{h_\alpha}\right)=0. \tag{3.34}$$

A derivation of Equation 3.34 is given in Appendix 2.

For a dipole magnetic field and plasma oscillations in the azimuthal direction, $h_\alpha = r\cos\lambda$ and Equation 3.34 may be written without the rather complicated first derivative term.

The magnitude of Earth's dipole magnetic field is given by

$$B=\frac{M\sqrt{4-3\cos^2\lambda}}{L^3 R_E^3\cos^6\lambda}, \tag{3.35}$$

where $M \approx 7.8\times10^{22}\,\mathrm{A\,m^2}$, the dipole moment of Earth, $R_E = 6378\,\mathrm{km}$, λ is the latitude, and

$$r = LR_E\cos^2\lambda \tag{3.36}$$

for the radial distance r, measured from Earth's center to a point on the dipole field line. The distance from Earth's center to a field line in the equatorial plane is LR_E. The elemental distance along a field line, $\mathrm{d}s$ is

$$\mathrm{d}s = LR_E\cos\lambda\sqrt{4-3\cos^2\lambda}\,\mathrm{d}\lambda. \tag{3.37}$$

Equation 3.34 may be simplified by writing it in terms of λ rather than s, and Equation 3.37 provides the link between these two variables. Set $y=\xi/h_\alpha$ and since we are solving in one dimension,

$$\frac{\mathrm{d}^2 y}{\mathrm{d}s^2}+\frac{\mathrm{d}y}{\mathrm{d}s}\frac{\mathrm{d}}{\mathrm{d}s}\left[\ln\left(h_\alpha^2 B_0\right)\right]+\frac{\omega^2}{V_A^2}y=0. \tag{3.38}$$

Using the chain rules for the first- and second-order derivatives gives

$$\frac{\mathrm{d}^2 y}{\mathrm{d}\lambda^2}\left(\frac{\mathrm{d}y}{\mathrm{d}s}\right)^2+\frac{\mathrm{d}y}{\mathrm{d}\lambda}\left[\left(\frac{\mathrm{d}}{\mathrm{d}\lambda}\frac{\mathrm{d}\lambda}{\mathrm{d}s}\right)\frac{\mathrm{d}\lambda}{\mathrm{d}s}+\left(\frac{\mathrm{d}\lambda}{\mathrm{d}s}\right)^2\frac{\mathrm{d}}{\mathrm{d}\lambda}\ln\left(h_\alpha^2 B_0\right)\right]+\frac{\omega^2}{V_A^2}y=0 \tag{3.39}$$

and Equation 3.37 defines $\mathrm{d}\lambda/\mathrm{d}s$.

The other terms in Equation 3.39 are

$$\frac{\mathrm{d}}{\mathrm{d}\lambda}\ln\left(h_\alpha^2 B_0\right)=\frac{3\sin\lambda\cos\lambda}{4-3\cos^2\lambda} \tag{3.40}$$

and

$$\frac{\mathrm{d}}{\mathrm{d}\lambda}\frac{\mathrm{d}\lambda}{\mathrm{d}s}=\frac{\sin\lambda}{LR_E}\left[\frac{1}{\cos^2\lambda\sqrt{4-3\cos^2\lambda}}-\frac{3}{\left(4-3\cos^2\lambda\right)\sqrt{4-3\cos^2\lambda}}\right]. \tag{3.41}$$

Substituting Equations 3.40, and 3.41 into Equation 3.39 gives the toroidal wave equation for azimuthal oscillations:

$$\frac{d^2 y}{d\lambda^2} + \frac{dy}{d\lambda}\tan\lambda + \frac{\mu_0 \rho \omega^2 L^8 R_E^8 \cos^{14}\lambda}{K^2} y = 0. \tag{3.42}$$

The shooting method or Runge–Kutta integration scheme is often used to numerically solve the toroidal wave equation. Equation 3.42 is also in a form suitable for solution using matrix methods, providing both the eigenvalues and eigenvectors from a relatively simple computer code.

Approximating the derivatives by central finite differences, on an equally spaced grid in λ, separated by d, Equation 3.42 becomes (Price *et al.*, 1999)

$$y_{i-1}\left(\frac{1}{d^2} - \frac{\tan\lambda_i}{2d}\right) + y_i\left(\omega^2 q_i - \frac{2}{d^2}\right) + y_{i+1}\left(\frac{1}{d^2} + \frac{\tan\lambda_i}{2d}\right) = 0, \tag{3.43}$$

with $y_0 = y_{n+1} = 0$ and

$$q_i = \frac{\mu_0 \rho(\lambda_i) L^8 R_E^8 \cos^{14}\lambda_i}{K^2}. \tag{3.44}$$

For a nontrivial solution, the coefficient matrix from Equation 3.43 must have a zero determinant. An IDL code that solves Equation 3.42 using the formulation of Equation 3.43 is available (see Appendix 1). The parameters in the code have been set to solve for an $L = 6.6$ dipole field line and the plasma mass density model of Cummings, O'Sullivan, and Coleman (1969). Harmonics of the resonant modes are a natural output of this process, with the eigenvalues giving the resonant frequencies as detailed in the computer code. The eigenvectors may also be plotted to show the plasma displacement along the resonant field line. The electric and magnetic field perturbations may be computed from the plasma displacement using Equations 3.29 and 3.32. This is illustrated in Section 3.5.

3.5 MHD Wave Mode Coupling in One Dimension

The wave equation (3.33) describes both the shear and fast Alfvén wave modes. Investigating the combined effects of mode coupling is straightforward in a one-dimensional Cartesian model. For solutions of the form $\exp[i(\boldsymbol{k}\cdot\boldsymbol{r} - \omega t)]$, Equation 3.28 becomes

$$-i\omega \boldsymbol{b} = \nabla \times (\boldsymbol{v}_1 \times \boldsymbol{B}_0). \tag{3.45}$$

The geometry and properties of the background (static) magnetic field $\boldsymbol{B}_0$ must be specified. The simplest geometry is a Cartesian box where a radial line in the equatorial plane on the dayside magnetosphere is mapped to Cartesian coordinates (e.g. Radoski, 1971; Waters et al., 2000). Assigning the ambient magnetic field to be in the **z**-direction, the **x**-coordinate represents the radial direction and the y-coordinate maps to the azimuthal direction. Performing the cross-product operations on the right-hand side of Equation 3.26 (for $\nabla \times \mathbf{B}_0 = 0$) gives

$$-\mu\rho_0\omega^2\xi_x = \left(\frac{\partial b_x}{\partial z} - \frac{\partial b_z}{\partial x}\right)B_z, \tag{3.46}$$

$$-\mu\rho_0\omega^2\xi_y = \left(\frac{\partial b_y}{\partial z} - \frac{\partial b_z}{\partial y}\right)B_z, \tag{3.47}$$

$$-\mu\rho_0\omega^2\xi_z = 0. \tag{3.48}$$

From Equation 3.29, the component perturbation magnetic fields are

$$b_x = \frac{\partial\xi_x}{\partial z}B_z, \tag{3.49}$$

$$b_y = \frac{\partial\xi_y}{\partial z}B_z, \tag{3.50}$$

$$b_z = -\left(\frac{\partial\xi_x}{\partial x} + \frac{\partial\xi_y}{\partial y}\right)B_z. \tag{3.51}$$

Combining Equations 3.46–3.48 with Equations 3.49–3.51 gives

$$\left(\frac{\omega^2}{V_A^2} - k_z^2\right)\xi_x = \frac{1}{B_z}\frac{\partial b_z}{\partial x}, \tag{3.52}$$

$$\left(\frac{\omega^2}{V_A^2} - k_z^2\right)\xi_y = \frac{ik_y b_z}{B_z}, \tag{3.53}$$

$$b_z = -ik_y\xi_y B_z - B_z\frac{\partial\xi_x}{\partial x}. \tag{3.54}$$

Define the symbol ε as

$$\varepsilon = \frac{\omega^2}{V_A^2} - k_z^2. \tag{3.55}$$

Eliminating the azimuthal plasma displacement ξ_y using Equations 3.53 and 3.54 and then eliminating b_z from Equation 3.52 gives

$$\varepsilon B_z\xi_x = \frac{\partial}{\partial x}\left(\frac{\varepsilon B_z}{k_y^2 - \varepsilon}\frac{\partial\xi_x}{\partial x}\right), \tag{3.56}$$

which is the "box model" wave equation used by Radoski (1974) and Kivelson and Southwood (1986).

The $(k_y^2 - \varepsilon)$ in the denominator of Equation 3.56 has been discussed in the literature. The singularity that occurs when $k_y^2 = \varepsilon$ represents a resonance of the shear Alfvén mode, and an imaginary part to k_z is often used to avoid mathematical unpleasantness there. Physically, this imaginary component is attributed to Joule heating losses of the resonant shear mode in the ionospheres. Furthermore, if there is no azimuthal variation ($k_y = 0$), then any fast mode oscillation, characterized by the b_z component, does not transfer energy into the shear mode. In the real system, wave mode coupling also arises from gradients in $\boldsymbol{B}_0$ and the anisotropic ionosphere conductance.

The solution to this second-order differential equation (3.56) requires a specification of the wave medium, through the Alfvén speed V_A. The solution yields the perturbation displacement ξ_x, given the wave spatial variations k_y and k_z. The perturbation magnetic fields are then calculated from Equations 3.49–3.51 and simplify to

$$b_x = ik_z B_z \xi_x, \tag{3.57}$$

$$b_y = \frac{-k_y k_z B_z}{k_y^2 - \varepsilon} \frac{\partial \xi_x}{\partial x}, \tag{3.58}$$

$$b_z = \frac{\varepsilon B_z}{k_y^2 - \varepsilon} \frac{\partial \xi_x}{\partial x}. \tag{3.59}$$

The k_y is associated with the azimuthal wave number (Olson and Rostoker, 1978). The field-aligned wave structure describes the shear Alfvén resonance and depends on the spatial variation of V_A.

Information on how the Alfvén speed varies in near-Earth space is mostly limited to spacecraft measurements of the plasma mass density in the equatorial plane. Chappell, Harris, and Sharp (1970a, 1970b, 1971) showed a number of examples obtained from the mass spectrometer on board the Orbiting Geophysical Observatory (OGO) 5 spacecraft. Although these studies were primarily interested in understanding characteristics of the plasmasphere, these results have contributed to the common use of a power law model where the plasma mass density ρ as a function of radial distance r is given by

$$\rho = \rho_0 \left(\frac{r_0}{r} \right)^{\alpha}, \tag{3.60}$$

where ρ_0 is the plasma mass density at the radial location r_0. All distances are calculated by taking the center of Earth as the origin. Values of α from 1 to 4 have been used for the plasmatrough, while some researchers have set $\alpha = 3$ for a diffusive equilibrium plasma for the plasmasphere (Menk *et al.*, 1999; Orr and Matthew, 1971; Poulter *et al.*, 1984; Warner and Orr, 1979; see also review by Reinisch *et al.*, 2009). The mass density model of Equation 3.60 in addition to the dipole magnetic field expression (Equation 3.35) may be used to provide values for V_A in the equatorial plane. This process is described in more detail in Chapters 6 and 7.

The methods of magnetoseismology allow estimates of V_A to be obtained from measurements of the field line resonant, shear Alfvén wave mode. A typical Alfvén speed radial profile, derived from field line resonant frequencies, is shown in Figure 3.2. The travel time along the background field from one ionosphere to the other is approximated by

$$T = \int \frac{ds}{V_A}, \tag{3.61}$$

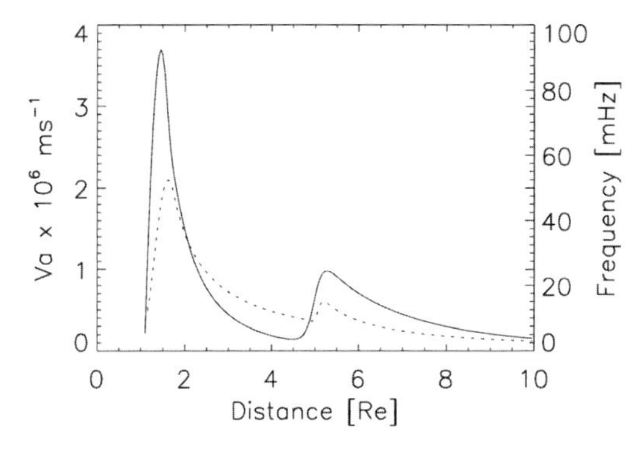

Figure 3.2 Typical radial dependence of the Alfvén speed (solid curve) and FLR frequency (dotted curve). The plasmapause is near $L = 5.0$. From Waters *et al.* (2000).

which gives estimates for the resonance frequencies ω_r. This is the Alfvén travel time expression for the fundamental period with an accuracy linked to the assumptions of the Wentzel–Kramers–Brillouin–Jeffreys (WKBJ) approximation, as discussed in Schulz (1996) and Sinha and Rajaram (1997).

The field-aligned wave number for harmonic n is

$$k_z = n\frac{\omega_r}{V_A}. \tag{3.62}$$

Including k_y in Equation 3.55 gives the condition for the location x_t of the fast mode cutoff (turning point):

$$\omega^2 - V_A^2(x_t)[k_y^2(x_t) + k_z^2(x_t)] = 0. \tag{3.63}$$

The frequency–spatial relationship between the ω in Equation 3.63 and the resonant frequencies in Figure 3.2 is shown in Figure 3.3. The relative positions of the resonance points, x_r and the turning points, x_t show whether the shear Alfvén resonance is coupled to a propagating or evanescent fast mode (in the equatorial

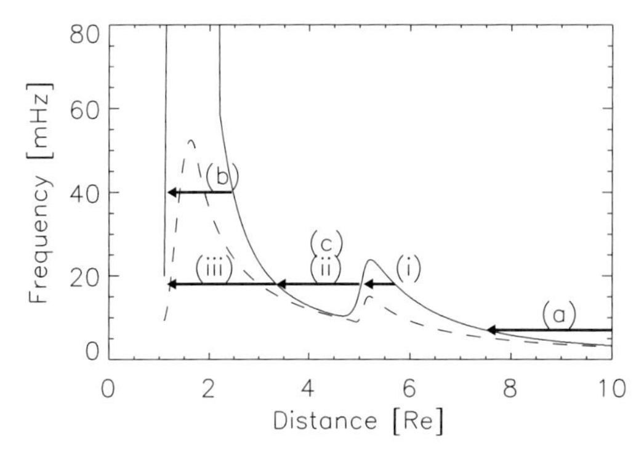

Figure 3.3 The fast mode turning points (Equation 3.63) (solid) and the FLR frequencies (dashed curve): (a) plasmatrough mode, (b) plasmatrough-plasmasphere mode, and (c) tunneling and inner magnetosphere trapped modes. From Waters *et al.* (2000).

plane). If the magnetopause is a strong wave reflector, then a cavity mode appears in the model between the magnetopause and a turning point in the plasmatrough, labeled (a). Another trapped cavity mode can exist between two turning points deep within the plasmasphere, labeled (b). A tunneling mode, labeled (c, i), can appear as a propagating mode in the plasmasphere, labeled (c, ii), and then appear as a plasmasphere trapped mode, labeled (c, iii).

Details of the wave structure may be obtained by solving Equation 3.56, which requires two boundary conditions. Set the inner boundary at $1.1 R_E$ to a strong wave reflecting barrier so that $\xi_x(\omega, x=1.1)=0$. The outer boundary at $x=10 R_E$ may be driven by a wave spectrum such as

$$\xi_x(\omega, x = 10R_E) = -\frac{1}{\omega}. \tag{3.64}$$

An example of the structure of a trapped plasmasphere mode is shown in Figure 3.4. A plasmasphere cavity resonance at 36.8 mHz couples with the FLR

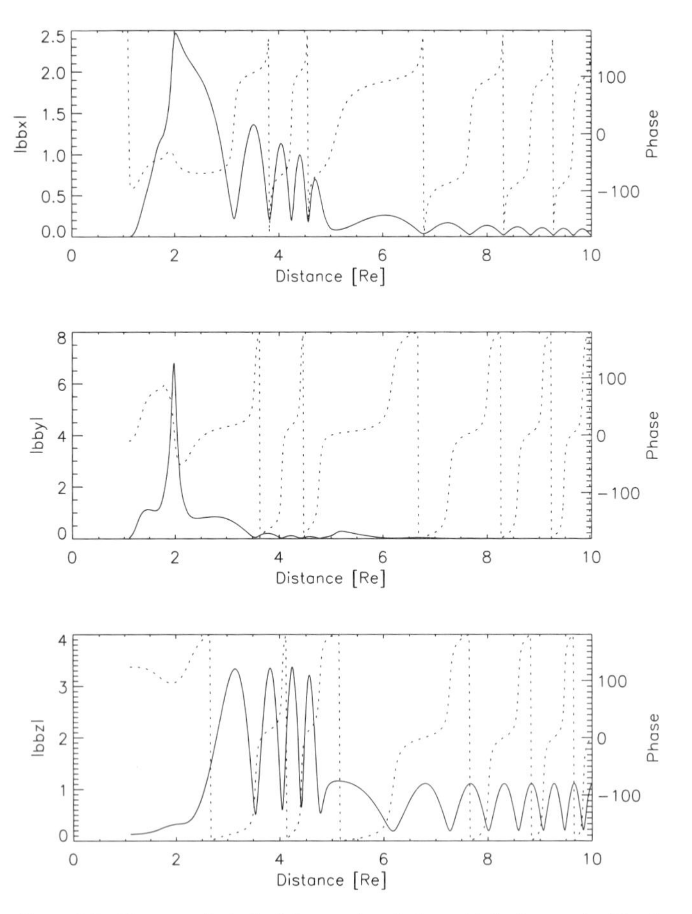

Figure 3.4 The magnitude (solid curves) and phase (dotted curves) of the three magnetic field components for the trapped plasmasphere mode.

shown by the peak in amplitude and characteristic change in phase in the b_y component at $2R_E$.

3.6 An Alternative Derivation of the Plasma Wave Equation, from Electromagnetism

The basic equations used in the previous sections keep the time derivative in the momentum equation while ignoring the displacement current in Ampère's law. A simpler formulation that only involves the familiar equations of electromagnetism is possible. This makes the resulting equations more suitable for numerical solution by computer for two- and three-dimensional studies in complicated geometries using the popular finite-difference time-dependent algorithms. The key to deleting the momentum equation from the analysis is in finding a suitable expression for the permittivity. A detailed explanation of this process was given by Lysak (1997).

The essential assumption is that the frequencies of interest are much smaller than the ion gyrofrequency Ω_i. The perpendicular components of the momentum equation are processed by the operator $\partial/\partial t + \nu$, where ν is the collision frequency (with neutrals) to give

$$\left(\frac{\partial}{\partial t}+\nu\right)^2 \boldsymbol{v}_\perp = \left(\frac{\partial}{\partial t}+\nu\right)\boldsymbol{v}_\perp \times \Omega_i + \frac{q}{m}\left(\frac{\partial}{\partial t}+\nu\right)\boldsymbol{E}_\perp. \tag{3.65}$$

The cross-product with the gyrofrequency can be eliminated and for wave frequencies $\omega^2 < \Omega^2$, the time derivatives with respect to the plasma velocity $\boldsymbol{v}$ may be neglected. Combining with Ampère's law and including the displacement current term gives

$$\varepsilon\frac{\partial \boldsymbol{E}_\perp}{\partial t} + \sigma_P \boldsymbol{E}_\perp - \sigma_H \boldsymbol{E}_\perp \times \hat{\boldsymbol{B}}_0 = (\nabla \times \boldsymbol{b})_\perp, \tag{3.66}$$

where σ_P and σ_H are the Pedersen and Hall conductivities and ε is an effective dielectric constant,

$$\varepsilon = 1 + \frac{c^2}{V_A^2}\sum_s \frac{m_s/M}{1+\nu_s^2/\Omega_s^2}, \tag{3.67}$$

with the sum over electrons and all ion species, c is the speed of light, and M is the average molecular mass of the ions. For collisionless MHD, $\varepsilon = 1 + c^2/V_A^2$ and the wave equation for low-frequency plasma waves follows the same procedure as for electromagnetic waves, using the Faraday and Ampère laws. The formulation of ULF wave propagation and coupling properties in the magnetosphere and ionosphere using this approach is taken up again in Chapters 6 and 8.

4
Sources of ULF Waves

4.1
Introduction

This book focuses on the use of naturally occurring plasma waves to remote sense properties and dynamics of the magnetosphere. This requires an understanding of the energy sources, generation mechanisms, and propagation of these plasma waves throughout the magnetosphere. The waves are mostly in the millihertz frequency range and the physics of their formation and propagation have already been discussed in Chapter 3. This chapter reviews observational results in the context of this theoretical background.

Chapter 1 introduced the widely used Pc1–Pc5, Pi1, Pi2 classification system for geomagnetic pulsations. It is now known that these perturbations measured by ground-based and satellite platforms are the signatures of ultralow-frequency (ULF) plasma waves propagating through the magnetosphere. Note that the descriptor "ULF" is the standard terminology in this field of science, but does not fit with the internationally agreed nomenclature for bands of the electromagnetic spectrum (which is ordered in units of 3×10^n and does not extend below 3 kHz), as defined by the International Telecommunications Union.

These ULF plasma waves are broadly of two types, depending on whether their energy source originates in the solar wind (exogenic) or from processes within the magnetosphere. Over the years ample evidence has accumulated that properties of Pc3–Pc5 waves depend on both the solar wind and the interplanetary magnetic field (IMF). For example, correlations have been demonstrated between wave power at the ground and the solar wind speed V_{SW}, the IMF magnitude B_{IMF}, and the cone angle θ_{Bx} defined in Equation 1.2 (Engebretson *et al.*, 1987; Greenstadt and Russell, 1994; Kessel *et al.*, 2004; Odera, 1986; Saito, 1969; Troitskaya, Plyasova-Bakunina, and Gul'yel'mi, 1971; Wolfe, Lanzerotti, and Maclennan, 1980) and with the density of the upstream solar wind (Heilig *et al.*, 2010). Figure 4.1 shows an early report of a relationship between Pc3 and Pc4 activity and solar wind speed. In contrast, Figure 4.2 represents the dependence of pulsation period (ordered in bands where P1 = 1–5 s period and P12 = 5–10 min period) over the year 1972 at a midlatitude observatory, on the IMF magnitude B_{IMF} (Verö and Holló, 1978). The curve corresponds to T (s) = $160/B$ (nT), that is, f (mHz) $\approx 6.3\ B_{IMF}$ (nT).

Magnetoseismology: Ground-based remote sensing of Earth's magnetosphere, First Edition. Frederick W. Menk and Colin L. Waters.

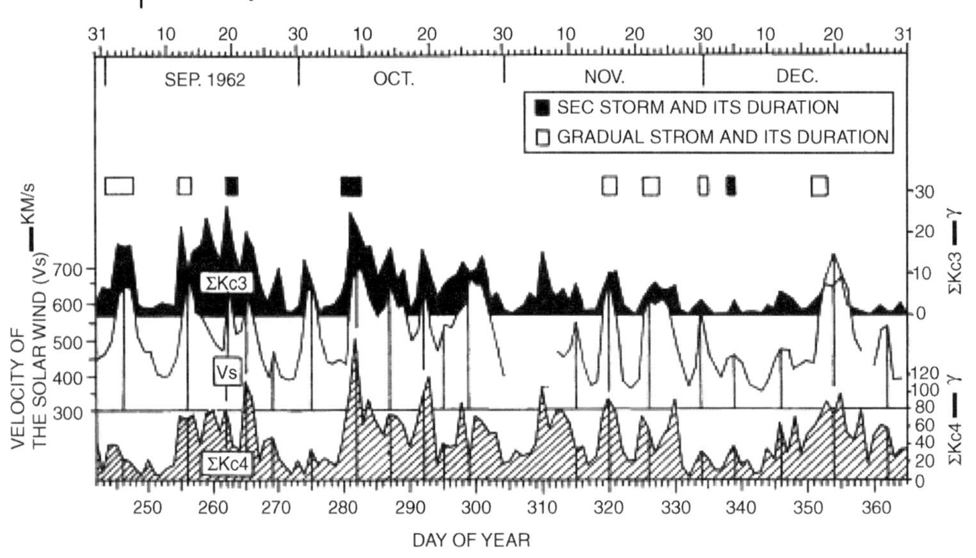

Figure 4.1 Correlation between Pc4 pulsation activity at a high-latitude station (bottom trace, shaded), Pc3 activity at a low-latitude site (top, solid), and solar wind velocity measured simultaneously by the Mariner II spacecraft (middle, unshaded). Open and filled boxes represent magnetic storm intervals. From Saito (1964).

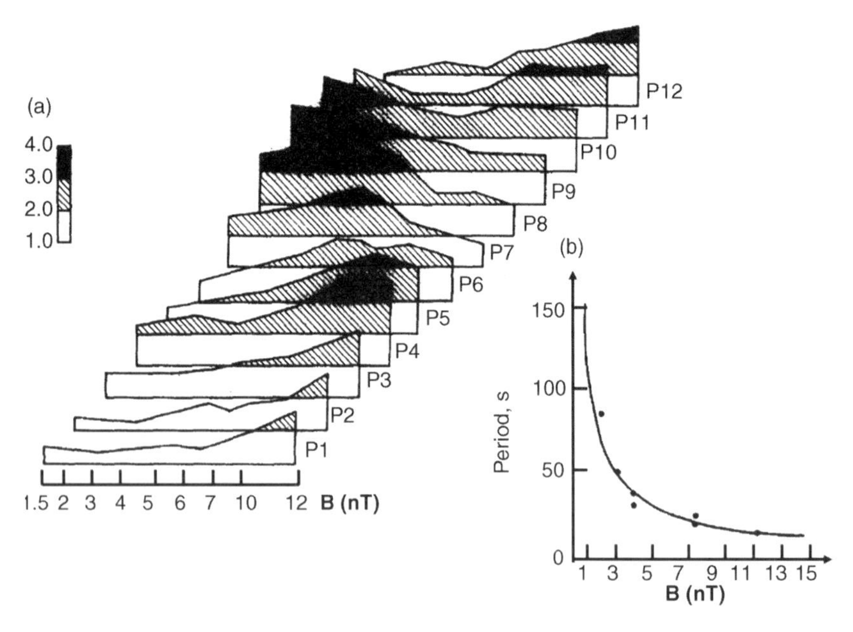

Figure 4.2 Variation of magnetic pulsation period over 1 year at a midlatitude ground station against magnitude of the upstream IMF, B_{IMF}. (a). Pulsation period ordered in bands (P1–P12) of increasing period, where shading represents level of activity. (b). Summary of results, where the continuous curve represents T (s) = 160/B (nT). From Verö and Holló (1978).

Geomagnetic substorms and other instabilities in the tail also form an important source of ULF waves on the nightside of Earth, but are transient phenomena and are not considered in this book. Magnetospheric remote sensing relies mostly on plasma waves on the dayside, since this is where the predominant source regions are located, and field line resonances (FLRs) are difficult to establish in the night hemisphere where the ionospheric conductance is low.

Wave sources internal to the magnetosphere involve wave–particle interactions. Some types of Pc4–Pc5 waves experience bounce and drift–bounce resonance interactions with warm or hot particle populations, possibly including hot ring current O^+ ions (Ozeke and Mann, 2008). Electromagnetic ion cyclotron waves (EMICWs) in the Pc1 and Pc2 frequency ranges are generated through an ion cyclotron resonance with ring current protons (Fraser and McPherron, 1982). In this case, the presence of "slots" in the Pc1 spectrum indicates the presence of thermal He^+ and warm O^+ ions in the outer magnetosphere (Fraser *et al.*, 1992).

4.2
Exogenic Sources

There are two main types of exogenic sources of magnetospheric ULF plasma waves: (i) wave–particle instabilities in the upstream solar wind, and (ii) periodic fluctuations in solar wind pressure or density. Both of these sets of processes result in compressional mode waves entering the magnetosphere.

A rich variety of plasma waves occurs in the magnetosheath, in particular Alfvén/ion cyclotron and mirror modes under low and high plasma β-conditions, respectively (Narita *et al.*, 2006; Schwartz, Burgess, and Moses, 1996). The region upstream of the bow shock is also home to a rich variety of waves, including relatively large amplitude waves with frequency around 30 mHz, upstream propagating whistler waves with frequency around 1 Hz, and 3 s period waves (Fairfield, 1969; Greenstadt, Le, and Strangeway, 1995; Le and Russell, 1992). Here, we focus on the first type because it has been shown that they drive field line oscillations within the magnetosphere.

The upstream wave region is called the foreshock, and of particular interest is the ion foreshock. This region is connected to the bow shock by the IMF field lines and is characterized by beams of field-aligned backstreaming ions under conditions when the IMF is uniform and steady and at a moderate angle to the solar wind flow. Solar wind ions gyrating around field lines and convecting antisunward undergo specular reflection at the bow shock. When the IMF field lines are nearly parallel to the tangent to the bow shock, called a quasi-parallel shock, the ions are reflected into the upstream foreshock region.

The geometry of the situation is illustrated in Figure 4.3. The angle between the IMF and the solar wind velocity vector is the cone angle, θ_{Bx}, defined in Equation 1.2, while the angle between the IMF and the normal to the shock boundary is θ_{Bn}. The foreshock boundary is defined by the tangential line indicated and is determined by the guiding center trajectory of the backstreaming ions. This is downstream from

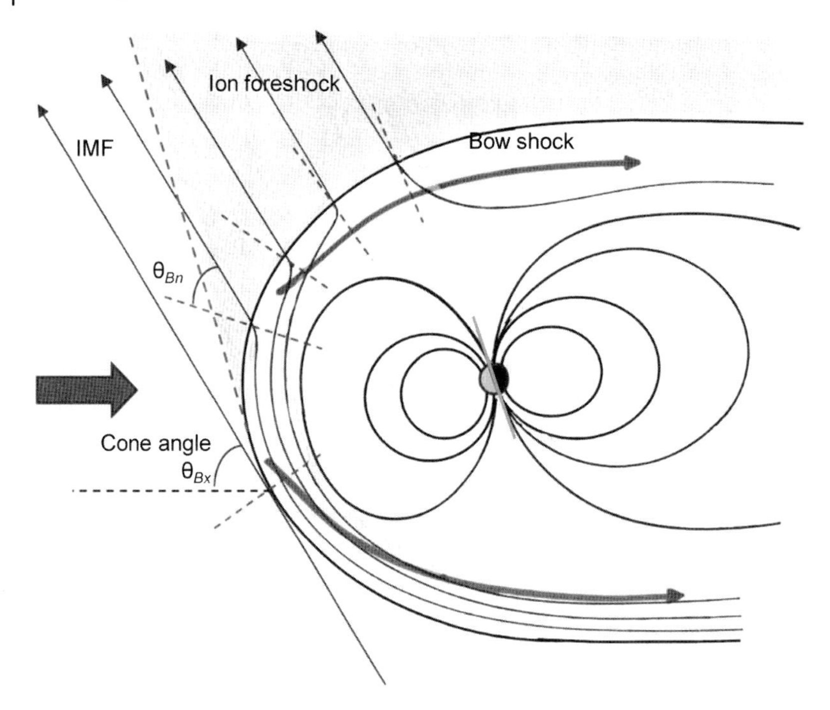

Figure 4.3 Schematic representation of the magnetopause, bow shock, and the ion foreshock region (shaded) where ULF waves are likely generated. The IMF is shown northward, and thick arrows represent plasma streamlines. Field lines (solid) map around the magnetopause and the plasma convects antisunward. (For a color version of this figure, please see the color plate at the beginning of this book.)

the tangent field line due to solar wind convection effects. The particles are confined to a plane containing the solar wind flow velocity and the IMF, the ***V*–*B*** plane, as shown in the figure. Of course, the bow shock is three dimensional and the IMF direction is variable, and although for specular reflection the guiding center motion is in the ***V*–*B*** plane, the ions will be reflected out of this plane and have large pitch angle (Le and Russell, 1992).

ULF waves with frequency in the range 10–100 mHz but typically near 30 mHz are common in the foreshock under quasi-parallel ($\theta_{Bn} \lesssim 35^\circ$) and oblique ($25^\circ \lesssim \theta_{Bn} \lesssim 50^\circ$) conditions (Russell *et al.*, 1983), at times of medium Mach numbers ($M_A \approx 2.3$–3) (Krauss-Varban, 1994). These waves are due to Alfvén/ion cyclotron waves excited by the resonance instability between the sunward streaming ion beams and the antisunward streaming solar wind plasma (Narita *et al.*, 2004). These are transverse waves with a nearly monochromatic spectrum.

Global two-dimensional hybrid (kinetic ions and fluid electrons) simulations and spacecraft observations for radial IMF conditions reveal the formation of a very perturbed foreshock region within which a slightly smaller ULF wave foreshock is embedded (Blanco-Cano, Omidi, and Russell, 2009; Eastwood *et al.*, 2005a, 2005b).

Under quasi-parallel conditions, weakly compressive sinusoidal waves are generated via the right-hand resonance instability with field-aligned ions, although the waves are left-hand polarized in the spacecraft frame. These waves can propagate at angles up to 30° to the magnetic field. Additional highly compressional fast magnetosonic linearly polarized waves occur close to the bow shock. The extent of the foreshock and the wave characteristics change with changing cone angle.

Under more quasi-perpendicular conditions, relatively narrow and sometimes energetic field-aligned ion beams occur in the foreshock, often in association with ~1 Hz whistler-like waves (Meziane *et al.*, 2011).

Spatial properties of ULF waves in the foreshock have been determined using multispacecraft observations. The wave dimensions are of order 1–3 R_E along the wave vector, and 8–18 R_E perpendicular to this, with a wave front shape similar to an oblate spheroid (Archer *et al.*, 2005). The waves can convect downstream to the subsolar region of the magnetopause and into the magnetosphere without significant change in their spectrum (Greenstadt *et al.*, 1983). The wave frequency depends on the strength and cone angle of the IMF, but is typically in the Pc3 range. The relationship has been variously expressed as follows (Heilig, Lühr, and Rother, 2007; Le and Russell, 1996; Ponomarenko *et al.*, 2002; Takahashi, McPherron, and Terasawa, 1984):

$$f(\text{mHz}) = 6B_{\text{IMF}}(\text{nT}), \tag{4.1}$$

$$f(\text{mHz}) = (0.72 + 4.67\cos\theta_{B_x})B_{\text{IMF}}(\text{nT}), \tag{4.2}$$

$$f(\text{mHz}) = (4.42 \pm 0.25)|B_{\text{IMF}}|(\text{nT}), \tag{4.3}$$

$$f(\text{mHz}) = 7.6|B_{\text{IMF}}|(\text{nT})\cos^2\theta_{B_x}, \tag{4.4}$$

$$f(\text{mHz}) = (0.708M_{\text{A}} + 0.64)(\text{mHz nT}^{-1})B_{\text{IMF}}(\text{nT}), \tag{4.5}$$

where $M_A = V_{SW}/V_A$ is the Alfvénic Mach number.

Multisatellite observations have identified upstream waves entering and propagating through the magnetosphere as compressional waves (Clausen *et al.*, 2008, 2009; Constantinescu *et al.*, 2007; Goldstein *et al.*, 2005; Sakurai *et al.*, 1999). Interestingly, the wave front curvature and propagation properties for Pc3 waves recorded during an outbound magnetosheath crossing by the four Cluster satellites show that these waves mostly originate from the cusp and electron foreshock, but not especially from the ion foreshock (Constantinescu *et al.*, 2007). This suggests that small Alfvén/ion cyclotron and mirror mode waves are initially stimulated in the electron foreshock and then couple to and are amplified by ion beam instabilities in the slightly downstream ion foreshock region.

Figure 4.4 summarizes results from a detailed statistical survey of Pc3 and Pc4 compressional mode wave power at ~350 km altitude using the CHAMP spacecraft (Heilig, Lühr, and Rother, 2007). Panels (a) and (b) show that compressional mode wave power increases with increasing solar wind speed, panels (c) and (d) demonstrate that wave power peaks at low IMF cone angle, and panels (e) and (f) indicate that normalized Pc3 wave frequency increases with increasing Alfvénic Mach

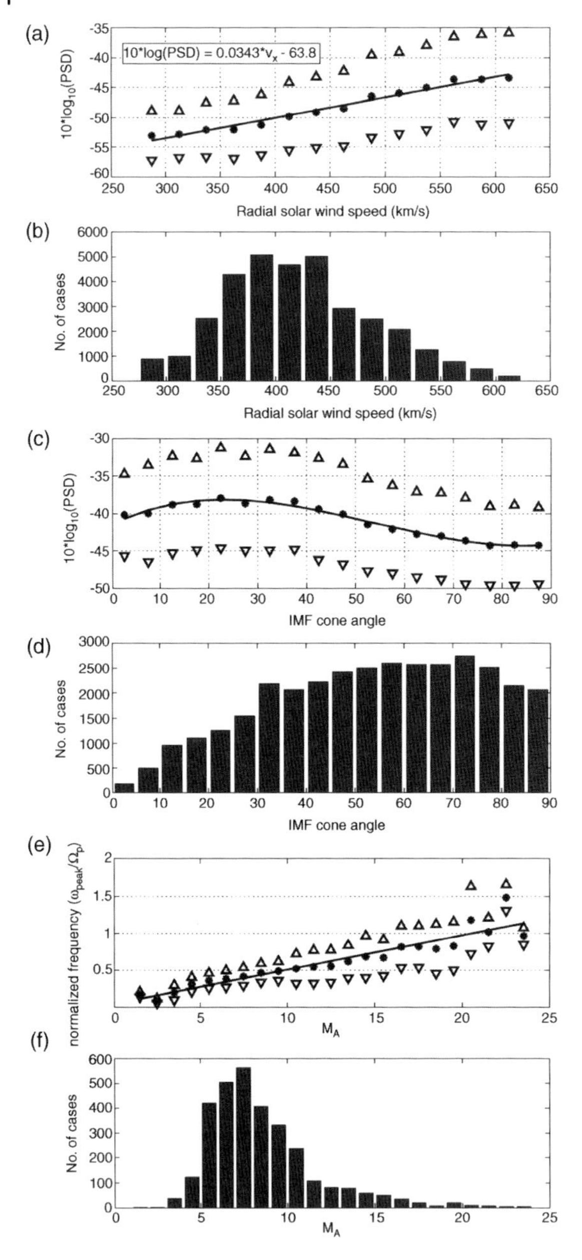

Figure 4.4 Dependence of power and frequency of upstream compressional waves recorded by the low-altitude CHAMP spacecraft during local daytime and between $\pm 60^\circ$ magnetic latitude. (a) Log wave power versus solar wind speed, where triangles represent one standard deviation from the linear fit line (correlation coefficient $C = 0.99$). (b) Solar wind speed distribution. (c and d) Log wave power versus IMF cone angle, and distribution of cone angles. (e and f) Normalized Pc3 wave frequency versus Alfvénic Mach number M_A ($C = 0.94$). Adapted from Heilig, Lühr, and Rother (2007).

number M_A (correlation coefficient = 0.94). This dependence arises because the upstream wave frequency is Doppler shifted during convection downstream. While the wave frequency was found to depend on IMF strength, as shown in Equation 4.5, cone angle did not seem to affect wave frequency.

In fact, the relationship between ULF waves in the magnetosphere and solar wind parameters in the foreshock is best described using multiple linear regression and artificial neural network analyses (Heilig *et al.*, 2010). These demonstrate that wave power depends on a number of factors, including the solar wind density in the foreshock and the Mach number, since when these quantities are low, the back-streaming ion density or the downstream ion convection rate is insufficient to excite the ion cyclotron resonance that generates the waves. An empirical model has been developed to predict the RMS amplitude of 22–100 mHz activity ($\mathrm{Pc3_{ind}}$) in the H magnetometer component at a midlatitude ground station:

$$\begin{aligned}\mathrm{Pc3_{ind}(pT)} &= 4.064 \times 10^{-5} V_{\mathrm{SW}}^{1.650} \times (\cos\chi + 2)^{1.946} \times P_{\mathrm{dyn}}^{0.540} \\ &\quad \times (\cos\theta_{B_x} + 2)^{2.675} - 16\,\mathrm{pT}, \qquad (4.6)\end{aligned}$$

where χ is the solar zenith angle and $P_{\mathrm{dyn}} = N_i m_p V_{\mathrm{sw}}^2$ is the dynamic solar wind pressure, in nPa.

Periodic compressional or Alfvénic fluctuations in the solar wind may also directly drive discrete frequency ULF waves in the magnetosphere (Kepko and Spence, 2003; Menk *et al.*, 2003; Potemra *et al.*, 1989; Stephenson and Walker, 2002). Multipoint studies provide strong evidence of correlated wave power in the Pc5 range in the upstream solar wind, near the magnetopause, at geostationary orbit, over the polar regions, and on the ground near the spacecraft footpoints, during high-speed streams and coronal mass ejections. Observational studies have shown that solar wind pressure variations are a major driver of Pc5 activity (including propagating compressional waves, field line resonances, and global modes) in the magnetosphere and on the ground (Kessel, 2008; Takahashi and Ukhorskiy, 2007, 2008), with correlation coefficients up to $\sim$0.7 between wave power on the dayside and dynamic solar wind pressure. It is likely that the compressional oscillations ("breathing mode" oscillations driven directly by pressure variations) then couple to toroidal mode field line resonances where the incoming wave frequency and field line eigenfrequency match. This may occur across a range of latitudes since the source waves may be impulsive and relatively broadband.

An interesting topic is the possible existence of magnetospheric Pc5 waves, standing global cavity modes, and field line resonances at stable discrete frequencies, called "magic" frequencies: 1.3, 1.9, 2.6, 3.4, and 4.2 mHz (Francia and Villante, 1997; Harrold and Samson, 1992; Mathie *et al.*, 1999a; Samson *et al.*, 1992; Ziesolleck and McDiarmid, 1994). Figure 4.5, for example, shows the distribution in Pc5 frequency recorded across a high-latitude array of ground stations for over 3 months. A tendency for selected frequencies is evident. Recent event-based and statistical studies suggest that these modes may be driven by periodic variations in solar wind dynamic pressure (Eriksson, Walker, and Stephenson, 2006; Fenrich and Waters, 2008; Villante *et al.*, 2007). While some workers examine the correlation

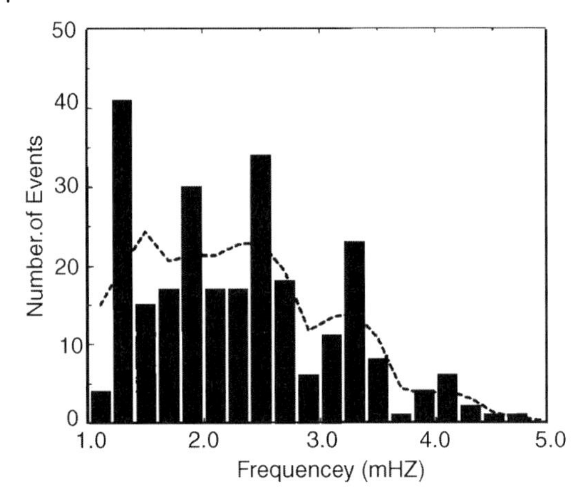

Figure 4.5 Distribution of Pc5 frequency observed over 3 months across a high-latitude ground station array, in 2 mHz bins. Dashed line shows occurrence averaged over 3 bins. From Mathie *et al.* (1999a).

between spectra from ground magnetometers or radars and solar wind pressure, others scrutinize the phase coherence between specific events, taking into account suitable propagation delays from the solar wind to the ground. There is even a suggestion that auroral luminosity variations are modulated at "magic" frequencies (Liou *et al.*, 2008). An alternative source mechanism for such discrete frequency signals is discussed in Section 4.5.

A striking result is summarized in Figure 4.6. This represents the occurrence of discrete spectral peaks in the 0.5–5.0 mHz range for over 11 years of measurements of number density in the upstream solar wind and for 10 years of magnetic field data from geostationary orbit near local noon (Viall, Kepko, and Spence, 2009). Horizontal bars indicate statistically significant occurrences of discrete frequencies in the solar wind at the 3σ level and in the magnetosphere at the 1σ level. Occurrence enhancements appear at discrete frequencies similar to the "magic" frequencies mentioned earlier in 87% of the solar wind records, and corresponding spectral peaks in the magnetosphere 54% of the time they occurred in the solar wind. These results suggest that specific frequency features in the solar wind number density often directly drive the magnetosphere to oscillate at repeatable discrete frequencies, but also a subset of repeatable frequencies are observed in the magnetosphere due to other physical processes.

In summary, multipoint observations show that Pc3 and Pc4 ULF waves generated in the foreshock region may be observed in the magnetosphere and on the ground, while there is mounting evidence that magnetospheric Pc5 waves may be directly driven by pressure oscillations in the solar wind. The first point means that ground observations of Pc3 and Pc4 pulsations may be used as a diagnostic monitor of conditions in the upstream solar wind, while the second point is important because Pc5 waves may be involved in the energization or loss of high-energy particles in the

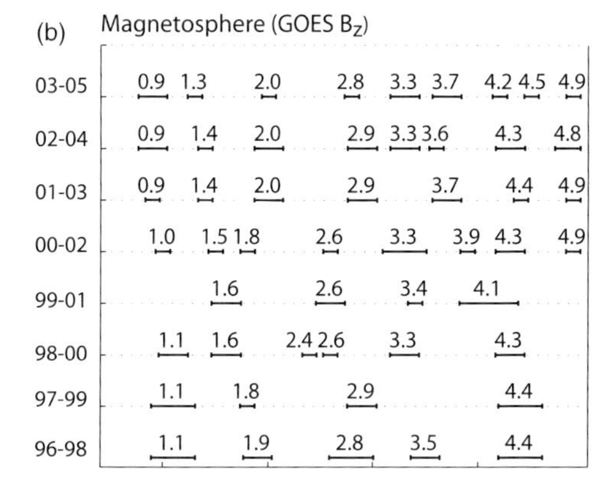

Figure 4.6 Summary of statistically significant occurrence of distribution enhancements (a) in the solar wind and (b) in the magnetosphere, over 3 year intervals. Width of each horizontal bar represents the approximate width of the frequency band enhancement. Corresponding peaks occur in both data sets for $f \approx 1.5$, 1.9, 2.3, 3.3, and 4.8 mHz. From Viall, Kepko, and Spence (2009).

radiation belts. However, it is not clear why these discrete frequencies should be present in the solar wind in the first place.

4.3 Boundary Instabilities

The observed correlation between solar wind speed and ULF power in the magnetosphere suggests that the Kelvin–Helmholtz instability (KHI) at the magnetopause may be a significant source of ULF wave energy (Pu and Kivelson, 1983; Southwood, 1968; Walker, 1981). This instability arises when the velocity shear between the relatively calm magnetospheric plasma and the streaming solar wind plasma exceeds a threshold value, exciting surface waves that propagate antisunward along the boundary and are evanescent within the magnetosphere.

Mathematical descriptions of the KHI often assume that the instability occurs on a tangential discontinuity, with the magnetic field and plasma flow direction parallel to the boundary. For a compressible plasma, fast and slow mode surface waves may be excited by the KHI, each with a lower and upper critical velocity. Such surface modes may couple to field line resonances within the magnetosphere (Southwood, 1974). The thickness of the magnetopause boundary layer may determine the frequency at which the linear growth rate of the instability is a maximum (Walker, 1981). The period of the pulsation is related to the scale thickness of the boundary as described in Equation 2.13. For example, if $V_0 = 200\,\text{km s}^{-1}$ and $d = 1\,R_E$, then $T = 320\,\text{s}$, in the Pc5 range.

Figure 4.7 illustrates the general situation. The shocked solar wind flowing at speed V_{SW} past the magnetopause flanks generates shear flow instabilities, which may excite fast mode waves that couple to field line resonances at different latitudes with frequency-independent or common phase speed. KHI and such "CPS" events are more likely under relatively high solar wind flow conditions (Engebretson *et al.*, 1998) and may include events excited by shocks and other large-scale solar wind discontinuities. In addition, incoherent solar wind buffeting and impulses may produce local magnetopause indentations and waves with independent phase speeds (IPS) that couple to poloidal mode field line oscillations (Liu *et al.*, 2009b; Mathie and Mann, 2000b).

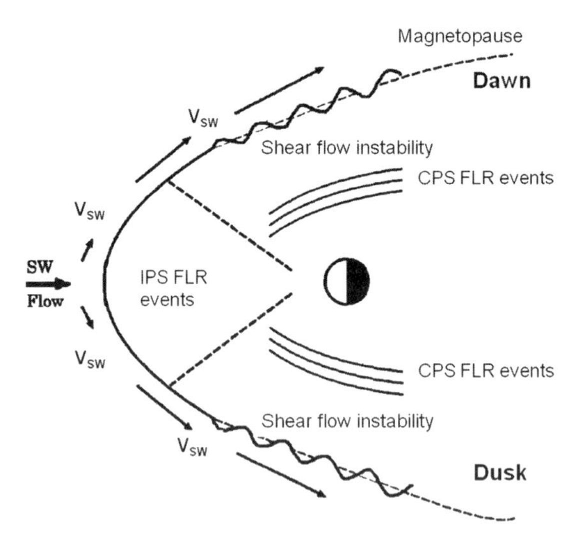

Figure 4.7 Illustration of the formation of shear flow instabilities under high solar wind conditions at the magnetopause flanks, which couple to FLRs with common azimuthal phase speeds (CPS) in the magnetosphere. Incoherent solar wind buffeting and impulses produce local magnetopause indentations and waves that couple to FLRs with independent phase speeds (IPS). Adapted from Mathie and Mann (2000b).

Numerical models now permit studies of the entire magnetosphere system under various conditions. These include global three-dimensional MHD simulations of the solar wind–magnetosphere interaction, which allow parameter studies of solar wind drivers (Claudepierre, Elkington, and Wiltberger, 2008). Coupled ULF surface modes are excited by the KHI near the dawn and dusk magnetopause, one mode propagating tailward along the magnetopause boundary and another along the inner edge of the boundary layer. In both cases, the preferred wave number is related to the boundary thickness, so that the KH waves are monochromatic.

Observational evidence of the KHI includes (i) a switch in propagation direction and polarization for FLRs at high latitudes, indicating propagation away from local noon with frequency-independent phase speed characteristic of flow speeds at the boundary (Dunlop *et al.*, 1994; Olson and Rostoker, 1978; Samson, Jacobs, and Rostoker, 1971) and (ii) a rapid decrease in amplitude with increasing distance into the magnetosphere, since the waves decay away from the boundary according to Equation 2.12.

Multispacecraft observations have provided new opportunities for *in situ* studies of wave distributions and properties. For example, electric and magnetic field data from the THEMIS satellite constellation confirm the importance of the KHI in exciting toroidal mode Pc5 waves in the outer magnetosphere, especially near the flanks, during solar minimum years (Liu *et al.*, 2009b). The excitation of the KHI and resultant MHD-scale vortices has been directly observed (Hasegawa *et al.*, 2009) and is facilitated by magnetic reconnection and compressional fluctuations of the dayside magnetosheath, although the KHI wavelength may exceed predictions based on linear theory.

It has also been suggested that the KHI can be excited across the boundary of flow channels in Earth's plasmasheet in the magnetotail. Multipoint spacecraft observations have detected ~5 min period KHI-produced magnetic oscillations in a flow shear, propagating earthward at about half the plasma flow speed. Such results suggest that the KHI may play a role in the braking of fast flows in the magnetotail.

The previous section highlighted recent results indicating that solar wind perturbations drive ULF waves at selected frequencies. Recent multisatellite observations of magnetopause motions find evidence of magnetopause oscillations at the "magic" frequencies described above (Plaschke *et al.*, 2009a, 2009b), as shown in Figure 4.8. These are believed to be due to Alfvén surface waves on the dayside magnetopause generated by mode coupling from compressional disturbances in the magnetosheath, that develop into standing Alfvén modes (i.e., Kruskal–Schwarzschild modes) (see Hasegawa and Chen (1974)) due to reflection from the conjugate ionospheres via field-aligned currents. The eigenfrequencies of the standing modes are determined by the magnetopause geometry.

In summary, simulation and observational studies show that the KHI at the magnetopause is probably an important source of long-period ULF wave activity. However, further work is needed to clarify to what extent this process is responsible for the range of Pc3–Pc5 pulsations that occur throughout the magnetosphere, and also on the significance of Kruskal–Schwarzshild mode surface oscillations.

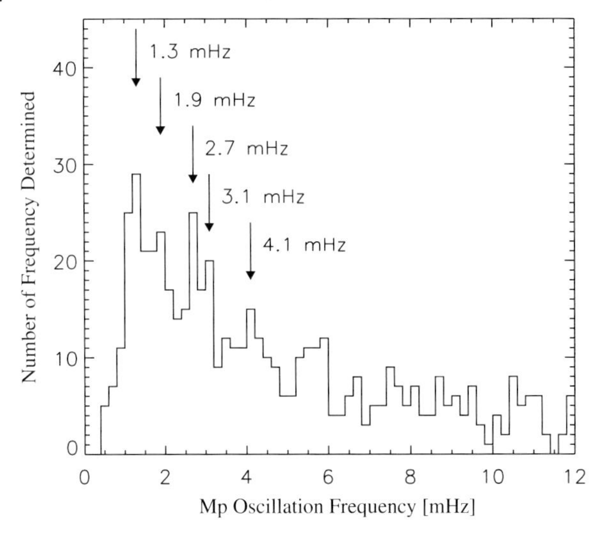

Figure 4.8 Magnetopause oscillation frequencies determined from 6697 magnetopause crossings, in 0.2 mHz bins. Maxima of the distribution are marked with arrows. From Plaschke *et al.* (2009a).

4.4 Field Line Resonances

The propagation of ULF plasma waves through the magnetosphere and down to the ground has been discussed by many workers (e.g. Allan and Poulter, 1992; Walker, 2005). Usually it is assumed that compressional mode waves enter or are otherwise present in the magnetosphere and propagate inward until they couple energy to standing shear Alfvén mode field line oscillations (Chen and Hasegawa, 1974; Menk *et al.*, 1994; Orr, 1984; Tamao, 1965; Southwood, 1974). The coupling is most efficient for relatively low azimuthal wave number (Allan, White, and Poulter, 1985), so the field line eigenoscillations – field line resonances – usually have relatively large longitudinal extent. On the other hand, field lines are assumed to be radially decoupled, leading to the notion of oscillations of shells of field lines. As discussed in Section 3.5, the resonant frequency varies with Alfvén speed and hence latitude, and we may expect that sufficient broadband hydromagnetic noise is always present in the magnetosphere to allow resonances to exist on all field lines, thus forming a continuum spectrum of oscillations (Lee and Lysak, 1991; Mathie *et al.*, 1999b; Waters, Menk, and Fraser, 1991). Azimuthal wave number is commonly described in terms of a dimensionless quantity (Olson and Rostoker, 1978):

$$m = \frac{2\pi R_E \Delta\phi}{360 S} \cos\lambda, \qquad (4.7)$$

where S is the station separation, $\Delta\phi$ is the estimated wave phase difference between these stations, and λ is the geomagnetic latitude. FLRs are most effectively driven by fast mode waves when $m \approx 3$ (Allan, White, and Poulter, 1985).

The physical principles of FLRs are well known and have been described in Chapter 3. Field line resonances are readily identified in ground data by a number of features (Orr and Hanson, 1981; Ziesolleck *et al.*, 1993): (i) a peak in wave power or amplitude at the resonant latitude, where the half-width is related to the Q of the resonance; (ii) a high degree of polarization or coherency at the resonant latitude; (iii) nearly linear wave ellipticity and a reversal of the sense of polarization across the resonance latitude with apparent propagation toward the resonance; (iv) large polarization azimuth angle at the resonant latitude; and (v) the azimuthal magnetic and radial electric field components in space being 90° out of phase at the resonant frequency. These properties are consistent with toroidal field line oscillations in space (see Figure 1.7) due to transverse standing waves with the Poynting flux directed along the field line and power predominantly in the H component on the ground (after 90° rotation by the ionosphere). Further details on resonance detection techniques such as analysis of the cross-phase spectrum between spaced stations are provided in Chapter 5.

How does the resonant frequency vary with latitude? Figure 4.9 shows an idealized radial mass density profile through the plasmasphere, across a steep plasmapause $0.2R_E$ wide, and in the outer magnetosphere, with the corresponding resonant frequency, determined using Equation 1.4. Dashed lines represent a (hypothetical) density biteout, while horizontal dotted lines indicate resonances at the same frequency on either side of the plasmapause boundary. Although this is an idealized representation, it resembles empirical models of the radial variation in equatorial electron density (e.g. Carpenter and Anderson, 1992). However, at low L-values, field lines have a significant portion of their length in the ionosphere and the mass density is additionally influenced by mass loading from ionospheric heavy ions and diurnal winds (Poulter, Allan, and Bailey, 1988; Waters *et al.*, 2000). This profile will be referred to again in Section 7.1.

Figure 4.10 illustrates the modeled variation in toroidal mode eigenperiod within the plasmasphere as a function of colatitude, 6 days after a simulated magnetic storm. Above 40° latitude (i.e., below 50° colatitude), the eigenperiods increase with latitude and are dominated by the behavior of H^+ ions near the equatorial plane. The diurnal variation results from refilling of depleted flux tubes from the underlying ionosphere, influenced by $E \times B$ drifts of the flux tubes due to azimuthal electric fields. Below 40° latitude, the O^+ ions become more important and result in a reversal of the eigenperiod profile near 20° latitude. The diurnal period variation follows the behavior of ionospheric O^+ ions, responding to neutral winds. The low-latitude reversal in the frequency of FLRs due to ionospheric mass loading has been detected observationally near $L = 1.6$ (Menk, Waters, and Fraser, 2000; Ndiitwani and Sutcliffe, 2010).

There is now abundant evidence of the existence of FLRs throughout the magnetosphere from high to low latitudes. As pointed out in Chapter 3, mathematical descriptions often assume a simple dipolar geometry that is not appropriate to

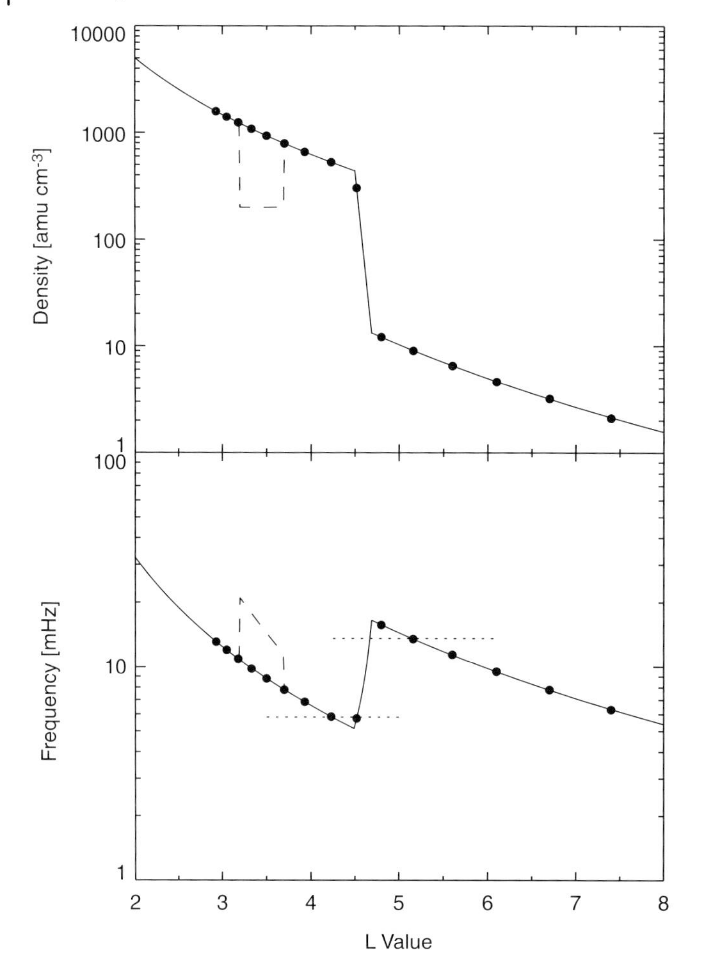

Figure 4.9 Model profiles of plasma density (top) and field line resonant frequency (bottom), including a typical plasmapause profile. Filled circles denote *L*-values of hypothetical ground stations each 110 km apart in latitude. Dashed lines represent density biteout. Horizontal dotted lines indicate resonances at the same frequency on either side of the plasmapause. From Menk *et al.* (2004).

high latitudes where field lines experience significant temporal distortion. This affects the frequency (Waters, Samson, and Donovan, 1996; Wild, Yeoman, and Waters, 2005) and polarization properties of the FLRs (Kabin *et al.*, 2007a). This is important because wave–particle energy transfer involves the wave electric field component parallel to the drift velocity of charged particles, usually regarded as the azimuthal field (poloidal mode), although this only applies to a dipole magnetic field. In a more realistic field, the polarization of Alfvén modes can no longer be described as poloidal or toroidal, as it becomes increasingly mixed and changes with local time. This happens because the contours of constant magnetic field no longer coincide

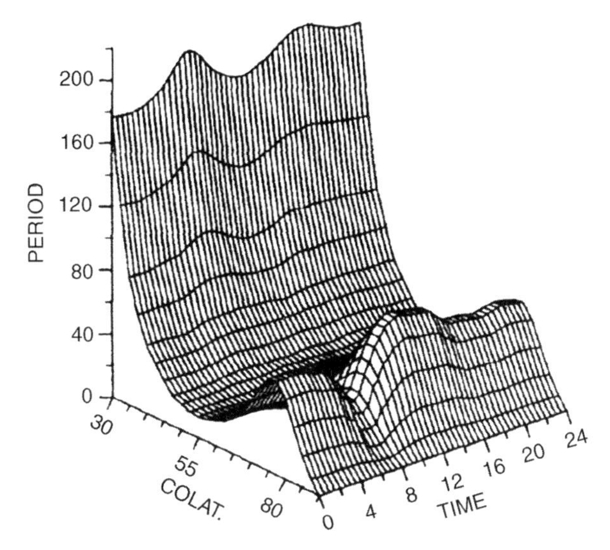

Figure 4.10 Modeled local time variation of fundamental toroidal mode eigenperiod in the plasmasphere as a function of colatitude. From Poulter, Allan, and Bailey (1988).

with contours of constant wave period for either mode in the equatorial plane, and it means that at high latitudes different Alfvénic modes may contribute to particle acceleration in different MLT sectors. The relevant equations were presented in Section 3.7. Observational confirmation of this effect has come from THEMIS multisatellite data (Sarris *et al.*, 2009b).

Recent modeling of MHD wave propagation and the formation of FLRs in response to a fast mode driver in a compressed dipole field geometry reveals that the spatial characteristics of high-latitude FLRs depend strongly on the source location at the magnetopause boundary (Degeling *et al.*, 2010). This is a consequence of the accessibility of fast mode waves from the magnetopause source to distant locations within the inhomogeneous magnetosphere, and it suggests that observed FLR structures may provide information on spatial characteristics of the wave source at the magnetopause.

A further complication arises when the magnetospheric plasma is in relative motion, such as azimuthal rotation near a KHI site. Model results suggest that monochromatic fast magnetosonic waves can excite harmonics of standing Alfvén waves simultaneously on different resonant surfaces (Kozlov and Leonovich, 2008). The plasma motion effect is greatest near strong velocity gradients such as the magnetopause and across a thin plasmapause, causing distorted phase and amplitude profiles near the resonant shell. This suggests that ground observations could be used to detect such high-velocity regions.

Multipoint satellite and ground measurements show that in the outer magnetosphere, just inside the cusp, Pc3 pulsations have dominant transverse toroidal and poloidal components, with wavelength $\sim 10^3$ km, transverse scale size $\sim 0.14 R_E$, and phase velocity $\sim 10^2$ km s^{-1} earthward (Liu *et al.*, 2008, 2009a). The Poynting flux is field aligned and away from the equatorial plane. These waves likely arise from incoming compressional mode waves coupling to guided Alfvén waves on the last closed field lines, exciting FLRs at lower latitudes. On the ground at high latitudes, Pc3 pulsations are often observed with FLR characteristics, although the field line eigenfrequency is actually much lower, in the Pc5 range. One possibility is that these are higher harmonics of FLRs excited at the upstream wave driving frequency (Howard and Menk, 2005; Ponomarenko *et al.*, 2005).

Field line resonances observed at low latitudes are most likely driven by fast mode (compressional) waves propagating from the upstream solar wind deep into the magnetosphere and coupling to discrete frequency toroidal and poloidal standing field line oscillations (Menk *et al.*, 2006; Menk, Waters, and Fraser, 2000; Ndiitwani and Sutcliffe, 2009; Waters *et al.*, 2000; Yumoto *et al.*, 1985).

How much energy does an FLR deposit into the ionosphere? Estimates for large Pc5 FLR events range from $\sim 6 \times 10^9$ W (Greenwald and Walker, 1980) to 10^{14}–10^{15} W for high solar wind speed global Pc5 events (Rae *et al.*, 2007b) and 10^{10}–10^{11} J for high-m particle-driven FLRs (Baddeley *et al.*, 2005). These values range up to 30% of a substorm budget, so clearly Pc5 FLRs may dissipate significant energy into the ionosphere.

Many observations of toroidal FLRs report wave propagation away from local noon, usually interpreted as evidence of a solar wind source. However, coherent low-m waves may also be found propagating sunward from the midnight sector. These have been observed in low- and midlatitude ground data (Mier-Jedrzejowicz and Southwood, 1979; Ostwald *et al.*, 1993) and at large L with spacecraft (Eriksson *et al.*, 2008). Numerical simulations of MHD wave coupling in the magnetotail waveguide (Wright and Allan, 2008) suggest that 5–20 min fast mode waves generated in the magnetotail waveguide by substorms may couple to earthward-propagating Alfvén waves and produce field-aligned currents resulting in narrow auroral arcs that move equatorward at ~ 1 km s^{-1}. The Alfvén waves phasemix as they propagate earthward, resulting in a rapid variation of wave fields perpendicular to B. The predicted wave properties agree with observations of Alfvén waves with local standing wave signatures in the plasma sheet boundary layer and on the ground. However, it is not clear whether this mechanism can explain the coherent sunward-propagating waves in the tail, or how common such waves might be.

In summary, although the fundamental properties of FLRs are well known, some key questions still exist: (i) How do the properties of FLRs change at high latitudes in realistic field geometries, and how do these affect particle acceleration? (ii) How significant is the effect of plasma motion on the properties of FLRs? (iii) How often do large FLRs deposit significant energy into the ionosphere, and what are typical values of energy deposition? (iv) What is the contribution of FLRs to the ULF wave spectrum at low latitudes, for example, near and inside the FLR cutoff latitude? (v) Are the FLR-like Pc3 that are detected in the outer magnetosphere harmonics of FLRs, and if so what is the contribution of these to general ULF activity at high latitudes? (vi) What is the origin of sunward-propagating waves, are they FLRs, and how common are they?

4.5 Cavity and Waveguide Modes

Under suitable conditions, the magnetosphere may form a resonant cavity for incoming millihertz-frequency fast compressional mode waves, between the reflecting magnetopause and the turning point where the radial wave number is zero. In an axisymmetric magnetosphere, the radial wave number is quantized and normal modes of the system can occur at discrete, quantized frequencies. In the presence of plasma inhomogeneities, these eigenmodes damp through coupling to transverse field line resonances whose spectrum is therefore dominated by the fast mode eigenfrequencies (Allan, White, and Poulter, 1985; Kivelson and Southwood, 1985). In essence, compressional mode energy present in the magnetosphere tunnels earthward past the turning point and couples to discrete frequency, driven FLRs that are embedded within the continuum spectrum. Cavity modes may also exist within the plasmasphere (Allan, Poulter, and White, 1986; Takahashi *et al.*, 2010a; Zhu and Kivelson, 1989), as illustrated schematically in Figure 3.3.

Since the magnetosphere is in reality not axisymmetric, compressional energy propagates from the dayside to the magnetotail, and so a waveguide model is more appropriate. However, cavity and waveguide modes are difficult to detect with spacecraft (Samson *et al.*, 1995; Waters *et al.*, 2002) and supporting evidence comes mostly from observational studies suggesting the preferential existence of discrete, quantized frequencies in the magnetosphere, especially at the so-called magic frequencies. These are often referred to as Cavity Mode Samson (CMS) frequencies, after their discoverer J.C. Samson (Samson *et al.*, 1991). Section 4.2 outlined observations that conversely suggest that discrete frequency modes may exist in the solar wind.

The discrete frequency CMS modes have been most clearly identified with ground-based HF radars that detect motions of the ionospheric plasma driven by the ULF wave fields (Harrold and Samson, 1992; Samson *et al.*, 1992, 1995; Walker *et al.*, 1992). High-latitude ground magnetometer observations also show evidence of discrete frequency FLRs, predominantly at the CMS frequencies (Figure 4.5) (Harrold and Samson, 1992; Mathie *et al.*, 1999a; Villante *et al.*, 1997).

Within the magnetosphere, spacecraft data also suggest the existence of discrete frequency modes. Figure 4.11 shows a 1 year statistical survey of ULF wave data recorded by the Cluster spacecraft mostly near 5 R_E (Clausen and Yeoman, 2009). The CMS frequencies are identified by vertical dashed lines, while dotted and solid sloping lines represent linear fits and 95% confidence levels for each wave component. While CMS frequencies do not stand out in all three magnetic field components, peaks in both the poloidal and toroidal components occur at 6.5, 8.0, 11–11.5, and 13 mHz. The observations come from locations significantly inward of the cavity/waveguide turning point for driving Pc5 FLRs and were therefore attributed to higher harmonics of waveguide/cavity modes.

Evidence is also accumulating that dayside waveguide and cavity modes may be observed within the plasmasphere. Since the cavity resonances are expected to be a fraction of an R_E apart, they will be difficult to detect with spacecraft but will be

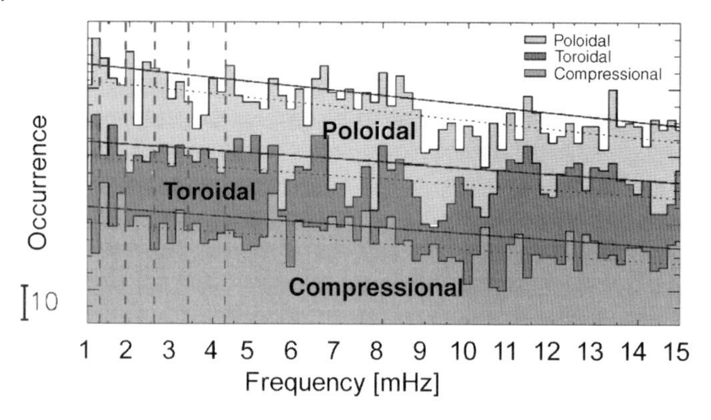

Figure 4.11 Occurrence of spectral peaks in 1 year of Cluster spacecraft data for the poloidal, toroidal, and compressional wave modes. Adapted from Clausen and Yeoman (2009).

manifested by a suite of closely spaced frequencies in ground magnetometer data (Samson *et al.*, 1995). The structure of such trapped plasmaspheric modes was predicted using a simple 1D waveguide model (Waters *et al.*, 2000) and confirmed observationally (Menk, Waters, and Fraser, 2000). However, if the plasmapause boundary is regarded as an imperfect reflector, then virtual resonance modes may form (Lee and Lysak, 1999). These can exist even in the absence of a clear plasmapause boundary and extend beyond the plasmasphere. Wave power within the plasmaspheric resonator may also escape through the plasmapause, producing field line oscillations at high latitudes (Teramoto, Nosé, and Sutcliffe, 2008).

Global Pc3 and Pc4 eigenmode oscillations have been observed throughout the dayside plasmasphere ($L \approx 1.7$–3.1) under conditions favorable for the propagation of broadband compressional mode power from the solar wind into the magnetosphere, but in the absence of a distinct plasmapause density signature (Takahashi *et al.*, 2009). These have been attributed to virtual cavity resonances in the inner magnetosphere, which may also account for the observed spectral properties of nighttime Pi2 pulsations (Kim *et al.*, 2005; Lee and Lysak, 1999; Lee and Takahashi, 2006; Teramoto, Nosé, and Sutcliffe, 2008). Pc4 pulsations recorded on the nightside at geomagnetically quiet times may also result from plasmaspheric cavity modes, which are excited by waves generated in the upstream solar wind (Takahashi *et al.*, 2005). Similar Pc4 pulsations have been reported in HF radar signals (Ponomarenko, Waters, and St-Maurice, 2010).

The shear flow between plasma in the magnetosheath and magnetosphere controls the reflection condition at the magnetopause, and when taking into account the boundary layer thickness may result in the formation of overreflection modes at the magnetopause (Mann *et al.*, 1999; Walker, 2000). Overreflection occurs when the characteristic spatial scale sizes of the wave and the inhomogeneity are comparable, and the wave and magnetosheath flow have similar propagation vectors. A magnetospheric fast mode incident at the magnetopause has a component of group velocity in the direction of magnetosheath flow, from which it may gain energy through the work done by the Maxwell and Reynolds stresses on the velocity gradient at the boundary (Walker, 2000). This offers an efficient process for the extraction of energy from the magnetosheath to magnetospheric waveguide modes on the flanks during fast solar wind speed intervals, and it may explain the production of discrete frequency ULF waves in the magnetosphere and statistical correlations between Pc5 power on the ground and solar wind velocity (Mann, O'Brien, and Milling, 2004; Mathie and Mann, 2001; Pahud *et al.*, 2009).

In summary, mathematical models suggest that fast mode eigenmodes of the magnetospheric cavity/waveguide and the plasmasphere should result in discrete frequency waves that may couple to FLRs. Evidence for these is provided mostly by radar data and statistical analysis of ground and *in situ* observations. However, the following difficulties remain: (i) Lack of clear evidence in spacecraft data for the existence of cavity modes; (ii) What is the significance of the CMS discrete frequencies? Are cavity/waveguide eigenmodes observed at other frequencies, including higher frequencies? (iii) Are there preferred conditions (e.g., magnetically quiet, high solar wind speed, and solar wind compressions) under which global cavity and waveguide modes preferentially exist? (iv) Under what conditions and at what frequencies do virtual cavity resonances exist compared to plasmaspheric cavity modes?

4.6 Spatially Localized Waves

Field line resonances observed with ground-based magnetometers are normally described as toroidal mode oscillations with low azimuthal wave number k_y, but since the MHD Alfvén and fast modes are coupled in the magnetosphere, spatially localized predominantly poloidal mode oscillations with high k_y values may be produced.

The conditions under which toroidal and poloidal Alfvén waves are generated were considered in detail by Klimushkin, Mager, and Glassmeier (2004). The eigenfrequencies of the two modes are different (Orr and Matthew, 1971), although nearly monochromatic guided poloidal modes occur in regions where the poloidal mode frequency reaches local minimum or maximum values, in the ring current and near the plasmapause (Denton and Vetoulis, 1998; Denton, Lessard, and Kistler, 2003; Klimushkin, Mager, and Glassmeier, 2004). While the frequency of a toroidal mode resonance ω_T is determined by the travel time of the shear Alfvén wave along a field line, the poloidal mode eigenfrequency ω_P is also affected by geometric and other factors, and in the WKB approximation is (Klimushkin, 1998; Schäfer *et al.*, 2007)

$$\omega_P = \omega_T - \omega_{geom} + \omega_\beta, \tag{4.8}$$

where ω_{geom} describes field line curvature effects and ω_β accounts for finite plasma pressure and perpendicular currents including ring currents.

The radial wave number for the poloidal mode, k_r, is given by

$$k_r^2 \propto \frac{\omega^2 - \omega_P^2}{\omega_T^2 - \omega^2}. \tag{4.9}$$

Waves propagate radially in the region where $k_r^2 > 0$ and are reflected at the point where $\omega = \omega_P$, where $k_r = 0$. When $\omega = \omega_T$, toroidal resonances occur and the poloidal mode is converted to a mainly toroidal one. Thus, wave behavior changes spatially and temporally. It is progressively easier to satisfy the poloidality condition with increasing difference between the toroidal and poloidal frequencies, as plasma pressure increases.

The existence of a peculiar type of large-amplitude, highly monochromatic wave mode called giant pulsations has been known for a long time (e.g. Glassmeier, 1980; Takahashi *et al.*, 2011). These are due to spatially localized poloidal mode oscillations resulting from resonant interaction between the wave electric field and energetic electrons that bounce and drift azimuthally around Earth. The resonance condition for a wave of frequency ω is

$$\omega - m\omega_d = N\omega_b, \tag{4.10}$$

where ω_d is the drift frequency of the particles around Earth, ω_b is the particle bounce frequency between conjugate points, m is the azimuthal wave number, and N is a positive or negative integer. Figure 4.12 is a schematic illustration of the resonance. Due to spatial integration effects (Ponomarenko *et al.*, 2001), these waves are heavily attenuated between the ionosphere and the ground, so most information on them has come from measurements of the wave fields in the ionosphere using HF sounders (Wright and Yeoman, 1999; Yeoman *et al.*, 2000), discussed further in Section 7.9, and fortuitous spacecraft conjunctions (Hughes, McPherron, and Barfield, 1978a). Evidence for poloidal field line eigenoscillations also comes from the detection of radial oscillations of plasmaspheric flux tubes from the Doppler shift of artificially produced VLF signals and simultaneous observation of FLRs at a similar frequency by ground magnetometers (Menk *et al.*, 2006). This is discussed further in Section 7.10.

Results from multipoint spacecraft missions show that spatially localized waves may exist for days in the outer magnetosphere during the recovery phase of storms (Eriksson *et al.*, 2005; Sarris, Li, and Singer, 2009a; Sarris *et al.*, 2007; Takahashi *et al.*, 1987). Figure 4.13 shows radial profiles of toroidal and poloidal mode eigenfrequencies calculated for conditions when a 23 mHz wave was detected by the Cluster satellites just outside the plasmapause, with maximum amplitude at $L = 4.56$, poloidal turning points $f_{obs} = f_P$ at $L = 4.4$ and $L = 5.0$, radial width of $0.6 R_E$, and azimuthal wave number $m \approx 155$ (Schäfer *et al.*, 2008). Similar trapped poloidal modes have also been observed just inside the dayside plasmapause (Schäfer *et al.*, 2007). Occasionally, such waves are seen at quiet times when no energetic particles are present and drift or bounce resonance is unlikely (Eriksson *et al.*, 2005).

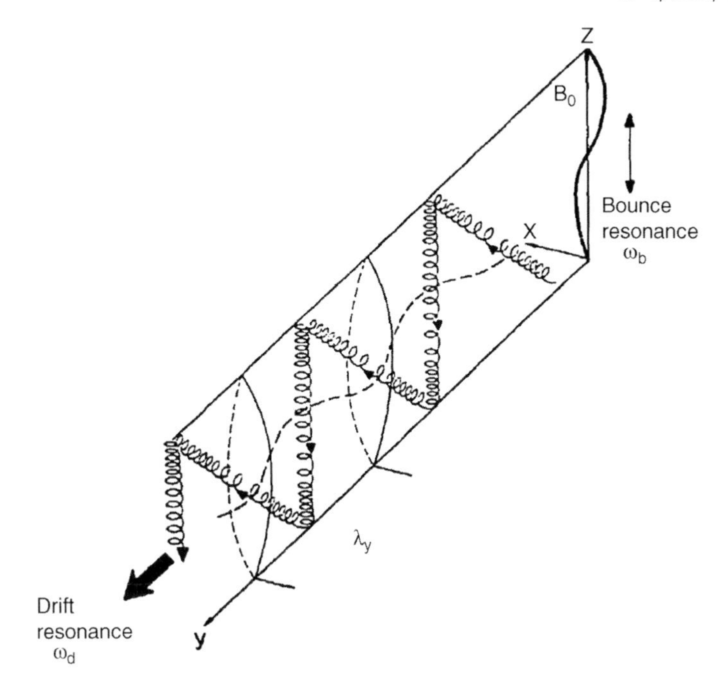

Figure 4.12 Schematic illustration of bounce and drift resonance interaction with high wave number ULF waves. Particles follow helical trajectories with changing pitch angle, executing bounce motion between conjugate points at frequency ω_b and azimuthal grad-B and curvature drift at frequency ω_d. Bounce resonance occurs when a multiple of ω_b matches the eigenfrequency of an even-mode standing wave. Drift resonance occurs when ω_d matches the wave's azimuthal wavelength. From Allan and Poulter (1992).

In addition to bounce and drift resonance, compressional high-m waves may also be generated by the drift mirror mode instability under high β-conditions when there is significant perpendicular pressure anisotropy. The frequency of the growing mode depends on the diamagnetic drift frequency, but the instability condition is affected by field line curvature and coupling to transverse shear Alfvén waves. The drift mirror waves will propagate slowly with the Larmor drift frequency. Suitable conditions were identified during about 30% of the Equator-S satellite orbits in the morning plasmasheet (Haerendel *et al.*, 1999). The observed plasma "blobs" drift slowly eastward and sunward and have lifetimes of 15–30 min. Storm-time Pc5 waves in the outer magnetosphere typically have azimuthal wave number in the range $|m| = 20$–100, limited radial extent, and a characteristic antisymmetric field-aligned nodal structure (Takahashi *et al.*, 1987). Long-lived high-m compressional Pc5 waves satisfying the drift mirror criterion have been observed under average magnetic activity conditions in the outer magnetosphere on the dawn side (Korotova *et al.*, 2009; Rae *et al.*, 2007a) and dusk side (Constantinescu *et al.*, 2009).

In summary, trapped spatially localized poloidal mode oscillations occur in the ring current region and near the plasmapause, and poloidal mode eigenoscillations

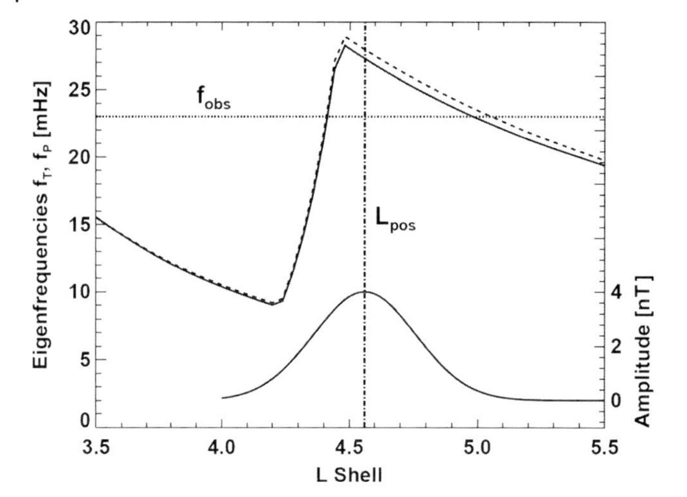

Figure 4.13 Location of poloidal mode resonance region observed at the plasmapause on August 8, 2003. Upper solid and dashed curves show radial profiles of toroidal and poloidal mode eigenfrequencies (assuming a power law distribution of the form $(r/R_E)^{-2}$ and a heavy ion mass loading factor of 2.0), respectively; horizontal dotted line shows observed frequency $f_{obs} = 23$ mHz. Lower curve represents amplitude profile. From Schäfer *et al.* (2008).

may also be detected within the plasmasphere. The following questions arise: (i) How frequently do high-m poloidal mode particle-generated waves occur? (ii) Can such waves be generated at quiet times in the absence of energetic particle distributions? (iii) How common are high-m poloidal mode waves at the plasmapause and are they a signature of the plasmapause? (iv) What is the significance of these waves in the energization of ring current particles (Ozeke and Mann, 2008)? (v) Techniques such as VLF sounders and HF radars provide the possibility of ground-based monitoring of high-m poloidal mode waves. What new results could emerge?

4.7
Ion Cyclotron Waves

Until recently, Pc1 and Pc2 (~0.2–5 Hz) plasma waves were generally believed to be generated in the equatorial region of the magnetosphere by ion cyclotron resonance with unstable distributions of energetic ring current ions, especially during the recovery phase of magnetic storms. These are electromagnetic ion cyclotron waves (EMICWs). The appearance of repetitive modulated packets in ground data was thus attributed to dispersive field-aligned wave packet propagation in the left-hand ion mode on successive bounces between hemispheres – that is, hydromagnetic whistlers (Obayashi, 1965). Nonpropagation stopbands occur at the local bi-ion

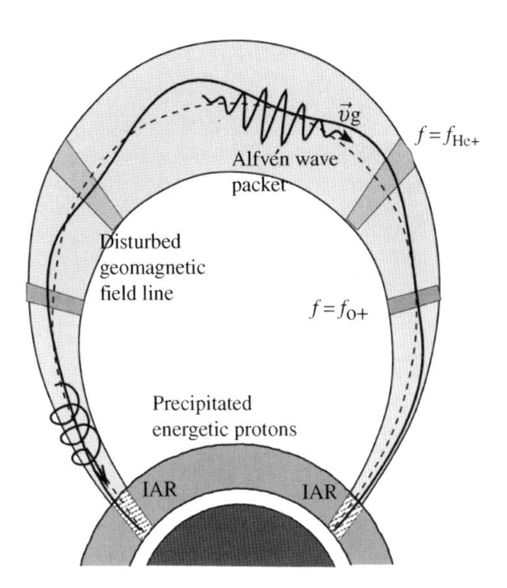

Figure 4.14 Schematic summary of Pc1 ion cyclotron wave generation and propagation. IAR refers to the ionospheric Alfvén resonator. From Demekhov (2007).

frequencies in He^+- and O^+-rich plasmas (Fraser and McPherron, 1982). On reaching the ionosphere, some of the wave energy couples to the right-hand mode and propagates in a horizontal waveguide (resonator) formed by the peak in the phase constant profile centered on the ionospheric F2 region electron density maximum (Fraser, 1975; Tepley and Landshoff, 1966) from source regions near the plasmapause (Fraser, 1976; Webster and Fraser, 1985). IPDP are an unstructured subtype of Pc1 and Pc2 pulsations generated by resonant interaction with westward drifting energetic protons near the plasmapause in the evening sector (Horita *et al.*, 1979). Figure 4.14 gives a schematic overview of the generation and propagation of structured Pc1 emissions.

A key difficulty with the above scenario is lack of evidence of wave packet bouncing between conjugate points (Demekhov, 2007). *In situ* observations show that EMICW propagation is almost exclusively away from a narrow zone near the equator (Fraser *et al.*, 1996; Loto'aniu, Fraser, and Waters, 2005) with minimal reflection at the ionosphere. This is supported by a model of wave generation within a narrow "backward wave oscillator" regime into which waves are reflected by mirrors just off the equator (Trakhtengerts and Demekhov, 2007). In addition, the packet appearance may result from modulation of the growth process by compressional Pc5 waves (Loto'aniu, Fraser, and Waters, 2009).

It now seems likely that Pc1 and Pc2 EMICWs are generated in a variety of regions under conditions that lead to the formation of suitable particle temperature anisotropies. The waves are often observed after magnetic storms (Engebretson *et al.*, 2008; Halford, Fraser, and Morley, 2010) near the plasmapause, and in the outer magnetosphere in connection with solar wind pressure fluctuations (Arnoldy *et al.*, 2005; Hansen *et al.*, 1992; Usanova *et al.*, 2008). They are also generated by interactions with cold plasma density gradients within plasma drainage plumes (Morley *et al.*, 2009). Observations of Pc1 and Pc2 waves therefore provide information on (i) the location of magnetospheric structures, as discussed further in Section 7.8, (ii) the presence of ring current particles, and (iii) He^+ concentrations (Fraser and McPherron, 1982).

5
Techniques for Detecting Field Line Resonances

5.1
Introduction

The ability to remote sense properties of near-Earth space using ULF waves depends on the excitation of the low-frequency, natural magnetized plasma oscillations that appear in the magnetosphere, described in Chapter 4. ULF resonance structures appear in the magnetosphere as a result of simple constraints: a bounded, magnetized plasma and a source of excitation energy. The combination of the nonlinear spatial variation of both the magnetic field and plasma mass density coupled with boundary conditions that change with time, latitude, and longitude yields rich resonant structures and dynamics. This chapter describes various experimental data analysis techniques that have been developed to identify signatures of ULF field line resonances. This is the first step toward developing the capability to remote sense the plasma mass density in the magnetosphere.

ULF waves in the magnetized plasma of near-Earth space appear as perturbations in the magnetic and electric fields and in the plasma density. In principle, ULF wave data may be obtained from any instrument sensitive to electric or magnetic fields or their effects on the plasma. The development of a ground-based magnetospheric diagnostic capability depends on being able to detect FLR signatures in data recorded by ground-based sensors such as vector magnetometers, scanning meridian photometers, and ionosphere-probing instrumentation such as Doppler sounders and HF radars. Although data from spacecraft-borne sensors have also been successfully used to detect ULF resonances, most of the experimental data have come from extensive magnetometer networks on the ground.

The most important property for the detection of FLRs in experimental data is the radial variation of the magnetic field and plasma properties, as discussed in Chapter 3. The first ULF resonance structure identification strategies relied on the variation of amplitude and wave polarization with latitude in addition to wave polarization properties at conjugate locations (Nagata, Kokubun, and Iijima, 1963; Sugiura, 1961). Some early attempts to identify FLRs used the magnetometer data time series directly (Ellis, 1960; Siebert, 1964), while others selected frequency domain parameters such as power density and wave polarization (Samson, Jacobs,

Magnetoseismology: Ground-based remote sensing of Earth's magnetosphere, First Edition. Frederick W. Menk and Colin L. Waters.

and Rostoker, 1971; Sugiura and Wilson, 1964) as a function of latitude. For magnetoseismology, we require accurate estimates of the FLR frequencies as a function of location on Earth, and these early attempts had limited success. For example, considerable scatter in the FLR frequency with latitude may be present in ground data (Obayashi and Jacobs, 1958), while some early workers found no evidence for FLR characteristics (Ellis, 1960).

Comprehensive studies of wave polarization at high latitudes by Samson (Samson, 1973; Samson, Jacobs, and Rostoker, 1971) resulted in a suitable MHD explanation for the appearance of ULF waves in the magnetosphere (Chen and Hasegawa, 1974a, 1974b; Southwood, 1974), confirming many features of the earlier model developed by Tamao (1965). A review of early observations of Pc4–Pc5 waves interpreted as resonant oscillations is given by Saito (1969). These studies required extensive latitudinal magnetometer arrays and the associated coordinated data analysis. More recently, a number of additional techniques and refinements of previous techniques have become available. These exploit the characteristics of both the wave amplitude and horizontal wave polarization properties with latitude and frequency for the detection of FLRs in ground-based magnetometer data. Furthermore, the routine detection of FLRs as a function of latitude without the need for large magnetometer arrays and with a low data processing burden is essential for magnetoseismology. Such techniques include polarization analyses (Samson, 1983; Samson and Olson, 1981), "gradient" analysis (Baransky *et al.*, 1985), and cross-phase (Waters, Menk, and Fraser, 1991).

In this chapter, a number of techniques that have been used to identify FLRs in ground magnetometer data are described. We assume the reader is familiar with basic digital signal processing techniques such as the fast Fourier transform (FFT) and spectral analysis. The methods used to detect FLRs in the data obtained from magnetometers and ionosphere probes such as Doppler sounders and HF radars were reviewed in Waters *et al.* (2006). That article also contains a computer code to compute the wavelet transform of ULF wave data if the reader is interested in digital signal processing using that alternative spectral estimation technique.

Magnetometer data are widely available from space physics World Data Centers and organizations devoted to the dissemination of geomagnetic field information (e.g., InterMagnet and SuperMAG). For digital signals, the bandwidth of the spectrum is set by the Nyquist criterion where the Nyquist frequency (maximum frequency) $f_N = 1/(2\Delta t)$ Hz, where Δt is the (constant) data sample interval in seconds. Data sampled at $\Delta t = 60$ s or longer from many stations around the globe are widely available. However, 1 min magnetometer data allow detection of FLRs only up to 8 mHz, limiting their application to high latitudes where ULF resonances in the Pc5 band are common. For low latitudes ($L < 3$), $\Delta t = 4$ s is generally suitable. With increasing capabilities in data storage, magnetic field data sampled at $\Delta t = 1$ s are becoming available, allowing the full range of ULF resonances and their harmonics to be detected at all latitudes.

The maximum FLR frequencies occur on magnetic field shells where the integrated Alfvén speed is the largest (see Equation 1.4). Since the magnetic field strength decreases with the cube of the radial distance (Equation 2.2), the Alfvén speed is larger closer to Earth. A typical radial profile for the Alfvén speed in the equatorial plane of the dayside magnetosphere is shown in Figure 3.2. Inside the plasmasphere the plasma mass density increases with decreasing radial distance and around $L \approx 1.5$–1.7, the Alfvén speed in the equatorial plane reaches a maximum. Closer to Earth the plasma mass density dominates and so both the Alfvén speed and the FLR frequencies decrease. The fundamental FLR around $L \approx 1.6$ is in the Pc3 range and may be 80 mHz or greater. The lower latitude limit for the detection of FLRs appears to be around $L \approx 1.3$ (Menk, Waters, and Fraser, 2000) where ionosphere damping processes appear to inhibit FLR oscillations.

The high-latitude limit for FLRs occurs at the outermost closed field line of Earth's magnetosphere where the fundamental frequencies are ~1–3mHz. An example of GEOTAIL spacecraft data from the dayside, outer magnetosphere is shown in Figure 5.1. For this interval, GEOTAIL was near 11 MLT in the subsolar

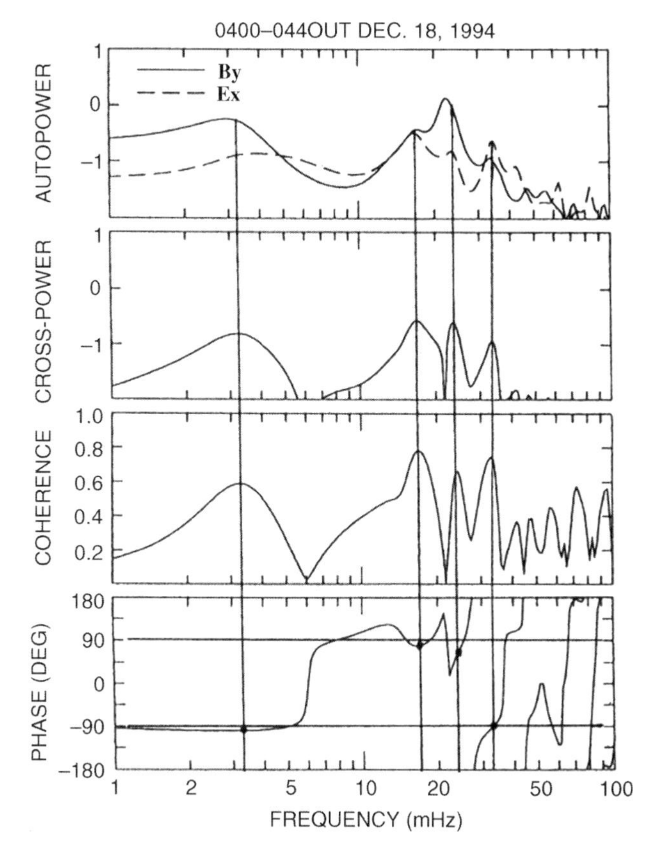

Figure 5.1 Spectra of the azimuthal magnetic field B_y and radial electric field Ex for 0400–0440 UT, recorded on December 18, 1994, from the GEOTAIL spacecraft in the subsolar region. From Sakurai *et al.* (2006).

region, $10.5R_E$ from Earth. Near the subsolar region of the magnetosphere, the higher frequency Pc3 band is enhanced. Although oscillations of the magnetic and electric fields were seen on all sensor components, FLRs are most prominent on the e_{radial} and b_{azi} components. The shear mode resonance has 90° phase relationship between the e_x and b_y signals. The enhanced power of the Pc5 resonance ~3 mHz is seen in Figure 5.1, followed by higher harmonics in Pc3 and Pc4 bands.

5.2
Variation in Spectral Power with Latitude

The variation of the Alfvén speed in the magnetosphere leads to an increase in the FLR frequency with decreasing radial distance except across the plasmapause region and at very low latitudes ($L < 1.4$) (Figure 3.2). For shear Alfvén FLRs that have largely an azimuthal (b_φ) magnetic perturbation in space (Section 3.4), transition through the ionosphere rotates the field vector by ~90°, as discussed in Section 5.6 and Chapter 8. Therefore, if FLRs have sufficient amplitude, they may be identified in the power spectra of data obtained from the magnetometer time series from a north–south-oriented sensor. A typical example from the research literature can be found in Miletits *et al.* (1990).

The process begins with an examination of the baseline removed, magnetic field time series from the north–south (x-component) sensors of a collection of stations that are located at different latitudes but at similar longitudes. For intervals that show ULF wave activity, the data are transformed into the frequency domain. We identify these spectra by $X(f)$ and $Y(f)$ for data obtained from the north–south [$x(t)$] and east–west [$y(t)$] magnetometer sensor time series, respectively. The common method is to use the FFT. However, the maximum entropy method (MEM) and wavelet and Hilbert transforms have also been applied. The FFTs of the time series segments are then represented as a stack plot of spectral amplitude or power density versus frequency. An example from the Churchill line of the Canadian magnetometer array for 16 UT, recorded on February 9, 1995, is shown in Figure 5.2. The spectra have been ordered with the highest latitude data (Rankin Inlet; 73.2° AACGM latitude) at the top to the lowest latitude (Pinawa; 61.0° AACGM latitude) at the bottom. Since the amplitude of ULF waves decreases for lower latitudes, identification of the FLR peak with latitude is facilitated by normalizing the spectra, as shown in Figure 5.2. The FLRs for each latitude obtained using the cross-phase method (described subsequently) are shown as asterisks.

The identification of spectral peaks can be tricky. The enhanced power just above 2 mHz for the Rankin Inlet and Eskimo Point (71.5°) data in Figure 5.2 are quite clear. However, at lower latitudes, the identification is more challenging, particularly for Pinawa. Additional resonance information, such as the phase, is therefore often included to improve FLR identification.

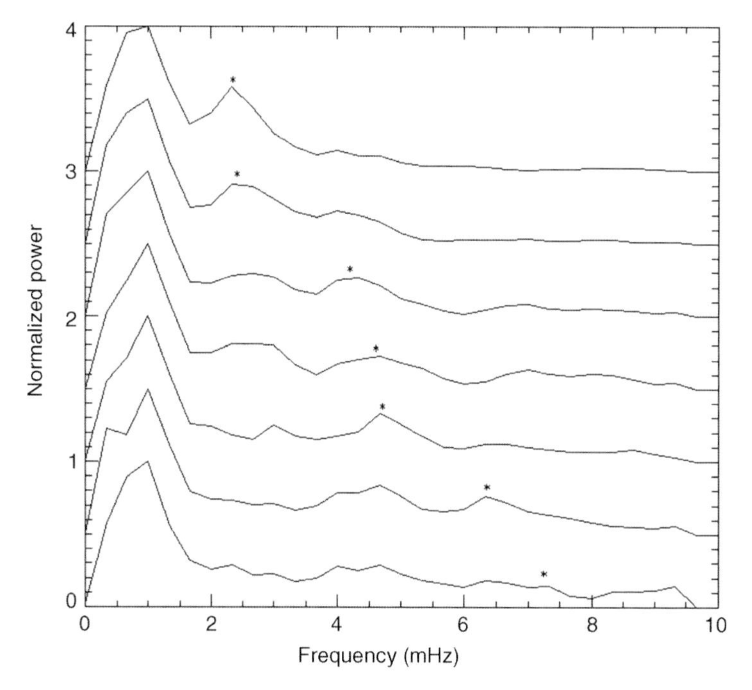

Figure 5.2 Normalized, stacked power spectra of the north–south magnetic field component data recorded by the Canadian magnetometer array 16 UT on February 9, 1995. From top to bottom, the data are from magnetic latitudes of 73.2°, 71.5°, 69.3°, 68.3°, 67.0°, 64.6°, and 61.0°. The asterisks (∗) indicate the FLR frequency identified using the cross-phase method.

5.3 Variation of Phase with Latitude

Waves contain both amplitude and phase information. As with other wave types, the phase properties of ULF waves also provide a rich source of information, a fact that was used in early attempts to use wave polarization information to identify FLRs (Sugiura and Wilson, 1964). There are several ways in which the phase information may be used to identify FLRs. The first process is similar to the spectral power with latitude already discussed: (i) Obtain ULF wave time series data from a latitudinal distribution of sensors (e.g., magnetometers, photometers, and HF radar beam). (ii) For a given time interval, calculate the FFT of these data. (iii) Plot the phase as a function of latitude for a given frequency.

The phase data are usually combined with the spectral amplitude with latitude plots as shown in Figure 5.3. For the selected frequency of 5 mHz, a resonance occurs at 66.5° AACGM latitude, identified as a peak in the power and decreasing phase with increasing latitude. The technique allows us to identify the latitude of FLR frequencies between those identified in Figure 5.2. An interesting and as yet

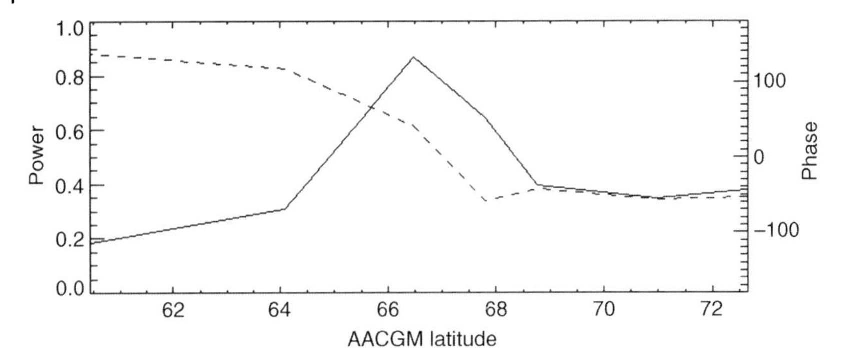

Figure 5.3 Power (solid curve) and phase (dashed curve) with magnetic latitude at $f = 5$ mHz, for the data in Figure 5.2.

unexplained feature of FLRs at high latitudes has been identified using this method. The magnetometer time series were replaced by Doppler velocity variations obtained by HF radars to investigate the signatures of FLRs in the ionosphere. An example from Fenrich *et al.*, (1995) is shown in Figure 5.4. The usual peak in power and decrease in phase with increasing latitude identifies the FLR in Figure 5.4a. However, for the event in Figure 5.4b, the phase variation with latitude goes the opposite way, earning these events the title "reverse-phase FLRs."

5.4
Wave Polarization Properties

A monochromatic wave is usually characterized by amplitude, frequency, and phase. An alternative representation is to consider wave polarization properties such as azimuth and ellipticity. For example, circular polarization has unit ellipticity, and azimuth in this case is meaningless. The sign of the ellipticity determines the left- or right-handedness of the wave polarization. A common notation from optics is to denote a pure polarized (no noise) wave by the Jones vector (Collett, 1993). For example, a right-hand circularly polarized wave in a 2D plane has the Jones vector [1,i]. Partially polarized signals can be described using the Mueller matrix or Stokes vector formulation from optics, and many of the ULF wave analysis methods use this formulation, particularly the pure state methods (Samson, 1973), discussed in Section 5.8.

The expected polarization of the Alfvén wave mode as a function of latitude (or frequency), including effects from the ionosphere, was described in Hughes and Southwood (1976). The representation of polarization properties in the time domain is the hodogram, and this has been used throughout ULF research history (e.g. Sugiura and Wilson, 1964). Hodograms provide useful information if the signals are noise free and narrowband. However, either or both of these constraints are often not satisfied for magnetometer time series. The combined power and wave polarization characteristics with latitude and frequency may be used to identify

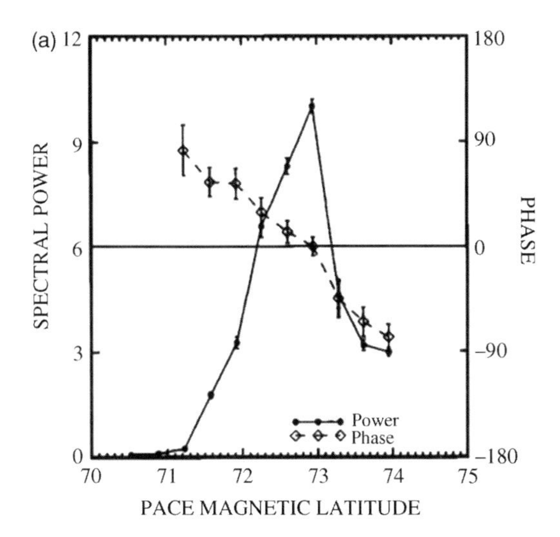

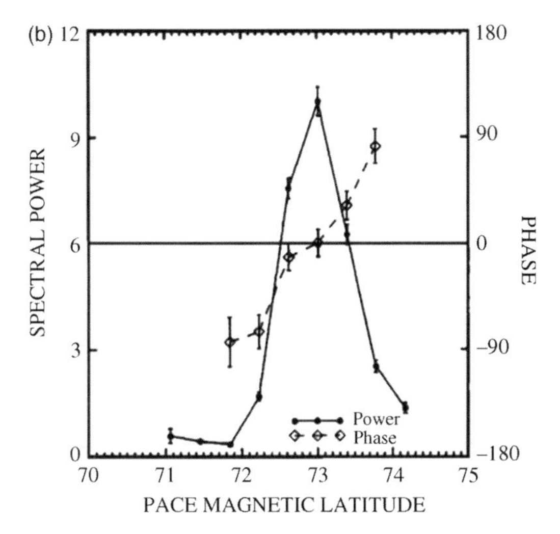

Figure 5.4 (a) Variation of power and phase at 1.9 mHz with latitude for a low-m event from the Goose Bay radar data recorded from 0735 to 0835 UT on September 28, 1988. (b) Variation of power and phase at 1.3 mHz with latitude for a high-m event from the Saskatoon radar data recorded from 2120 to 2220 UT on October 18, 1993. From Fenrich *et al.* (1995).

FLRs. An example from a low-latitude magnetometer array is shown in Figure 5.5. The peak in power and change in ellipticity with latitude identify an FLR at $L \approx 1.9$. The frequency of interest may be removed from the processing by the process of demodulation. The methods for FLR identification are similar, but now use the demodulated spectra. An example from the literature using demodulated data is Figure 3 in Beamish, Hanson, and Webb (1979). These methods have provided estimates for resonance widths but resonance widths measured in the ionosphere

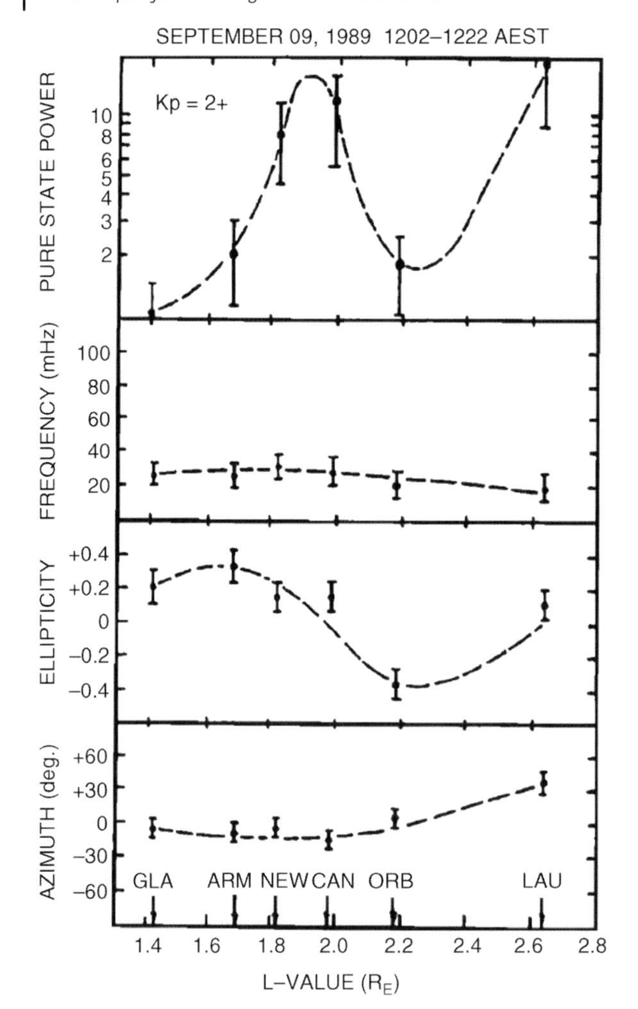

Figure 5.5 Latitude variation (*L*-value) of the wave pure state power (top), frequency, and polarization for data recorded over the Australian low-latitude magnetometer array. The peak in power and reversal in the sense of polarization near $L = 1.9$ identify the resonance location, while the resonance width is estimated to be about $0.2R_E$ from the -3 dB width of the peak in power. From Ziesolleck *et al.* (1993).

are smaller than those measured on the ground (Ziesolleck *et al.*, 1998). This is discussed further in Chapter 8.

These techniques assume that spectral peaks that change with latitude correspond to the FLR, which requires resonant excitation over the sensor array with a relatively high-quality resonance (Q-factor). However, resonance effects may be masked by the spectrum of the source energy (Kurchashov *et al.*, 1987). In addition, these methods require data from many (typically greater than four) magnetometers arranged in latitude, with the associated data processing burden. The determination of temporal variations of FLRs using this format thus becomes tedious.

5.5 Spectral Power Difference and Division

These methods are identified as the "gradient" method in the literature (Baransky *et al.*, 1985, 1990) and depend on magnetic field measurements of the ULF fields at closely spaced, latitudinal sites. What does "closely spaced" mean? This question was discussed in Waters, Samson, and Donovan (1996). The required latitudinal magnetometer separation depends on the quality of the resonance and the variation of FLRs with latitude, which is directly related to the radial variation of the Alfvén speed in the magnetosphere. Essentially, we need to obtain sufficient difference in FLR frequency at the two sites while keeping the stations close enough to ensure coherent signals. For $L < 2.8$, site separations of $\sim$80 km have been successfully used while at higher latitudes ($L > 5$), separations of the order of $\sim$200 km have been used (Green *et al.*, 1993; Waters, Menk, and Fraser, 1991; Waters, Samson, and Donovan, 1995).

The process first requires that a suitable time interval of magnetometer data from two latitudinally separated, north–south oriented sensors is transformed into the frequency domain, usually by the FFT and the researcher's preferred windowing function. We denote these as the equatorial and poleward spectra ($X_E(f)$ and $X_P(f)$) to avoid confusion with data recorded in the northern and the southern hemispheres. These amplitude or power spectra are then either subtracted or divided and plotted as a function of frequency. The idea is illustrated in Figure 5.6, using the driven, damped simple harmonic oscillator equations. An example using magnetometer data from the Fort Churchill and Gillam magnetometers for 16 UT, recorded on February 9, 1995, is shown in Figure 5.7. The FLR is identified as the frequency where $X_P(f) - X_E(f) = 0$ and $X_P(f)/X_E(f) = 1$ (Baransky *et al.*, 1985) shown by the asterisk. The "gradient" aspect of the calculation (division by station separation) allows interpolation of the variation of FLR frequencies with latitude (Kawano *et al.*, 2002). Further examples appear in Menk, Waters, and Fraser (2000) and in Pilipenko and Fedorov (1994), which is a review of ULF resonance parameters that may be derived from measurements with closely spaced, latitudinal magnetometers.

5.6 Single Station H/D

FLR signatures measured on the ground have small azimuthal wave numbers since the smaller spatial scale perturbations attenuate in amplitude from the ionosphere to the ground, as discussed in Chapter 8. This selects those FLRs in the magnetosphere that have a predominantly azimuthal (east–west; y) variation in the perturbation magnetic field. In the magnetosphere, the FLR exhibits properties of the

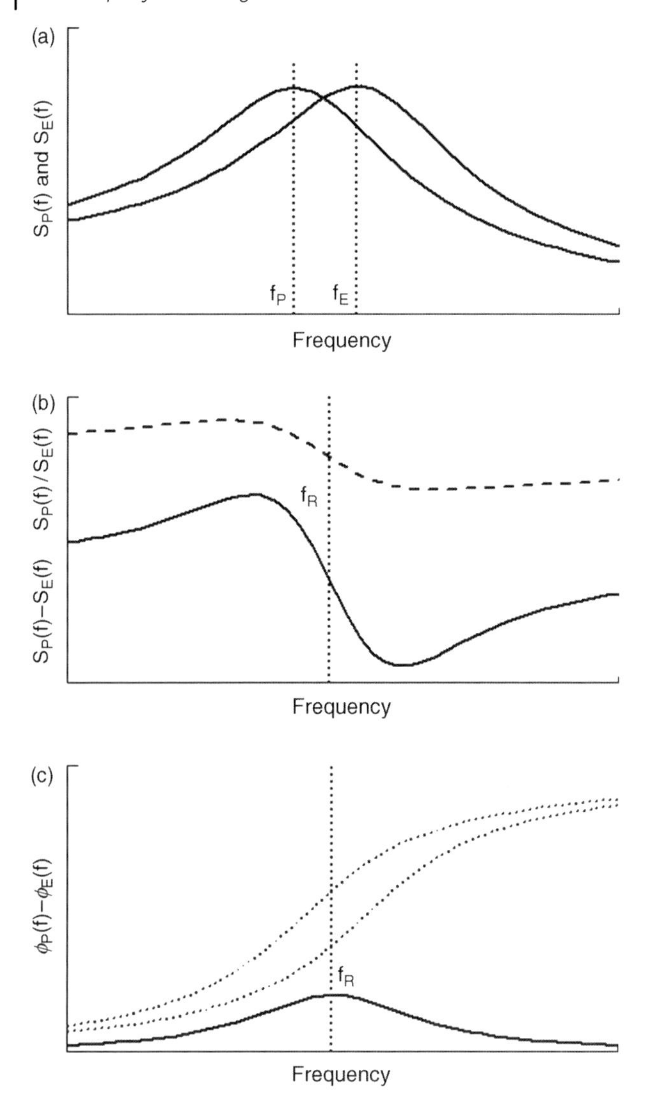

Figure 5.6 Variation with frequency of two simple harmonic underdamped oscillators. (a) Amplitudes with resonant frequencies f_E and f_P. (b) Subtraction (solid) and division (dashed) of these amplitude functions. (c) Phase response (dashed curve) and cross-phase (solid curve).

shear Alfvén wave mode where the perturbation magnetic field and the associated field-aligned current are related by $\nabla \times \mathbf{b} = \mu \mathbf{j}$. However, in the atmosphere, $\nabla \times \mathbf{b} = 0$, which implies a rotation of the perturbation magnetic field in the north–south (x) direction for detection by ground magnetometers. If the x-component magnetometer sensor data contain the FLR signal and the y-sensor data are dominated by the source spectrum, then a two-sensor, horizontal and orthogonal arrangement should be suitable for FLR detection.

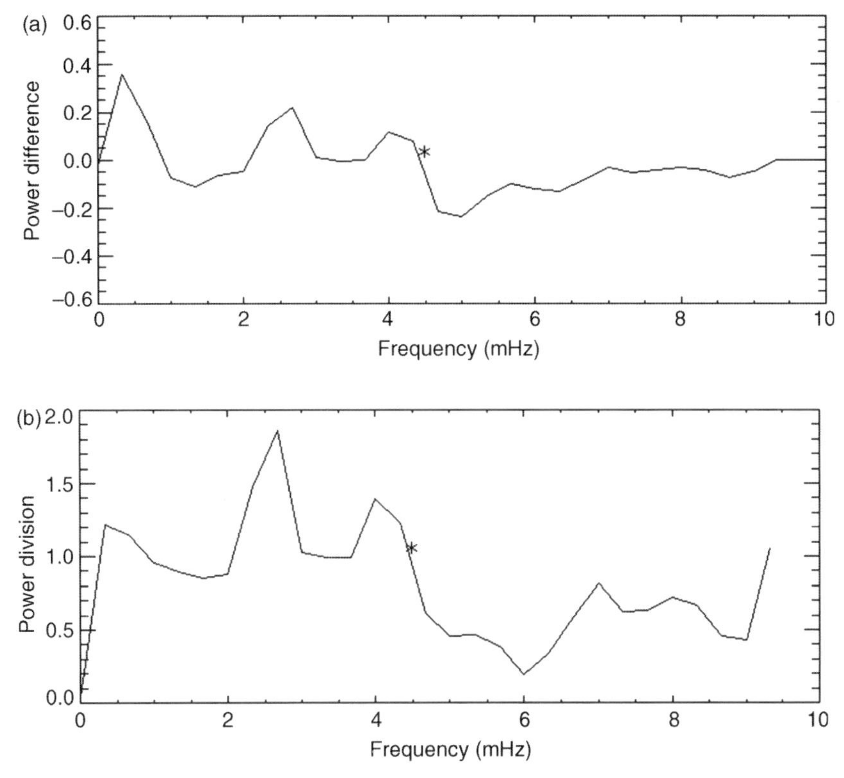

Figure 5.7 (a) Spectral power difference and (b) division for data recorded by the Fort Churchill (69.3° AACGM latitude) and Gillam (67.0°) magnetometers from 1600–1630 UT on February 9, 1995.

A method that has been used is to calculate the spectral power using data from the two orthogonal sensors and plot the ratio $X(f)/Y(f)$. This is also known as the H/D technique (Pilipenko and Fedorov, 1994; Vellante *et al.*, 1993b). The x-component spectra should peak at the FLR frequency. The $X(f)/Y(f)$ results for the Churchill line magnetometer data for February 9, 1995 as a function of latitude are shown in Figure 5.8. On comparing with the spectral power with latitude in Figure 5.2, we see some improvement in the FLR signature at the lower latitudes as the y-component spectrum $Y(f)$ has a normalizing effect on the x-component data. However, $Y(f)$ is not necessarily flat and depending on the variation of power with frequency in the y-component data, the FLR frequency identified by $X(f)/Y(f)$ may be shifted to lower or higher frequencies. Single station power ratio measurements were described in Menk *et al.* (2004), Pilipenko *et al.* (1999), and Vellante *et al.* (1993a).

A driven harmonic oscillator passes through a phase difference of 90° at resonance between the resonance oscillation and the driver term. Therefore, a cross-phase analysis using the $X(f)$ and $Y(f)$ spectra may yield the FLR where the phase difference is 90°. The frequencies where the cross-phase between $X(f)$ and $Y(f)$ is between 85° and 95° for each station are shown as asterisks in Figure 5.8. In general,

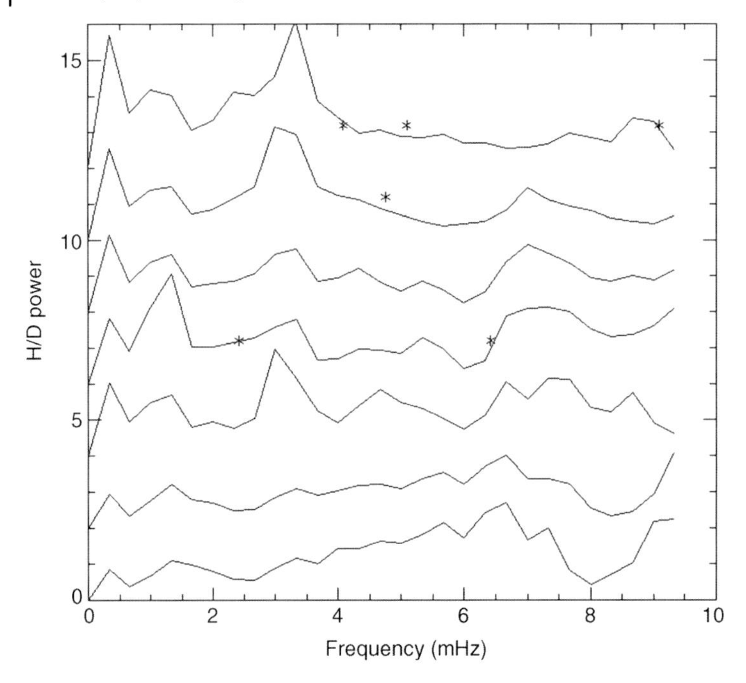

Figure 5.8 Spectral power division $X(f)/Y(f)$ from each of the seven stations in the same order as Figure 5.2. Asterisks (*) show frequencies where the phase difference is between 85° and 95°.

we have found that the $X(f):Y(f)$ cross-phase spectra at a single site are not useful for FLR identification. The $X(f):Y(f)$ cross-phase spectra appear to be dominated by wave polarization variations most probably from variations in the $Y(f)$ from the energy excitation source. For high latitudes and FLRs in the Pc5 band, these would arise from properties of the Kelvin–Helmholtz instability active at the magnetopause.

5.7 Cross-Phase from Latitudinally Separated Sensors

So far, the detection of FLRs in experimental data has focused on selected time series intervals. Multiple spectral plots become tedious to examine in this form and so spectra stacked by time on the horizontal axis with frequency on the vertical axis are used. These "dynamic" spectra are generated by computing successive FFTs, sliding the FFT analysis window along the time series. One spectral parameter that provides immediate FLR identification in the dynamic spectrum form is the cross-phase.

Figure 5.6c illustrates the principle of the cross-phase method for identifying FLRs in ULF data obtained from two latitudinally spaced sensors (Waters, Menk, and Fraser, 1991). For sensors spaced of order 100 km apart, it does not appear to matter what wave processes are present at frequencies other than the FLR (including any

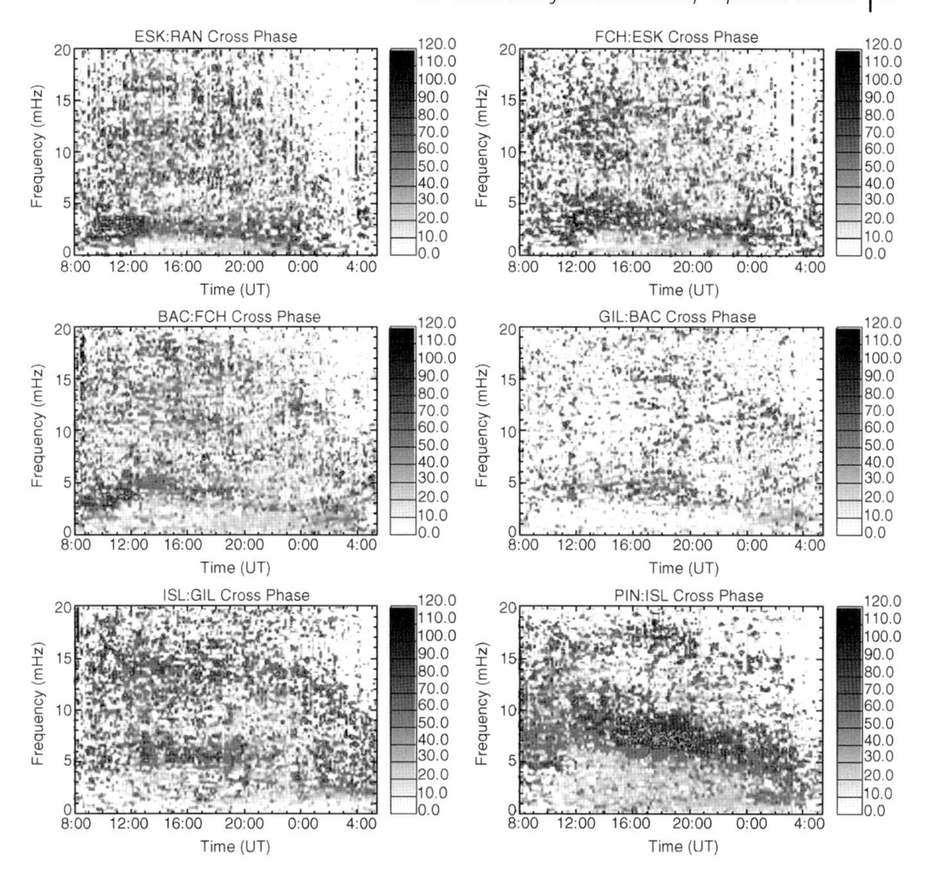

Figure 5.9 Dynamic cross-phase spectra data recorded on February 9, 1995 by the Churchill line of magnetometers of the Canadian array. Time axis is from 0800 to 0530 UT and local noon is at 1800 UT. Cross-phase scale is from 0° to 120°. (For a color version of this figure, please see the color plate at the beginning of this book.)

additional harmonics). The cross-phase function shows an extreme value at the resonant frequency. This is a positive value for the phase difference given by $\varphi_P - \varphi_E$ at most latitudes. Around plasmapause latitudes, the gradient in the Alfvén speed reverses and the values for $\varphi_P - \varphi_E$ at the resonant frequencies are minima (Dent *et al.*, 2006; Waters, 2000).

The cross-phase technique applied to data obtained from the Churchill line magnetometers recorded on February 9, 1995 is shown in Figure 5.9. The seven stations are taken as six consecutive latitude pairs decreasing in latitude from left to right and moving down the figure, and the FLRs, including harmonics, may be identified from the peaks in the cross-phase as a function of time of day. The changes in resonant frequency throughout the day reflect variations of the magnetic field and plasma mass density, weighted by the inverse of the Alfvén speed along the field from one ionosphere to the other. For the higher latitudes (topmost panels), the temporal variation shows an arch shape, mostly due to changes in the magnetic field, where the field is compressed around noon and more stretched at dawn and dusk. There are also magnetic field topology effects that alter the wave polarization and frequency in the more distorted field geometry (Rankin *et al.*, 2005). These are discussed in Chapter 7. At lower latitudes, the FLR frequencies increase. Generally, the magnetic field is well modeled at the lower latitudes and changes in the FLR frequencies are more directly related to plasma mass density variations in the magnetosphere.

5.8 Using ULF Wave Polarization Properties

In the spectral domain, ULF data may be processed to yield the rich mix of polarization properties with frequency. General techniques for the analysis of pure states and polarized signals applied to ULF data have been discussed in a series of papers during the 1970s–1980s (Arthur and McPherron, 1977; Samson, 1973; Samson and Olson, 1981). These techniques involve the eigenvalues and eigenvectors of the complex spectral matrix, computed from multichannel time series. Finding the eigenvalues and eigenvectors of the complex matrix is not required if we wish to search only the spectral information for specific polarization states (Olson, 1987).

The "vector" that describes a digitally recorded time series from m different data channels is

$$x^T = [x_1(t), x_2(t), \ldots, x_m(t)]. \tag{5.1}$$

The $x_j(t)$ $[j = 1 \cdots m]$ may be time series from sensors at the same site $[x(t), y(t), z(t)]$, sensors at different locations, or any combination of components and instruments. The spectral matrix S is formed by the outer product of the FFT of these time series:

$$S = \begin{bmatrix} Z_1 Z_1^* & Z_2 Z_1^* & Z_3 Z_1^* & \cdots & Z_m Z_1^* \\ Z_1 Z_2^* & Z_2 Z_2^* & Z_3 Z_2^* & \cdots & Z_m Z_2^* \\ \vdots & \vdots & \vdots & \vdots & \vdots \\ \vdots & \vdots & \vdots & \vdots & \vdots \\ Z_1 Z_m^* & \cdots & \cdots & \cdots & Z_m Z_m^* \end{bmatrix}, \tag{5.2}$$

where

$$Z_j(k) = \sum_{t=0}^{N-1} x_j(t) e^{-2\pi ikt}, \tag{5.3}$$

for $k = 0, 1, 2, \ldots, N-1$ and $*$ denotes complex conjugation. The spectral matrix S is square and for real input data $x_j(t)$, the diagonal elements are real and represent the autopowers of each time series. The off-diagonal elements of S are the cross-power spectra. The cross-phase is computed from the phase angle of the cross powers in the usual way for complex numbers. Equation 5.3 is usually implemented using the FFT. The spectra are smoothed by some window function that may be applied either in the time or frequency domain.

Given a polarization state ν, a normalized detector for this state may be defined (Olson and Samson, 1979):

$$D = \frac{\nu^{+} S \nu}{\mathrm{Tr}(S)}. \tag{5.4}$$

For real input data, the spectral matrix (Equation 5.2) is Hermitian and the trace of S ($\mathrm{Tr}[S]$) represents the total signal power (at some frequency) and therefore $0 \le D \le 1$ in Equation 5.4. In practice, the parameter D is raised to some power (usually 4) and this is analogous to the "order" of the polarization filter if the data are then filtered by this polarization state. This process has a geometric interpretation. From Equation 5.1, the spectral matrix spans an "m"-dimensional space and Equation 5.4 projects a particular vector ν onto this space (dot product). The detector D is large when the vector ν is close to a state in S that contains the outer product of ν with its complex conjugate, that is, the polarization state ν. Therefore, Equation 5.4 describes a parameter D that provides a measure of the presence of some polarization state ν in the data. From here on the discussion will be restricted to the detection of the FLR wave mode. The detection of other wave modes follows a similar development.

i) **Polarization from Orthogonal, Horizontal Magnetometer Sensors.** Suppose that time series data from only two orthogonal sensors from one magnetometer site are available. The polarization state should change with frequency if a resonance is present. Figure 5.6 shows that a resonance with large signal-to-noise ratio might be identified by constructing a polarization filter that describes a state ν that has most of the power in the $X(f)$ and minimal power in the $Y(f)$ channels. This process is equivalent to the normalized $X(f)$ component power described in Section 5.2. However, it has the flexibility to be able to easily alter the polarization state. An example was given by Olson (1987).
ii) **Polarization from Two Latitudinally Spaced Sensors.** Polarization properties (ellipticity and azimuth) are usually computed using data obtained from orthogonal sensors. There is no reason why the same process cannot be applied to the data from the same magnetometer components, spaced in latitude. The ideal amplitude and phasing for this case are shown in Figure 5.6. At the resonant frequency, the spectral amplitudes will be approximately equal and the phase difference is a maximum (or minimum). The "ellipticity" calculation from these data should show a trend toward circular at the resonant frequency, if the phase difference is not too large. Figure 5.10 shows an example using the same magnetometer x-component data as in Figure 5.9. The FLRs and harmonics are easily identified and follow the temporal and frequency properties of the cross-phase analysis shown in Figure 5.9. Monthly averaged dynamic ellipticity spectra for low-latitude stations were shown in Menk *et al.* (2012).

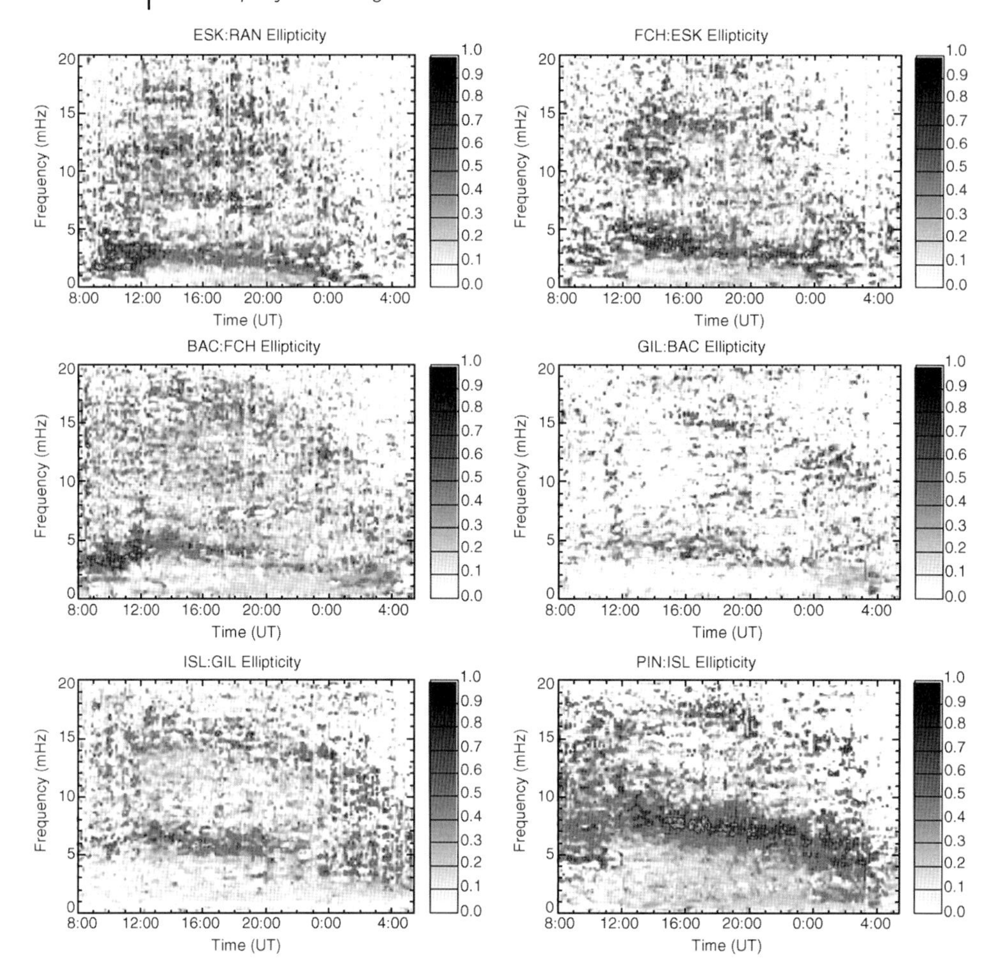

Figure 5.10 The "ellipticity" spectra computed from the north–south component magnetic field data from pairs of latitudinal spaced stations of the Canadian Churchill line. The processing used the same time series as Figure 5.9. (For a color version of this figure, please see the color plate at the beginning of this book.)

The pure state methods are not limited to two-station magnetometer time series data. The dimension of S can be as large as we need. The principles are easily extended to data from a latitudinal magnetometer array. All we need to do is define a "state" and then search S by the dot product process. An example of this type of multistation identification of FLRs was discussed by Plaschke *et al.* (2008). Using an analytic form for the amplitude and phase variations across an FLR, the spectral matrix was constructed from data obtained from multi-latitude sites in the Canadian magnetometer array. Further developments may provide robust techniques for searching large data sets for FLR signatures.

5.9 Automated Detection Algorithms

The ubiquitous nature of ULF waves and FLRs requires low-burden, data analysis methods. The detection of FLRs over long periods of time (days or years) has been solved by computing dynamic spectra. In particular, FLR signatures are easily identified by eye in cross-phase dynamic spectra. However, training the computer to identify the FLR frequencies in these spectra is a greater challenge. This section reviews the methods that have been developed to autodetect FLRs and describes some possible improvements.

The often-clear FLR signature in cross-phase spectra facilitates FLR frequency detection based on the statistics of cross-phase values (Berube, Moldwin, and Weygand, 2003). In summary, the method is as follows:

i) Compute the cross-phase and the $X(f)/Y(f)$ ratio dynamic spectra for magnetometer time series data obtained from two nearby, latitudinally spaced sites. The FFT length sets Δf and the number of frequency estimates in the Nyquist interval *n_freq*. The time step *t_step* determines the number of data points to step successive FFT windows through the data. Therefore, the spectra consist of cross-phase and power ratio values at "points" identified by each *t_step* and Δf.
ii) Define a range in time and frequency for computing the "local" statistics. For example, for $\Delta f = 1$ mHz, a 10 mHz range for the "local" frequency yields $n\_f = 10$ frequency estimates centered on the frequency of interest in the spectra. A similar process gives *n_t*, the number of time steps.
iii) Step through the spectra with this "local" analysis box and compute the average and standard deviations of the cross-phase values $(\Delta\varphi)_{av}$ and $(\Delta\varphi)_{SD}$.
iv) For each time step column in the spectra, select the cross-phase values $\Delta\varphi$ that have two standard deviations $(\Delta\varphi)_{SD}$ larger than the mean $(\Delta\varphi)_{av}$.
v) For these cross-phase values, develop a *t*-statistic, $(\Delta\varphi)_{av}/(\Delta\varphi)_{SD}$ and select the frequency (row in the spectra) that has the largest *t*-statistic (ensuring $t > 1$).
vi) For each frequency, examine the slope in the power ratio spectra. The criterion should be consistent with the power ratio variation in frequency described in Section 5.3.

The method depends on the selection of the "local" box and hence the values chosen for *n_t* and *n_f*. The user should experiment with these parameters for the particular data set of interest as the cross-phase statistics in the spectra depend on the latitude and station spacing. Berube *et al.* (2003) reasoned that cross-phase values "far" from the resonant frequency are random so $(\Delta\varphi)_{av}$ is close to zero. In reality, the method depends only on $(\Delta\varphi)_{av} \approx 0$ away from resonance, which does not require that the $\Delta\varphi$ values be random. In fact, the $\Delta\varphi$ values often appear constant and near zero with frequency and time for frequencies less than the resonant frequency.

The application of this auto-FLR detection algorithm is illustrated on the Canadian magnetometer data described previously, and the result is shown in Figure 5.11. These data represent a greater challenge for the method, which was originally described for low-latitude data. The Canadian data contain higher harmonics with

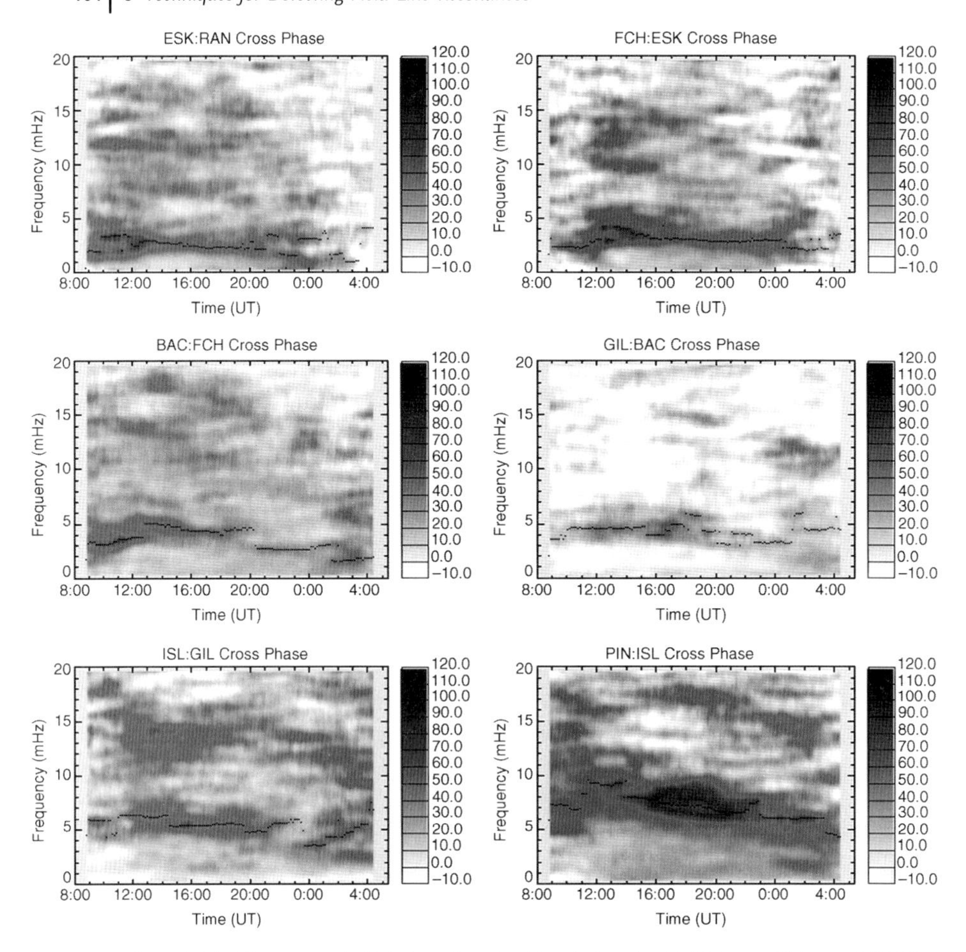

Figure 5.11 The automatic FLR detection algorithm in Berube, Moldwin, and Weygand (2003) applied to the cross-phase data in Figure 5.9. (For a color version of this figure, please see the color plate at the beginning of this book.)

varying clarity of the cross-phase FLR signature as seen in Figure 5.9. The method tracks the fundamental quite well when the cross-phase signature is clear, becoming less reliable as the signature fades.

There is no reason why the auto-FLR detection algorithm should be limited to cross-phase and power ratio spectra. Figure 5.12 shows the results when the method is applied to the ellipticity spectra shown in Figure 5.10. The algorithm seems to perform better than that shown in Figure 5.11. However, if we take the ellipticity or cross-phase spectra and run a standard image processing, median smoothing over the data, followed by a peak detection, then the FLR detection success rate is essentially the same.

The algorithm described in Berube, Moldwin, and Weygand (2003) has been used a number of times in the literature with varying degrees of success. Boudouridis and

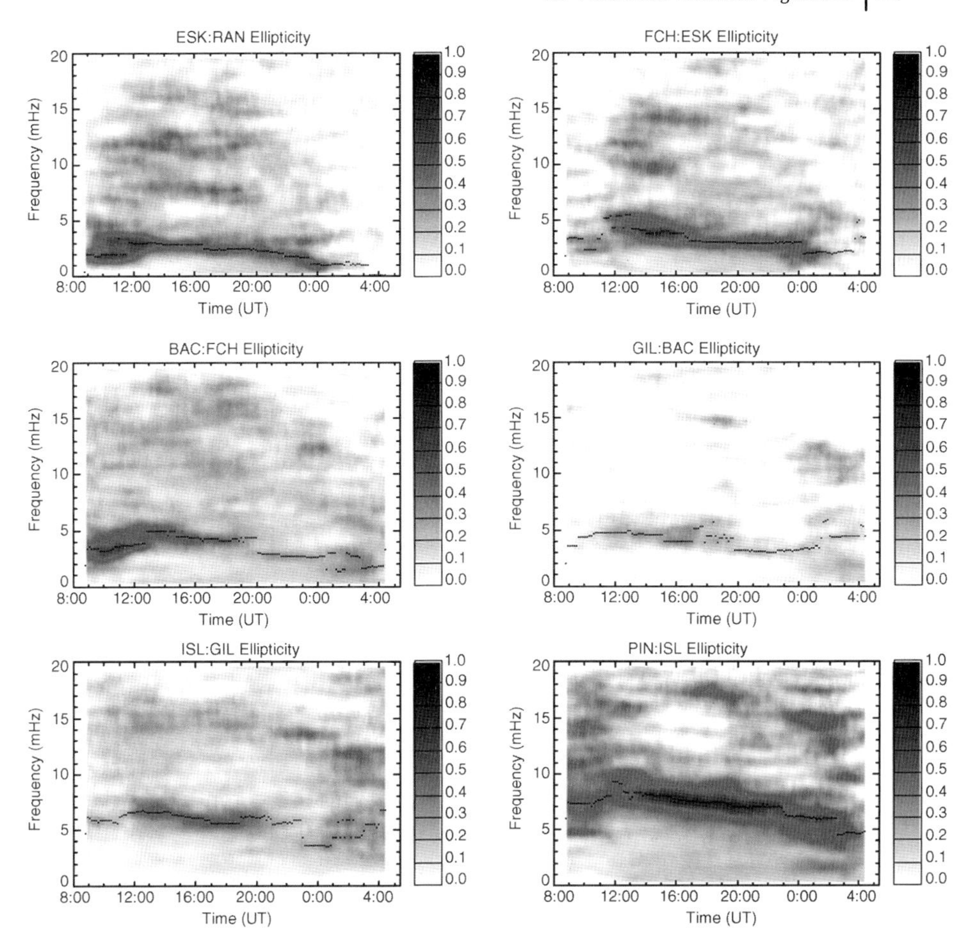

Figure 5.12 The automatic FLR detection algorithm in Berube, Moldwin, and Weygand (2003) applied to the ellipticity data in Figure 5.10. (For a color version of this figure, please see the color plate at the beginning of this book.)

Zesta (2007) compared FFT and wavelet transform implementations of the algorithm and found that the FFT gave the better results. In our experimentation with the method, the final check rarely required the power ratio step. This was also reported by Vellante *et al.* (2007). We could also eliminate the *t*-statistic step using a median smoothing. Clearly, more research is required in order to perfect techniques that reliably identify FLRs and harmonics in long data sets.

6
Ground-Based Remote Sensing of the Magnetosphere

The possibility of remote sensing the cold plasma mass density in near-Earth space using naturally occurring ULF waves has been recognized since the late 1950s. The relatively continuous ULF geomagnetic pulsations observed in magnetometer records were interpreted by Dungey (1954, 1955) as oscillations of magnetic flux tubes and it was soon recognized that these oscillations could provide information about near-Earth space (Gul'elmi, 1966; Obayashi and Jacobs, 1958). These early investigations showed the possibilities for remote sensing the radial plasma density profile in the equatorial plane using ground-based magnetometer records.

The techniques described in Chapter 5 provide experimental data that may be used by remote sensing algorithms to estimate the composition of the plasma in the magnetosphere. The most common constituent of space plasmas is hydrogen. In addition, the ionosphere contributes atomic oxygen, and helium ions are also detected. The remote sensing techniques based on experimentally determined FLR signatures provide the total plasma mass density. There are additional space- and ground-based methods that provide the electron concentration and ion-specific information, such as He^{+} from extreme ultraviolet emissions. These techniques are described in this chapter.

6.1
Estimating Plasma Mass Density

Resonances in the ULF (1 : 100 mHz) band are ubiquitous in Earth's magnetosphere. They arise from oscillations with wavelengths of size similar to the magnetosphere dimensions. The combined magnetic and fluid equations discussed in Chapter 3 show how the resonant frequency of the shear Alfvén mode in a cold plasma is related to the plasma mass density distributed along the ambient magnetic field, and the methods required to solve for the resonant sequence were also introduced. If the resonant frequency and a harmonic number are known for a field line from experimental data (Chapter 5), then we can estimate the plasma dynamics between conjugate ionosphere boundaries.

Magnetoseismology: Ground-based remote sensing of Earth's magnetosphere, First Edition. Frederick W. Menk and Colin L. Waters.

FLRs have been compared with resonances on a uniform stretched string (Sugiura and Wilson, 1964). Analogous parameters for the string tension, mass per unit length, and string length are the magnetic tension, plasma mass density, and the distance along the field between the ionospheres, respectively. In the magnetosphere, the plasma mass density is distributed along the field line in a nonuniform manner, so the shape of the oscillation along the field is nonsinusoidal compared with the uniform density, stretched string case. This has important implications for estimating the plasma mass density along the field, off the equatorial plane, and these are discussed later in this chapter.

Signatures of FLRs are seen in ground magnetometer data, mostly during the daytime. For a latitudinal array of magnetometers, the resonant frequencies may be determined over a large radial distance in the magnetosphere. An example using data from the Canadian magnetometer array is shown in Figure 5.9. Harmonics appear in the highest latitude data (RANK-ESKI) and we see the resonant frequency increases with decreasing latitude. The arch-shaped temporal variation is caused by a combination of plasma mass density in the magnetosphere and changes in the magnetic field (Mathie *et al.*, 1999b; Waters, Samson, and Donovan, 1995). The task is to extract the plasma mass density information from the data.

The process for estimating the plasma mass density from ULF resonances is as follows.

i) Obtain the FLR frequencies from the data (magnetometer, photometer, radar, etc.) as described in Chapter 5.
ii) Select a realistic magnetic field model.
iii) Choose a suitable functional form for the variation of the plasma mass density along the magnetic field (e.g., Equation 3.60).
iv) Solve the FLR wave equation for the plasma mass density as described in Chapter 3.

Since FLRs are detected for hours during the day, a magnetometer array represents a slowly rotating magnetosphere plasma mass density probe. Combining the FLR data from all station pairs in Figure 5.9, the temporal variation may be represented on a polar plot as shown in Figure 6.1. This format shows temporal and spatial estimates of magnetospheric plasma mass densities derived from FLRs measured on the ground at a range of latitudes.

There are a number of assumptions involved in producing plots such as Figure 6.1. The plasma is assumed to comprise solely of H^+ and so the number density units are given in H^+ cm^{-3}. If additional information regarding the ion composition is available, then the units may revert to kg m^{-3} or each ion number density may be specified. The plasma mass density along the field has the form specified (e.g., Equation 3.60) and the estimated values correspond to the plasma mass density in the vicinity of the equatorial plane, where the Alfvén speed is minimum.

The choice of the ambient magnetic field model affects the method of solution of the shear mode differential equation (Equation 3.34). For the inner magnetosphere, the dipole equations are often suitable, and the plasma mass density may be easily found by a small modification to the IDL code described in Section 3.4. For the outer regions, the ambient field responds to the impacts of solar wind energy on the

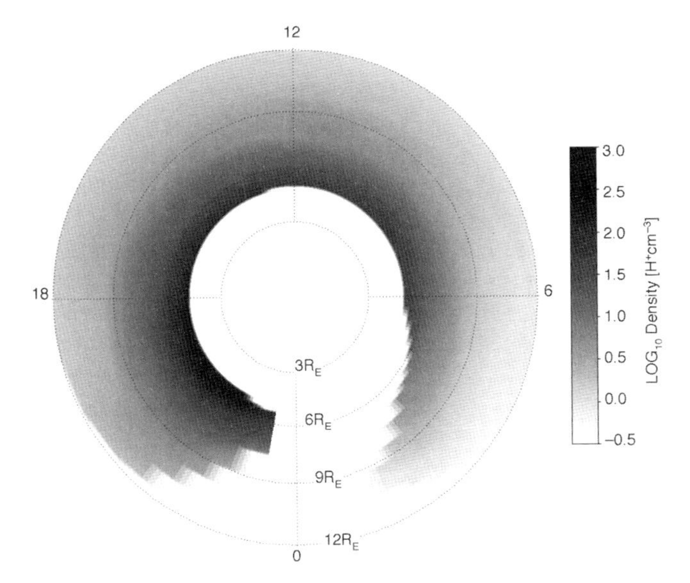

Figure 6.1 Logarithm of magnetosphere plasma mass density in units of $H^+ \, cm^{-3}$ as a function of radial distance and MLT, derived from FLRs detected with the CANOPUS magnetometer array on February 9, 1995. From Waters *et al.* (2006). (For a color version of this figure, please see the color plate at the beginning of this book.)

magnetosphere, and the associated magnetic field distortions are often modeled using the Tsyganenko descriptions. The shear mode differential equation must then include the geometric twists of the magnetic field. Finally, the differential equation to be solved assumes a pure shear mode oscillation. These assumptions are examined in detail in Sections 6.6 and 6.8.

6.2 Travel Time Method of Tamao

Seismology is the science of earthquakes, usually associated with the propagation of elastic waves within a planet. Helioseismology concerns acoustic pressure waves in the Sun and is discussed in Chapter 9. The Greek, "seio" means "to shake" to give "seismos," meaning an earthquake. Textbooks on Earth science describe three types of wave motions associated with earthquakes: body waves, surface waves, and normal modes. The longitudinal and transverse "body waves" are the P and S waves recorded by seismographs. For large earthquakes, normal modes appear with discrete frequencies, similar to the sound from a ringing bell. Since the early 1900s, studies of the time of arrival of S- and P-type seismic signals at multiple surface locations over Earth have revealed details of its internal structure. The essential features are an impulsive, point-localized source and wave propagation to convey the information away from the source to multiple detection infrastructure. A similar process for probing the properties of near-Earth space is discussed in this section.

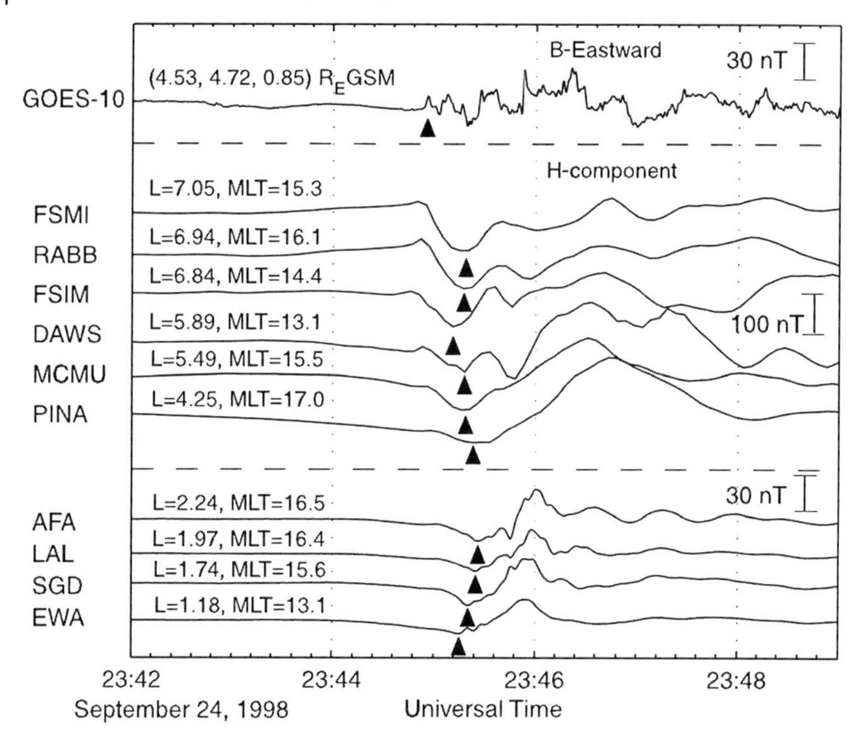

Figure 6.2 Magnetometer time series associated with a sudden impulse event on September 24, 1998. The FSMI-PINA data are from the Canadian sector, while AFA-EAW are from the SPMN and IGPP-LANL arrays. The arrival times are indicated by triangles. From Chi and Russell (2005).

The first requirement is for a spatially localized, impulsive source. For the magnetosphere, a promising candidate is the beginning portion of a sudden impulse (SI), known as a preliminary impulse (PI). Multiple data recording infrastructure is provided by surface magnetometer instrument arrays. This is a little different from the seismology case where the impulsive source and detection instruments are often located close to the same radial distance from Earth's center. For the magnetosphere, the detection instrumentation is located on Earth's surface, while the impulse source is many R_E further out in space. An example of multiple ground magnetometer records resulting from a shock in the solar wind is shown in Figure 6.2. The time series from a geostationary satellite (GOES-10) is also shown. The time of arrival at each location is indicated by triangles. From the arrival times and the location of the sensors, the plasma mass density in the magnetosphere can be inferred.

The principles of the method were described by Tamao (1964). Although he described in detail the fast and shear mode propagation properties for a uniform Alfvén speed in a cylindrical system, the application to the nonuniform Alfvén speed magnetosphere was also discussed. Given an impulsive disturbance near the magnetosphere nose, the method is illustrated in Figure 6.3. From the source to

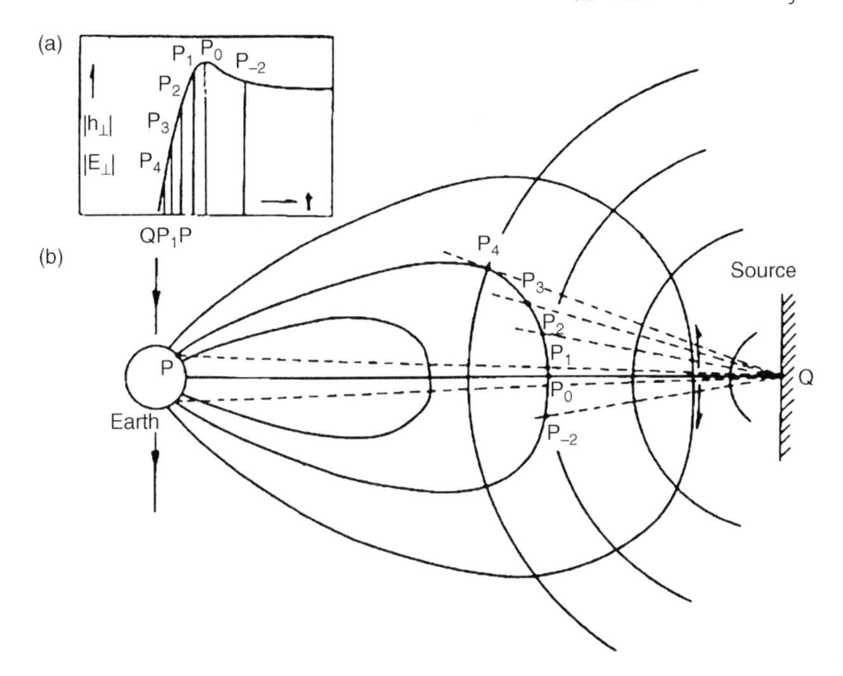

Figure 6.3 Illustration of propagation of hydromagnetic waves from a point source Q. The circles around Q indicate wave fronts of the fast (isotropic) mode. Conversion to the shear mode for a particular field line occurs at P4, P3, P2, and so on (b). The resulting time trace on the ground at point P is indicated in panel (a). From Tamao (1964).

the point P, a direct route via the fast mode is possible. However, a fast mode also converts to shear modes in the magnetosphere, as discussed in Chapter 3. In Figure 6.3, a path from $Q \rightarrow P_0$ as a fast mode followed by conversion to the shear mode and a route from $P_0 \rightarrow P$ along the field line to the ground are shown. The travel time is given by $\mathrm{d}s/V_\mathrm{A}$ (Equation 1.4) for each segment, so the travel times depend on the plasma mass density.

Taking the impulse arrival times and locations in Figure 6.2, the plasma mass density with radial distance may be estimated, using the assumptions described in Chi and Russell (2005). Values for the magnitude and field line geometry in the magnetosphere are required. If the location of the source impulse is known, then the number of parameters to be determined is reduced. Similarly, if the plasma mass density is described by some functional form, then the method is simplified. The R^α density model of Equation 3.60 is often used. Therefore, by minimizing the sum of squared differences between the observed and modeled travel times, the plasma mass density parameters are obtained. The inferred plasma mass density with radial distance using this method and for the travel time data in Figure 6.2 is shown in Figure 6.4. Time delays for Pc3–Pc4 signals measured across high-latitude ground magnetometer stations and with HF radars have been

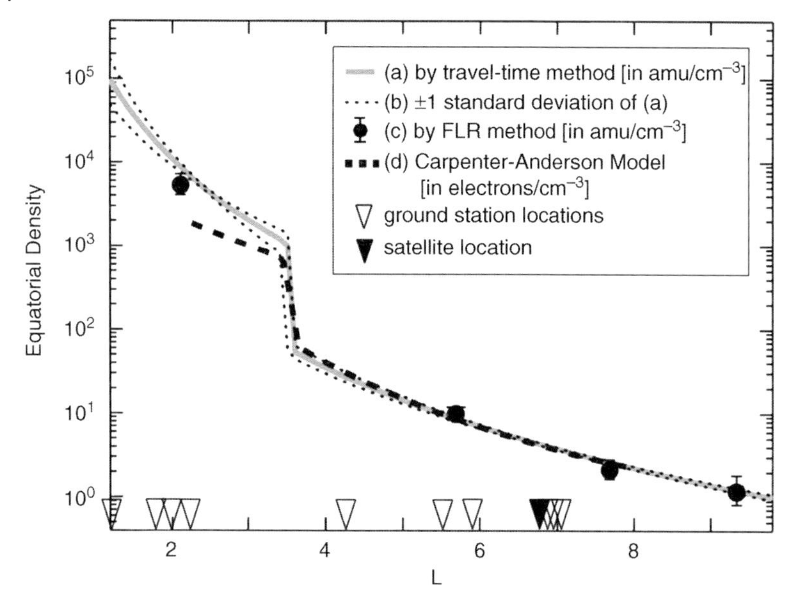

Figure 6.4 Equatorial plasma mass density with radial distance inferred from the impulse travel times of Figure 6.2 (shaded line) compared with empirical model prediction (dashed line) and FLR measurements (solid circles). From Chi and Russell (2005).

interpreted in terms of travel time along Tamao-type paths in Howard and Menk (2001, 2005) and Ponomarenko *et al.* (2005).

The travel time method uses MHD waves that have wavelengths much larger than Earth's radius. Therefore, any disturbance will quickly fill the magnetosphere volume, bounce around between leaky boundaries such as the plasmapause, ionosphere, turning points, and magnetopause, yielding complicated time series at ground sensors. This is why it is important to determine the travel times as soon after the impulse as possible. Tamao ignored effects of wave refraction on the travel time of the fast mode propagation portion. These might become significant for the inhomogeneous magnetosphere and were considered in Chi, Lee, and Russell (2006) using an impulse-driven, 3D MHD model described by Lee and Lysak (1999). For example, the incoming spherical wave fronts shown in Figure 6.3 might be distorted. This was the case in Chi, Lee, and Russell (2006), particularly near the plasmapause where the point source spatial disturbance on the magnetopause is not necessarily the same as that on the plasmapause, distorting the time delays within the plasmasphere. More research is required in order to determine the corrections required for the inferred plasmasphere densities using this method.

There are also questions related to the transfer of the signals from the magnetosphere through the ionosphere to ground sensors. Tamao (1964) recognized this and devoted a large portion of his mathematical derivation to estimates of the disturbance fields through the ionosphere. In his simplified treatment, the background magnetic field was vertical, perpendicular to a thin current sheet ionosphere. These approximations and the associated consequences for ULF wave signals detected by ground magnetometers are discussed in Chapter 8. Since the method depends on time delays and the steps described above consider travel time only in the magnetosphere, Kikuchi and Araki (2002) raised the question of an Earth–ionosphere waveguide travel path where the propagation speed would be that of light. For an impulse at the magnetopause equatorial plane, shorter travel times are expected for latitudes outside rather than inside the plasmasphere. If a signal travels a path at high latitudes to the ionosphere and then takes an ionosphere-assisted route to low latitudes, it will reach latitudes within the plasmasphere before a signal that took a magnetosphere–plasmasphere route. Recent MHD modeling efforts discussed in Chapter 8 are close to being able to investigate these effects.

In summary, the Tamao-based travel time method for estimating plasma mass densities in the magnetosphere has a number of requirements in common with the FLR-based method described in Chapter 5 and Section 6.1. The background magnetic field properties must be known. For the plasma mass density throughout the magnetosphere, a particular functional form simplifies both algorithms. Although the FLR-based method is best achieved using closely spaced pairs of latitudinally located magnetometers, the travel time method requires a spatial array. Examples of the travel time method in the literature have described events using latitude array data from similar local times. However, the travel time method can be extended to include a local time dependence in the functional form for the plasma mass density. The FLR method requires the excitation of shear mode resonances, which appear to be very common during the dayside magnetosphere. The travel time method requires impulsive excitation, which occur less frequently, and knowledge of the location of the "point source."

6.3 Determining Electron Density

Under appropriate conditions, circularly polarized very low-frequency (VLF) plasma waves interact with trapped magnetospheric electrons resulting in changes in particle pitch angle and energy. These waves are therefore important for the scattering of electrons into the loss cone and precipitation into the atmosphere, contributing to depletion of the radiation belts.

Consider a circularly polarized electromagnetic wave propagating parallel to the geomagnetic field. The wave $\boldsymbol{E}$ and $\boldsymbol{b}$ vectors are perpendicular to the field and rotate in a right-hand sense (relative to the propagation direction) at the wave frequency ω. The phase velocity is given by the dispersion relation:

$$\frac{\omega}{k} = v_{\mathrm{ph}} = \frac{c[\omega(\omega_{\mathrm{c}} - \omega)]^{1/2}}{\omega_{\mathrm{pe}}}, \tag{6.1}$$

where ω_{c} is the electron cyclotron frequency (Equation 3.5) and ω_{pe} is the plasma frequency (Equation 3.3). An electron moving at velocity v parallel to the field and toward and through the wave gyrates in the same way as the wave's electric field vector rotates, which relative to the electron is at Doppler-shifted frequency $\omega + kv$. If the electron gyrofrequency and Doppler-shifted wave frequency match, a resonance interaction occurs in which the electron (or wave) may gain energy at the expense of the wave (or electron) energy, depending on the phase angle between the particle and wave. The electron cyclotron resonance condition is given by

$$\omega = \omega_{ce} - kv \tag{6.2}$$

and typically occurs near the equatorial plane where the EM waves are amplified by the transfer of transverse energy from the spiraling electrons. The loss in electron energy decreases the pitch angle, causing the particles to mirror at lower altitudes or precipitate into the atmosphere. Ionospheric effects include absorption signatures on riometers and correlated optical auroral emissions (Hansen and Scourfield, 1990).

Naturally occurring lightning flashes couple energy to such right-hand polarized VLF electromagnetic waves, which propagate along dispersive paths between the hemispheres in field-aligned ducts. Frequency–time spectra from receivers near the conjugate point show characteristic falling tone traces that are historically called whistler signals because of their sound in audio recordings. The properties of whistlers were first described in detail by Storey (1953), who pointed out that measurements of the dispersion properties would provide information on the electron density at high altitudes.

If the highest frequency of the whistler is comparable to the minimum electron gyrofrequency along the path, then dispersion at the higher frequencies causes rising tones contiguously with the descending tones at lower frequencies. This results in a "nose" in the dynamic spectrum at the frequency of minimum group delay, which identifies the field line apex.

The group delay of a nose whistler of frequency f, integrated along the path, is (Carpenter, 1962; Smith, 1961b)

$$T = \frac{1}{2c} \int_{\text{path}} \frac{f_{pe} f_c}{f^{1/2} (f_c - f)^{3/2}} \text{d}s. \tag{6.3}$$

Thus, $T \propto \sqrt{N_e}$, and because of the gyrofrequency term in the denominator, the electron density at the apex of the flux tube has greatest effect on the whistler dispersion. The duct in which the trapped whistler propagates has an upper cutoff frequency of half the gyrofrequency and results from a field-aligned density enhancement of order 5–10% with a lifetime of a few hours (Smith, 1961a). Duct size is rather small, $0.035–0.07L$ in thickness and $\sim 0.3 R_E$ in longitude, and multiple ducts often exist simultaneously separated by $\sim 0.1 R_E$ (Angerami, 1970; Smith and Angerami, 1968). The physics of duct formation was described in detail in Cole (1971) and Walker (1978).

Whistlers are often scaled from the time delay measured at three frequencies to provide an equatorial electron density estimate with uncertainty of order 5% (Ho and Bernard, 1973). The limiting factor for the availability of whistler-based electron density data has been data storage requirements at the receiver and the human effort involved to scale the observations. The recent development of an automated whistler detector and scaling process (Lichtenberger *et al.*, 2008, 2010) is likely to revolutionize access to and the interpretation of magnetospheric electron density data.

Whistler signals were first described in 1894 (Preece, 1894) and have been used for systematic studies of the magnetospheric electron density for over 50 years (Al'pert, 1980). The first major discovery using whistlers was the existence of the "Carpenter knee" in the magnetospheric density profile, now known as the plasmapause (Carpenter, 1963). Since then, whistler observations have become one of the most important tools for remote sensing the spatial and temporal behavior of the magnetospheric electron density, in particular near the plasmapause (e.g. Carpenter, 1983, 1988; Carpenter and Park, 1973). New findings include the radial variation in density, the existence and properties of cross-L drifts, the penetration of storm-time electric fields and related convection, plasmapause dynamics and plasmaspheric refilling, diurnal and seasonal variations in plasma density, the existence of the plasmapause bulge in the dusk sector and its evolution to form plumes (e.g., Figure 3 in Carpenter, 1983), and modulation of the duct path length by magnetospheric ULF waves.

Whistler observations have also been compared with *in situ* measurements, confirming that the density distribution along plasmaspheric field lines follows a diffusive equilibrium model, although in the outer magnetosphere this is somewhere between diffusive equilibrium and an R^{-4} dependence (Carpenter *et al.*, 1981). This is illustrated in Figure 6.5, which compares electron density profiles measured from the upper hybrid resonance emission recorded by the ISEE-1 satellite within about 15° longitude of the ground observatory. Under these quiet conditions, the profiles follow an R^{-3} diffusive equilibrium dependence although at the plasmapause itself (not shown here), the power law dependence was in the range R^{-8}–R^{-10}.

Three limitations with whistler observations are the spatial and seasonal variation in lightning activity that triggers the whistlers, the duct cutoff frequency that restricts naturally occurring whistlers to $L \geq 3$, and absorption in the lower ionosphere that requires at least one end of the flux tube to be in darkness. The first two limitations are overcome by using whistler-mode signals produced by artificial VLF signals from high-power transmitters normally used for navigation and submarine communication (Andrews, Knox, and Thomson, 1978; Saxton and Smith, 1989). This opens up the possibility of measuring the Doppler shift of the received VLF signal, which may vary in response to changes in the electron density along the path or changes in path length, since phase path length

$$\phi \propto \int N_e^{1/2} ds. \tag{6.4}$$

Measurement of the group delay of the ducted whistler mode signal (by comparison with the subionospheric propagation time) yields an estimate of the equatorial electron density. In addition, the Doppler technique provides a sensitive way to measure radial motions of flux tube plasma in the equatorial plane of the

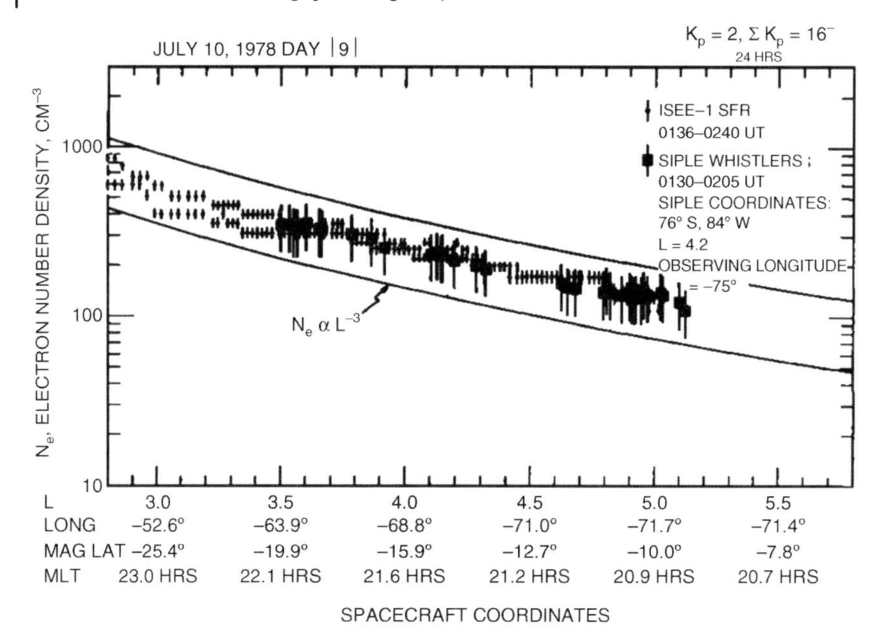

Figure 6.5 Comparison of electron density profiles in the magnetosphere determined from *in situ* measurements by the ISEE-1 spacecraft and from whistlers recorded at Siple, $L = 4.2$, on July 10, 1978. From Carpenter *et al.* (1981).

plasmasphere driven by azimuthal electric fields. Compressional mode ULF waves propagating from the solar wind through the magnetosphere may drive poloidal and toroidal mode field line resonances that are detected by VLF Doppler shifts and ground magnetometer cross-phase measurements, respectively. Radial velocities $\geq 10\,\mathrm{m\,s^{-1}}$, associated with electric fields of $\geq 0.1\,\mathrm{mV\,m^{-1}}$ and fluxes of order $10^{12}\,\mathrm{m^{-2}\,s^{-1}}$, can be detected at $L \approx 2.5$ (Andrews, Knox, and Thomson, 1978).

6.4 Verification of Ground-Based Mass Density Measurements

As we have seen, remote sensing of the magnetospheric equatorial plasma mass density may be achieved using measurements of ULF field line resonances obtained from cross-phase analysis of multistation ground magnetometer data. Although we can estimate the uncertainty in this approach, it is important to verify the technique. This is achieved by comparison of FLR-derived mass densities with independently obtained VLF whistler electron densities and *in situ* particle and wave observations. However, we need to be mindful that due to heavy ion mass loading, FLR-derived mass densities (expressed in $\mathrm{H^+\,cm^{-3}}$ or amu $\mathrm{H^+\,cm^{-3}}$) may at times be larger than electron number densities. This may provide additional information on the ion composition provided that the different techniques are suitably calibrated.

The first event-based comparison of VLF and ULF cold plasma densities was reported by Webb, Lanzerotti, and Park (1977) for $L \approx 4$, assuming electron density varies with altitude as $(r_0/r)^4$ (Equation 3.60), using a dipole magnetic field model, and assuming the FLR latitude occurs where ULF wave polarization reverses on the ground rather than at the location of the wave amplitude maximum. Although many of their VLF and FLR observations came from opposite sides of the plasmapause, when the observation locations overlapped the inferred electron and mass densities were the same within experimental error. This error arose mainly from uncertainty in determining the resonance location.

A detailed comparison of ULF-derived plasmaspheric mass densities and electron densities from both naturally triggered and artificially produced VLF whistlers was presented by Menk *et al.* (1999). Mass densities were calculated using the Taylor and Walker approximation (Taylor and Walker, 1984) for resonant frequencies obtained from cross-phase analysis of observations from the midlatitude SAMNET magnetometer array. Uncertainty in determining the resonant frequencies resulted in density error of ~15%, while uncertainty in determining the electron density was ~5%. Figure 6.6 compares results for October 16, 1990, when K_p was in the range 3+ to 2− and the plasmapause was near $L = 4.5$. The SAMNET mass densities at 08 UT (UT ≈ LT) are represented by filled circles, while solid and dashed lines represent values based on solutions of the wave equation for pure toroidal and poloidal mode oscillations in a dipole field and an $(r_0/r)^3$ density distribution (Orr and Matthew, 1971). Asterisks and squares denote VLF electron density

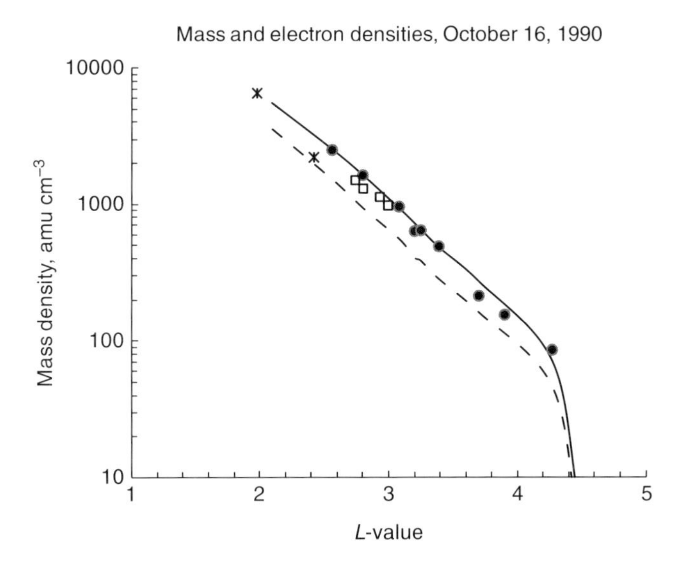

Figure 6.6 Equatorial plasma densities, recorded on October 16, 1990, based on SAMNET FLR measurements at 08 UT (UT ≈ LT; filled circles) and VLF electron density measurements at 07 UT (asterisks) and 20 UT (squares) compared with solution of the wave equation for pure toroidal and poloidal mode oscillations (solid and dashed lines, respectively). Adapted from Menk *et al.* (1999).

measurements at 07 and 20 UT from about 4 hours west in local time and agree well with results from other studies (Park, 1974). Clearly, the mass and electron densities are quite similar, and the FLRs are well described in terms of toroidal mode eigenoscillations. Similar comparisons of FLR mass densities and VLF electron densities (Clilverd *et al.*, 2003; Dent *et al.*, 2003; Menk *et al.*, 2004) confirm that under magnetically quiet conditions, the two techniques yield very similar density values.

We now consider the comparison of ground magnetometer FLR-based density determinations with *in situ* measurements by satellites. Loto'aniu *et al.* (1999) found that mass densities determined using cross-phase measurements from the CANOPUS magnetometer array spanning 61°–74° magnetic latitude, assuming an $(r_0/r)^4$ mass density distribution and a T89 magnetic field model, compared very well with CRRES electron density data when the satellite was passing over the ground stations under relatively quiet magnetic conditions.

Detailed comparisons have also been performed between ground magnetometer FLR-derived mass densities, VLF electron densities, and direct measurements of the electron density with the RPI payload on the IMAGE spacecraft (Clilverd *et al.*, 2003; Dent *et al.*, 2003). The very successful Radio Plasma Imager (RPI) experiment on the IMAGE spacecraft comprised an active radio sounder and passive receiver, which determined plasma wave modes and electron densities (Reinisch *et al.*, 2000, 2009). Figure 6.7 shows plasma density profiles determined using four different techniques for August 19, 2000, a geomagnetically quiet day ($K_p < 2$) in the recovery phase of a major storm that had occurred 7 days previously. Mass densities were calculated using FLR cross-phase observations from the SAMNET and IMAGE magnetometer arrays spanning the United Kingdom and Europe; VLF electron

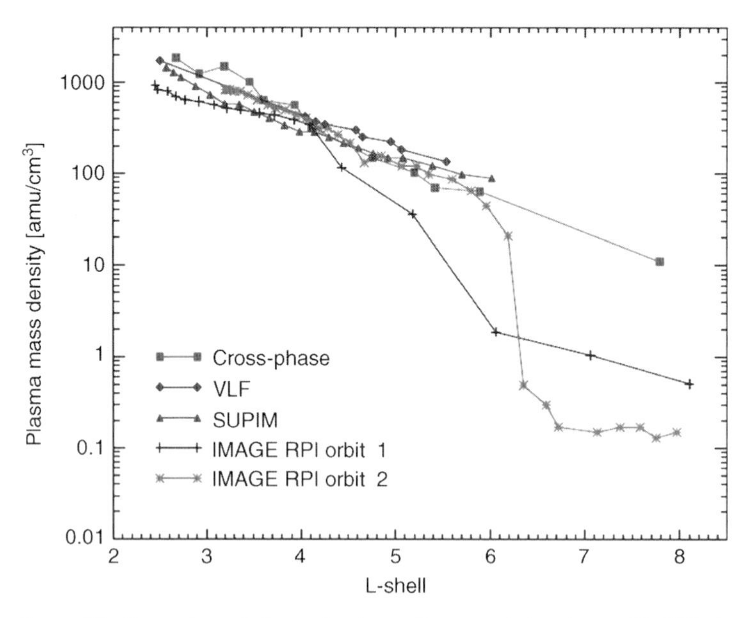

Figure 6.7 Comparison of plasma density profiles, recorded on August 19, 2000, determined using ground-based cross-phase VLF whistler and RPI *in situ* measurements and SUPIM model prediction. From Dent *et al.* (2003).

densities are based on natural whistler observations at Dunedin, New Zealand (~162°W geographic) and artificial Doppler whistler-mode observations at Halley (26°W), and the RPI observations are from passes over eastern Canada (orbit 1) and northern Russia (orbit 2). Also shown are predicted equatorial mass densities calculated with the *ab initio* SUPIM model (Bailey and Sellek, 1990). Despite the longitudinal separations between the measurement points, density profiles from all methods are very similar inside the plasmasphere, except for RPI orbit 1, when the IMAGE spacecraft was traversing an azimuthally localized outer plasmasphere density depletion. This excellent agreement validates the cross-phase technique and also demonstrates that there is little heavy ion mass loading on this occasion.

Figure 6.8 shows results from an independent but simultaneous study that also compared mass and electron densities using four separate techniques, for January 22, 2001, under moderately disturbed conditions ($K_p = 4$) (Clilverd *et al.*, 2003). In Figure 6.8a, magnetometer data are from the SAMNET ground array (solid line and error bars) and an $L = 2.5$ pair of low power magnetometers (LPM) in the Antarctic peninsula (circle), while electron densities are from ducted whistler-mode signals recorded by VLF receivers monitoring the same $L = 2.5$ flux tube (triangle). RPI data (dotted line) are from a nearby orbit with electron densities extrapolated to the equatorial plane assuming uniform field-aligned distribution. The dashed line represents the electron density distribution predicted by an empirical model (Carpenter and Anderson, 1992). It is clear that the cold plasma densities determined by all techniques agree within experimental error (represented by the error bars) over most of the plasmasphere. The study also compared the diurnal variation in FLR mass density and VLF and RPI electron densities with SUPIM model mass density predictions for this day at $L = 2.5$, shown in Figure 6.8b. The model has difficulty in representing temporal variations, possibly due to varying heavy ion concentrations during the day (Clilverd *et al.*, 2003).

We explore the topic of heavy ions by comparing in Figure 6.9 mass density measurements from the MPA instrument on board the 1989-046 LANL spacecraft with mass density determinations from the CANOPUS ground magnetometer array during February 1–19, 1995 (Waters, 2006). The D_{st} index is also shown. The ground magnetometers are 4.5 hours east of the spacecraft longitude, while the MPA data have been converted to equivalent H^+ number densities. The "background" trace provides an indication of the detection limit of the MPA instrument. Days 32–35, 39, and 42–46 were magnetically moderately disturbed, but conditions were quiet on days 50 and 51.

These observations are mostly from the plasmatrough region and show generally good agreement between the satellite and ground-based densities. The satellite data show an increase in density on most afternoons as it progresses through the plasmapause bulge. Increased densities on days 40–42 and 48–50 indicate active refilling periods. On some days (e.g., 38–39 and 42) the FLR-based mass densities are higher than the spacecraft values, most likely due to the presence of heavy ions.

Similar comparisons of FLR-derived mass densities with *in situ* electron density measurements in the plasmasphere and near the plasmapause show that increases in mass loading are most likely associated with enhanced O^+ populations in the inner plasmatrough following magnetic storms (Dent *et al.*, 2006).

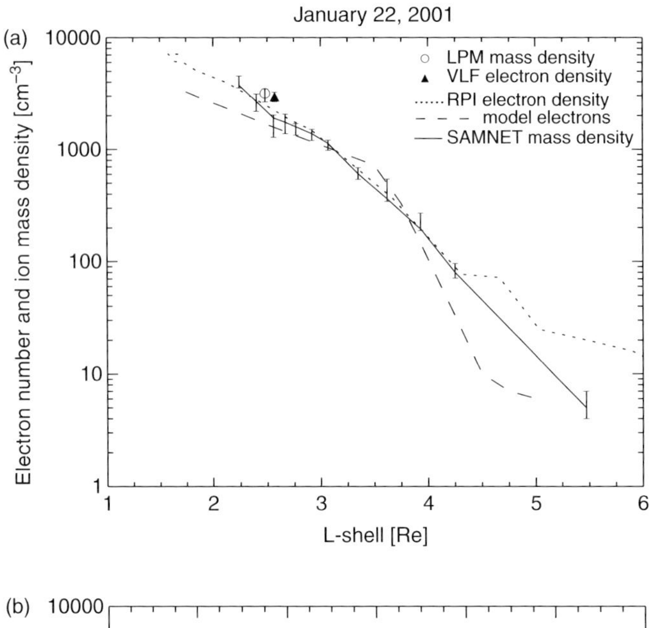

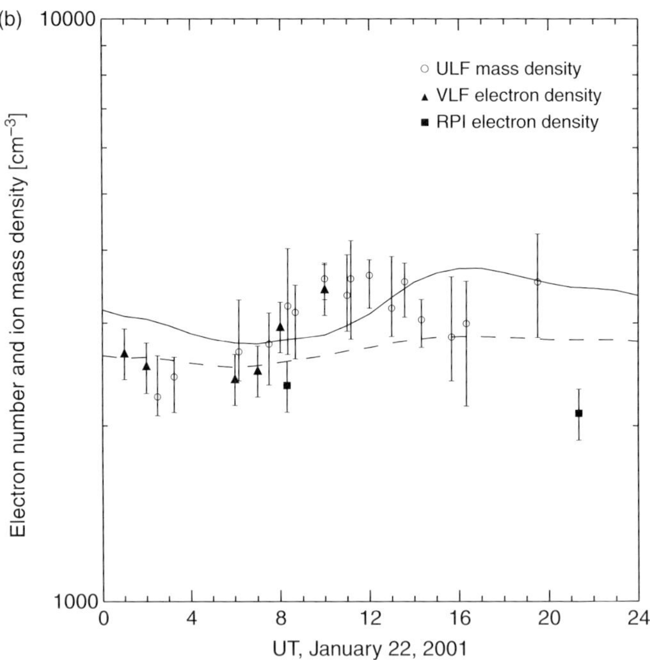

Figure 6.8 Variation in plasma mass density (from FLR measurements), electron density (from VLF measurements and RPI), and model predictions, on January 22, 2001 (a) as a function of latitude and (b) with UT. Solid line in (b) represents SUPIM model mass density prediction. Adapted from Clilverd *et al.* (2003).

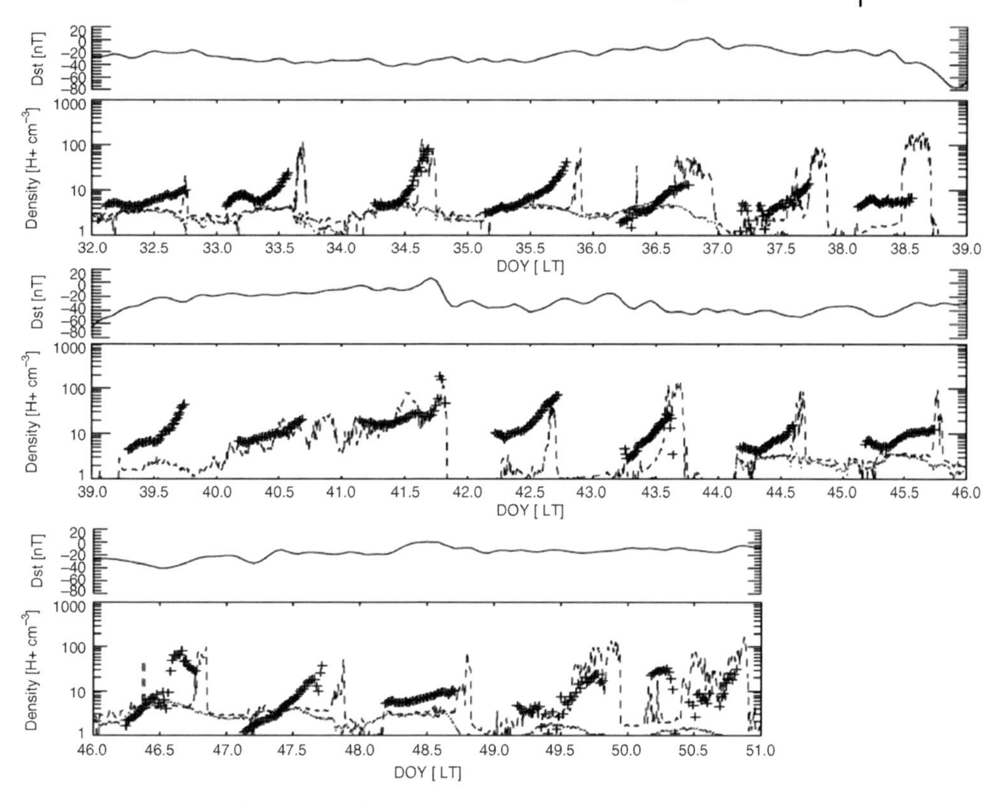

Figure 6.9 Comparison of equivalent H^+ densities measured *in situ* by the LANL 1989-046 satellite (dashed lines) with mass densities estimated using FLRs identified with the CANOPUS magnetometer array (plus symbols), during February 1–19, 1995. Dotted line shows the background spacecraft threshold. From Waters *et al.* (2006).

6.5 Determining Ion Concentrations

The availability of ion mass density and electron density data suggests the possibility of determining the plasma ion composition. This may be achieved using observations from the EUV imager experiment on the IMAGE spacecraft, which provides line of sight measurements of the intensity of resonantly scattered sunlight at 30.4 nm. This intensity is proportional to the He^+ concentration at the equatorial plane (Sandel *et al.*, 2003), and equatorial He^+ abundances are inferred from the measured brightness using the concept of effective path length (Gallagher, Adrian, and Liemohn, 2005) and the intercalibration process described by Clilverd *et al.* (2003). In this way it is found that for the plasma conditions described in Figure 6.8a, there was about 4% by number and 16% by mass of He^+ relative to H^+.

For a neutral plasma comprising $N = N_e$ electrons, x protons, and y He^+ and z O^+ ions, we have $N = x + y + z$ and $M\,(\text{amu cm}^{-3}) = x + 4y + 16z$. There are two equations and three unknowns. If the He^+ concentration is known from the above procedure, then the H^+ and O^+ abundances can also be found. By comparing the FLR-derived mass density estimates with VLF electron densities and IMAGE EUV-derived He^+ densities during a prolonged disturbed interval, Grew *et al.* (2007) were able to deduce the plasma composition and its variation near the plasmapause. They found that in the plasmasphere and plasmatrough, the $H^+ : He^+ : O^+$ composition by number was around 82 : 15 : 3, despite significant spatial and temporal changes in total mass density. However, within about 12 h of a strong decrease in D_{st}, the O^+ composition just outside the plasmapause rose to around 60%, suggesting the formation of an oxygen torus. The estimated ion concentrations did not agree with predictions from empirical models, which are based on statistical averages.

Using a similar process, Obana, Menk, and Yoshikawa (2010) showed that for upward flowing plasma during poststorm plasmaspheric refilling, the ion composition by number of $H^+ : He^+ : O^+$ was in the range 84–92 : 2–3 : 6–13, suggesting a somewhat rich heavy ion component. A further example is presented in Section 7.5 in the context of longitudinal variations in plasma density.

In summary, comparison of ground magnetometer FLR measurements with VLF whistler data and *in situ* particle measurements and imager data verifies the integrity of FLR mass density estimates and permits the plasma composition to be monitored. These are important results for magnetoseismology remote sensing.

6.6
Field-Aligned Plasma Density

The development of applications from basic research necessarily undergoes successive refinements as new techniques and information become available. The plasma mass density, remote sensing algorithms using FLRs is no exception. In this section, the assumptions of the method outlined in Section 6.1 are examined and various recent refinements are discussed. In particular, we examine whether the method can be extended to provide plasma mass density estimates off-equator, along the ambient magnetic field. Observations from the RPI experiment on the IMAGE spacecraft may be used to infer the field-aligned electron density, and are reviewed in Reinisch *et al.* (2009).

The power law described by Equation 3.60 originally described the radial variation in plasma mass density in the equatorial plane, but it has also been used to model the plasma mass density profile along magnetic field lines. Schulz (1996) found that for a dipole field, values of the power law exponent α, between 0 and 6, gave rise to evenly spaced harmonics. These eigenfrequencies form a linear harmonic series when plotted against integer harmonic number. For a dipole magnetic field model, setting $\alpha = 6$ makes the Alfvén speed, V_A independent of radial distance, that is, if the mass density ρ is proportional to $1/r^6$, then V_A varies as $(1 + 3\cos^2\theta)^{1/2}$. Harris (1974) proposed that $\alpha = 4$ be used for the outer magnetosphere (plasmatrough), while $\alpha = 3$ has been associated with a diffusive equilibrium plasma for the inner magnetosphere (plasmasphere) (Poulter *et al.*, 1984; Warner and Orr, 1979).

Typical spectra of ULF oscillations obtained from spacecraft data show multiple (up to eight) FLR harmonics (Takahashi and McPherron, 1982), which leads to the question of whether such rich resonant structures contain additional information on the plasma spatial distribution. Much effort has been focused on the information that might be contained in the harmonic ratios (e.g. Denton *et al.*, 2006). In particular, the data might provide information about the plasma mass density along the field and, therefore, the functional form might not always match a particular power law index. This has led to recommendations for $\alpha = 1$ or less, instead of 3 or 4. Negative values of α would imply a plasma mass density increase with increasing radial distance in space. Refinements to the algorithm for probing off-equator locations would provide much sought-after values for the Alfvén speed in the magnetosphere, particularly for MHD wave modeling researchers. However, before we attempt to estimate off-equator plasma mass densities using FLRs, an examination of the uncertainties in the method should be considered.

The experimental FLR data come with uncertainties in the frequencies, either from the Δf of the spectra or from the spread in frequencies in a statistical study of FLRs. If we assume the functional form of plasma mass density given by Equation 3.60, then estimates for ρ_0 and α are available from the data. The associated uncertainties in ρ_0 and α can be estimated given the observed harmonics and the frequency uncertainties Δf. A conservative value for the uncertainty in the harmonic frequencies would be $\Delta f = 0.2\,\text{mHz}$ for five observed harmonics in spacecraft data such as GOES at geosynchronous orbit. The process is identical if Δf is different for each harmonic.

Using the computer algorithm described in Section 3.4, the FLRs can be calculated over the ρ_0 and α space. If the calculated frequencies are within the experimental error, then these frequencies are assigned a value of 1. Therefore, for five observed harmonics, each (ρ_0, α) pair is assigned a number between 0 and 5, depending on whether the calculated values are within the range of the observed values $\pm\Delta f$. An example for a typical set of five harmonics seen at geosynchronous orbit is shown in Figure 6.10. The darkest section from $\rho_0 \approx 16$ down to $\rho_0 \approx 9$ shows where all five harmonics are within the experimental uncertainties. This gives a range in α of -2 to 3. Therefore, the experimental uncertainties in the observed harmonics do not really allow a distinction between α be it positive or negative (density enhancement) and do not have sufficient precision for ρ_0 or α to be expressed beyond the integer place.

A similar example was presented for midlatitude ground magnetometer data in Figure 10 in Menk *et al.* (1999) where it was noted that "For a constant second harmonic frequency and varying fundamental frequency, a variation in (power law index) α from 3 to 6 or from 1 to 3 at $L = 2.8$ requires a change in plasma mass density of only +2%. It is therefore not surprising that the mass density index is a highly variable parameter, and the actual value chosen in model calculations probably is not critical." The same point has been made by others (Maeda *et al.*, 2009; Vellante and Förster, 2006) and also applies to estimates of electron densities using whistler signals, as discussed in Section 6.3.

This insensitive nature of the "best" harmonic sequence to a range of α was recognized by Carpenter and Smith (1964), and is apparent from the tabulated values in Cummings, O'Sullivan, and Coleman (1969). This insensitivity must be related to the relatively small effective section of the field line that determines the harmonic sequence.

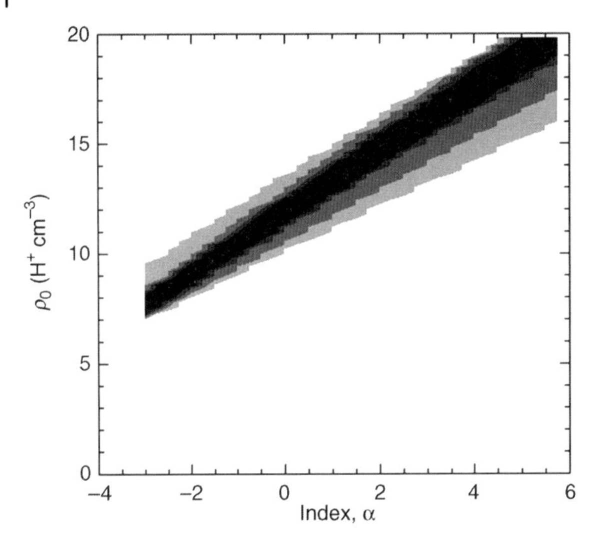

Figure 6.10 Range of mass density and power law exponent, given a five harmonic sequence and associated uncertainties of $\Delta f = 0.2$ mHz.

A discussion of the effective field line length that determines the ULF harmonic frequencies was given by Vellante and Förster (2006). Using the equations of Taylor and Walker (1984), they obtained an integral expression for the fundamental period T:

$$T \approx \frac{L^4}{\pi C}\left[4z_i \int_{-z_i}^{z_i} (1 - z^2)^6 \cos^2\left(\frac{\pi z}{2z_i}\right)\rho(z)\mathrm{d}z\right]^{1/2} \tag{6.5}$$

where z_i is the field line footprint in the ionosphere and $z = \cos\theta$ for θ the colatitude. The normalized integral in Equation 6.5 was termed the effective density ρ_{eff}/ρ_0 and this parameter indicates the portion of the field line that contributes to the resonant period.

In order to illustrate the implications for magnetoseismology, we have used plasma density values from the field line interhemispherical plasma (FLIP) model (Richards, 2002) for southern hemisphere winter (June 2001) conditions and have taken the plasma mass density data for local noon at various latitudes. In order to show the differences in the northern hemisphere compared with the southern hemisphere, we have modified the presentation of ρ_{eff}/ρ_0. Since the effective section of a resonating field line is in the vicinity of the Alfvén speed minimum (equatorial plane), the integration of Equation 6.5 was calculated from the equator to the ionosphere for each hemisphere, as shown in Figure 6.11. The normalization was over the full field line to give maxima of the normalized integral ~0.5 for each hemisphere. The difference in the plasma mass density in the north compared with

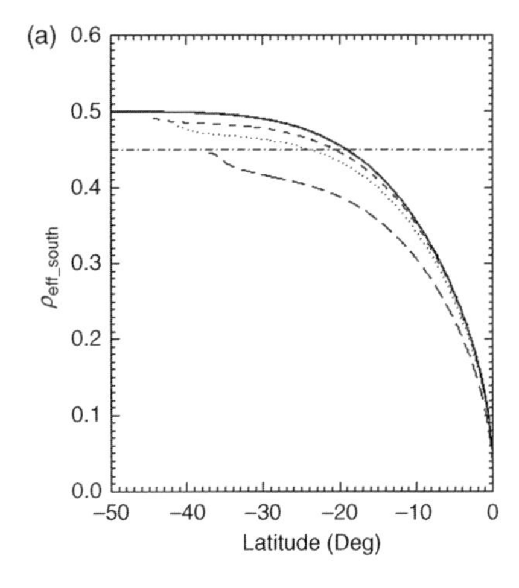

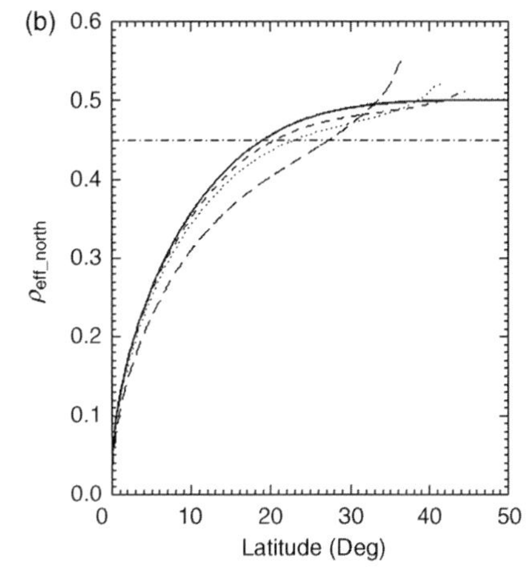

Figure 6.11 The fraction of effective field line length that determines the FLR harmonic sequence, calculated using Equation 6.5, as a function of latitude. The L values for each curve are $L = 1.6$ (long dash), $L = 1.8$ (dot), $L = 2.0$ (dash), $L = 4.0$, 5.0, 6.0 and 6.6 (solid). The south and north hemispheres are shown in panels (a) and (b), respectively.

the south is seen for the lower latitude L-values of 1.6 and 1.8. Winter for the southern hemisphere gives a smaller plasma mass density and hence a larger V_A at the lower altitudes. For field lines with $L > 3$, the nonequinox conditions appear to be irrelevant.

Figure 6.11 indicates that 90% of the information that determines the harmonic sequence for field lines with $L > 3$ is contained within $\sim 20^\circ$ of the equatorial plane. The effective field line length for a range of latitudes (L-values) may be determined using this 20° point for the $L = 6.6$ case in Figure 6.11, that is, $\rho_{eff} = 0.45$. We now evaluate the integral in Equation 6.5 for field lines over $L = 1.4$–6.6. The colatitudes for both the northern and southern hemispheres are calculated for the 0.45 point of the integration. These define the section of each field line that contributes to each harmonic series. The results are shown in Figure 6.12. For $L > 3$, only about 30% of a field line length contributes to the harmonics. Therefore, information on plasma mass densities beyond $\sim 20^\circ$ from the equatorial plane for $L > 3$ is unavailable using the ULF harmonic technique. For $L < 3$, a greater portion of the field is sensitive to the harmonics.

For $L > 3$, the ULF harmonic frequencies when plotted with (integer) harmonic number are linear within experimental uncertainty, which means the frequencies convey a limited amount of information about the plasma mass density. In principle, only two harmonics are sufficient to construct the line on which the harmonics will lie. Additional harmonics (if available) may assist in improving the accuracy of the slope and frequency intercept of the harmonic sequence.

A particular ULF harmonic sequence yields a unique α and ρ if we use Equation 3.60 or a unique coefficient set for any other model (e.g., polynomial) in the minimum least-squared sense. However, the relatively small effective field line length makes these parameters relatively insensitive to the harmonic sequence. We

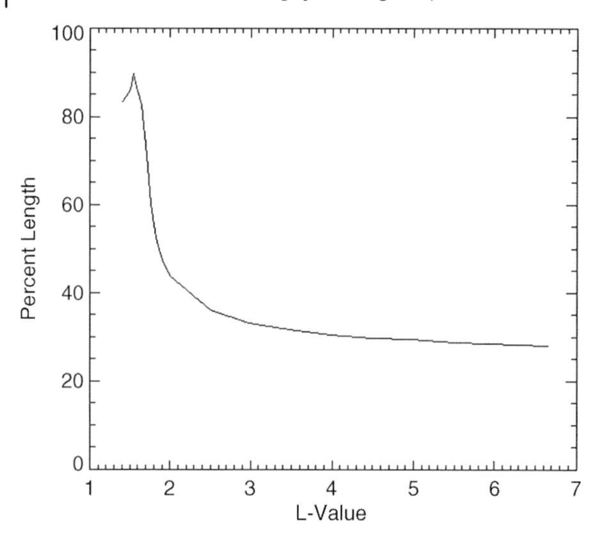

Figure 6.12 The percentage of field line length that contributes 90% to the integral in Equation 6.5, as a function of latitude (L-value).

can, for instance, use plasma mass density profiles that are wildly different at low altitudes for a given field line. If these different plasma mass density functions (along the full field line length) are actually similar within the effective near-equatorial region, then the harmonic sequences calculated for each density function will be the same, within experimental uncertainty.

6.7 Plasma Density at Low Latitudes

Figure 6.12 shows quite a distinct change in character for $L < 2$. The inner plasmasphere is a difficult region for obtaining cold plasma density information. Rocket and balloon-borne sensors do not reach sufficient altitudes and although there are many low Earth orbit (LEO) spacecraft around 1000 km altitude, there are few measurements between LEO and geosynchronous orbit. Some data are available from spacecraft with highly elliptical orbits that may give snapshots of information in the $L < 3$ region. Therefore, a ULF harmonic-based technique that could estimate the plasma mass density in the inner plasmasphere would be very useful, even if the plasma composition details were lacking. At present, the most convenient way to test the ULF harmonic method for estimating plasma mass density at low latitudes is to use a plasmasphere composition model.

Figure 6.12 indicates that the ULF harmonic approach for estimating plasma mass density in the $L = 1.5$–3 region requires consideration of a larger section of the field line than that required for $L > 3$. The toroidal harmonics depend on the integration involving V_A along the field line. The variation in the Alfvén speed along the geomagnetic field is different for low and high latitudes. For high latitudes, the Alfvén speed is minimum near the equatorial plane where the contribution to the ULF harmonic values is confined. For low latitudes, $L < 2$, the proportion of the field line that contributes to the harmonic sequence increases quickly. At $L = 1.8$ and $L = 1.6$, 54% and 86%, respectively, of the field line length are involved in determining the harmonics. For the smaller L-values, the large variation in the Alfvén speed near the ionosphere influences the eigenvalues (harmonic sequence). This was the interpretation for the unequally spaced harmonics shown in Figure 9 in Poulter, Allan, and Bailey (1988) for $L = 1.7$ and $L = 1.33$. In these results, the harmonic frequencies when plotted against harmonic number do not lie on a straight line.

For low latitudes, the contribution to the ULF harmonic sequence for a given L-value includes a larger proportion of the field line, which should provide the opportunity to remote sense the plasma mass density down to the topside ionosphere (Hattingh and Sutcliffe, 1987). The density model of Equation 3.60 is inadequate, being unable to model both the density variation along the outer field line distances and the sharp increase in the topside ionosphere. Therefore, the estimation of plasma mass density using FLR harmonics for $L < 3$ compared to $L > 3$ becomes much more challenging.

The Harmonic Derived Density (HARDD) method (Price *et al.*, 1999) uses observed multiple harmonics to estimate the plasma mass density at low latitudes without assuming any functional form. At the heart of this method is the formulation of the shear mode differential equation in matrix form, as discussed in Section 3.4. However, instead of solving for the eigenfrequencies, the (nonlinear) inverse problem is solved, given a harmonic sequence, to obtain plasma mass density estimates at specific locations along the field line. The procedure is as follows:

i) Obtain the FLR harmonics from the data.
ii) Provide an initial estimate of the plasma mass densities at the sample sites of Equation 3.44. These are usually obtained from computer simulations such as the FLIP model.
iii) Generate the matrix (Equation 3.44) for each measured harmonic and calculate the determinants.
iv) If the determinants are nonzero, use a Newton method iteration on the mass densities and recompute.

An example of results from this method, compared with three density model outputs, is shown in Figure 6.13. There were three observed harmonics at $L \approx 1.8$, so this provides plasma mass density estimates at three altitudes. The advantage of this method is that there is no need to superpose any functional form on the plasma mass density in order to obtain a solution. However, there are a number of assumptions. The method, at present, assumes hemispherical symmetry in the Alfvén speed. The most restrictive problem is the lack of harmonics seen in low-latitude ULF data. For the event in Figure 6.13, the third harmonic at $L \approx 1.8$ was $\sim$100 mHz. The bandwidth of incoming fast mode excitation energy often only reaches $\sim$70–80 mHz at this latitude.

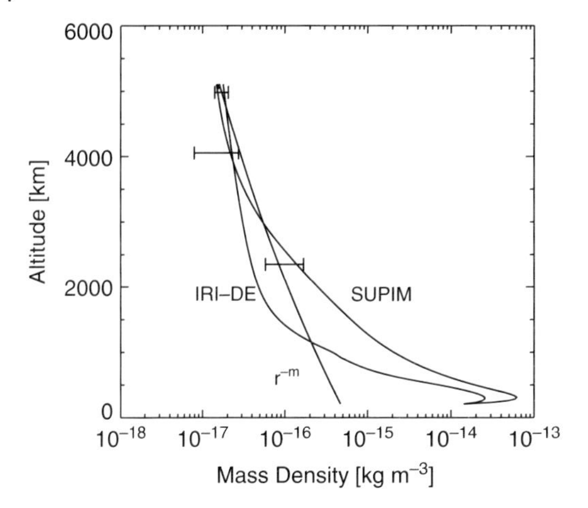

Figure 6.13 Comparison between plasma mass densities versus altitude determined using the International Reference Ionosphere plus diffusive equilibrium models (IRI-DE), an r^{-6} model, HARRD method (horizontal error bars), and Sheffield University plasma model (SUPIM). From Price *et al.* (1999).

6.8
Plasma Density at High Latitudes

The method of remote sensing magnetospheric plasma mass densities using ULF resonances requires solving the shear mode differential equation assuming some ambient magnetic field. A dipole magnetic field can be expressed in the right-hand orthogonal dipole coordinate system that gives the analytic expression $h_\alpha = r\cos\lambda$ in Equation 3.34. However, the magnetic field beyond about 5 R_E becomes less dipole-like and depends on local time and magnetic activity. A representation of the ambient magnetic field for average solar wind conditions with zero IMF B_y from the Tsyganenko04 and IGRF models was shown in Figure 2.6. The field lines are traced starting from a geographic latitude of 57°. The "stretched" field on the flanks and nightside decreases the resonant frequencies, while a compressed field on the dayside increases them (Waters, Samson, and Donovan, 1996). However, high-latitude ULF data show an additional feature superposed on the expected "arch" variation of FLR with time of day. An example of the average fundamental FLR with time for Davis, Antarctica (−74.5 AACGM latitude) is shown in Figure 6.14. Superposed onto the arch is a decrease near local noon. Clearly, the geometry of the ambient magnetic field must be included in any algorithm for magnetoseismology.

A first step away from a dipole description for the magnetic field is to use the formulation of Equation 3.34 and compute the h_α numerically. This was done by Singer, Southwood, and Kivelson (1981) for the Olson–Pfitzer magnetic field model and by Waters, Samson, and Donovan (1996) for the Tsyganenko96 model. For the "toroidal" ULF wave mode, the h_α are computed for two field traces, separated in longitude. For example, if we begin the calculation at 57° geographic latitude, then we first convert this start location to GSM coordinates. We then trace the field to the other hemisphere using the selected field model. For the second field line, we begin with the same GSM latitude

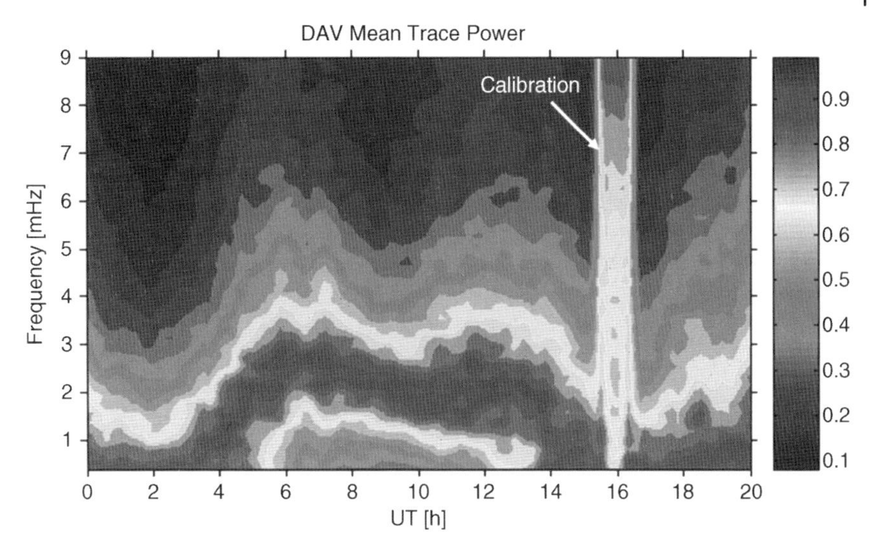

Figure 6.14 Normalized, mean trace spectral power over 0.1–9 mHz from magnetometer data recorded at Davis, Antarctica. The data are for the full year 1996 and local magnetic noon is near 0940 UT. (For a color version of this figure, please see the color plate at the beginning of this book.)

as the first, but separated by a degree or so in longitude. The h_α can be normalized (by the value at the start), as the derivatives are the important quantity in Equation 3.34 for determining the frequency. This approach works quite well, provided the ambient field does not twist so that the plasma displacement ξ always points azimuthally. However, we can progress one step further to extend the magnetic "ribbon" calculation of Singer, Southwood, and Kivelson (1981) and remove the assumption of a pure shear mode.

Torsion components of the ambient magnetic field become more pronounced for nonzero IMF B_y and depend on local time. In these cases, the two field line "ribbon" calculation is certainly possible, but the ribbon twists and turns in space as the field is traced between the hemispheres and the actual wave perturbation does not oscillate in the ribbon plane. We need a formulation that considers the properties of a flux tube. Rankin, Kabin, and Marchand (2006) and Kabin *et al.* (2007b) described refinements that do this. Their system uses nonorthogonal coordinates suitable for any magnetic field as the method numerically calculates the geometric tensor coefficients that describe a flux tube. The relevant equations in the contravariant (superscript) and covariant (subscript) form are

$$\begin{aligned}
\frac{\partial e^1}{\partial t} &= -i\omega(g^{11}e_1 + g^{12}e_2) = -\frac{V^2}{J}\partial_3 b_2, \\
\frac{\partial e^2}{\partial t} &= -i\omega(g^{12}e_1 + g^{22}e_2) = -\frac{V^2}{J}\partial_3 b_1, \\
\frac{\partial b^1}{\partial t} &= -i\omega(g^{11}b_1 + g^{12}b_2) = \frac{1}{J}\partial_3 e_2, \\
\frac{\partial b^2}{\partial t} &= -i\omega(g^{12}b_1 + g^{22}b_2) = \frac{1}{J}\partial_3 e_1,
\end{aligned} \tag{6.6}$$

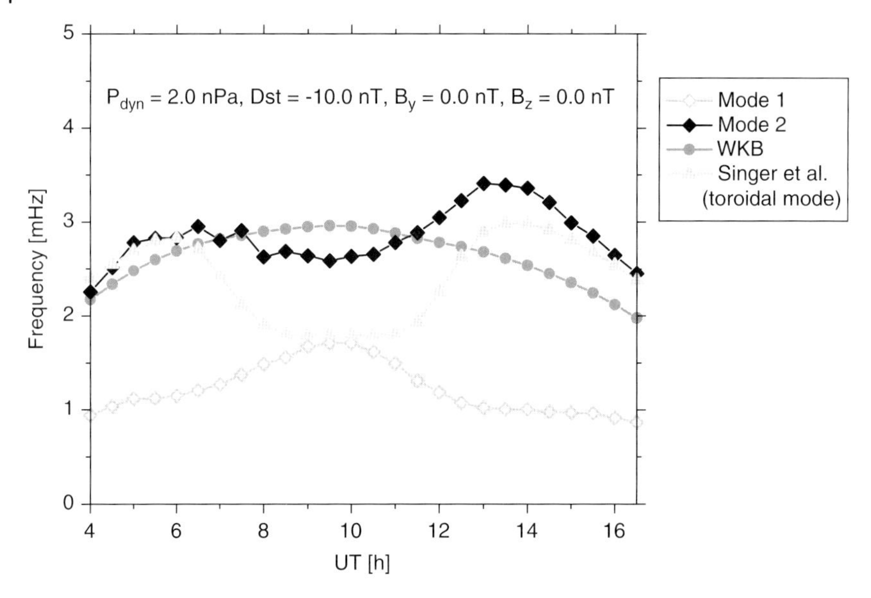

Figure 6.15 Various FLR calculations for the data in Figure 6.14: First eigensolution (open gray diamonds) and second eigensolution (Equation 6.6; filled black diamonds), WKB solution using Equation 1.4 (circles) and Equation 3.34 (triangles).

where $V^2 = 1/(\mu_0\varepsilon)$ with $\varepsilon = 1 + c^2/V_A^2$ and the g^{ij} are the components of the geometric tensor.

The solutions to Equation 6.6 yield eigenmodes given the boundary conditions that the electric fields are zero at the ionosphere. Figure 6.15 shows the first two modes, compared with the results from Equation 3.34 and also Equation 1.4 for the Tsyganenko field model traced from Davis, Antarctica. The ambient field has both curvature and torsion, which depend on the local time. Magnetic torsion τ as a function of distance s along the field is

$$\begin{aligned} \tau(s) &= \left[\hat{\mathbf{B}}(s) \times \mathbf{n}(s)\right] \cdot \frac{d\mathbf{n}(s)}{ds}, \\ \mathbf{n}(s) &= \frac{d\hat{\mathbf{B}}(s)}{ds}. \end{aligned} \tag{6.7}$$

For the ambient field traced from Davis, increased torsion tends to increase the resonant frequency in the morning and afternoon, while around noon the curvature pulls the resonant frequency down. The latitudinal extent of these effects may be predicted using the Tsyganenko model for various conditions. Figure 6.16a shows the torsion from Equation 6.7 calculated from a Tsyganenko96 model as a function of latitude. The noon decrease is certainly apparent over the range of latitudes in the Scandinavian magnetometer array, as shown in Figure 6.16b. Therefore, these properties of the magnetic field need to be included if details of the plasma mass density are to be extracted from ULF resonances at high latitudes.

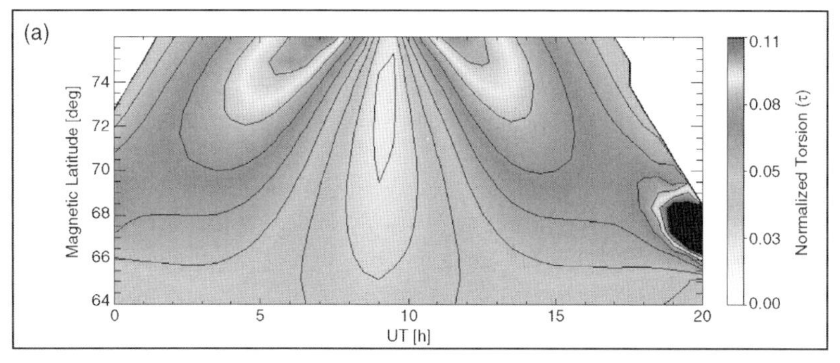

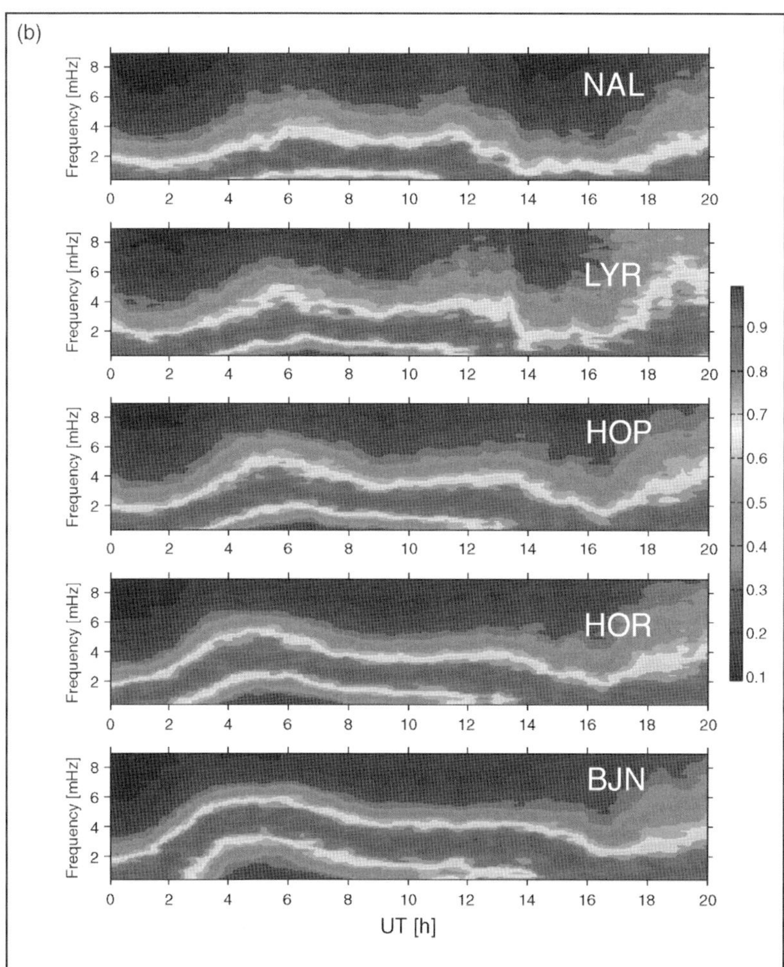

Figure 6.16 Extent in latitude of field line tension and torsion that affects FLR frequencies. (a) Estimates using the Tsyganenko 1996 model. (b) Normalized trace spectra of the horizontal components of magnetometer data from various stations in the Scandinavian IMAGE magnetometer array for the year 1996. (For a color version of this figure, please see the color plate at the beginning of this book.)

Finally, we return to the discussion of a ULF harmonic sequence plotted against (integer) harmonic number. The dipole magnetic field used with the power law plasma mass density model yields harmonic sequences that lie on straight lines. This is a two-parameter model and additional harmonics do not provide additional information. Harmonic sequences that do not lie on a straight line have been observed, even when experimental uncertainties have been considered. There are two possible sources; the plasma mass density and the magnetic field. For low latitudes, the nonuniform plasma mass density at the lower altitudes produces harmonics that do not lie on a straight line. For high latitudes, the geometry of the magnetic field can produce the same effect and these must be included in order to extract the effects caused by the plasma mass density variations.

7
Space Weather Applications

The magnetosphere is a dynamic region responding to the variability in solar wind fields and particle fluxes. The core cold plasma of the magnetosphere, with ionosphere–magnetosphere coupling, provides the base population for higher energy particle dynamics. This complex region is where we place expensive spacecraft technology. Magnetoseismology allows us to remote sense these areas. In this chapter, space weather applications of the remote sensing methods described in Chapters 5 and 6 are illustrated.

7.1
Magnetospheric Structure and Density

Measurements of field line resonances using arrays of ground-based magnetometers provide information on the plasma mass density distribution throughout the magnetosphere and its temporal variation, as shown in Figure 6.1. A model equatorial mass density profile, and the corresponding variation in FLR resonant frequency with latitude, was presented in Figure 4.9. Dashed lines illustrate a local density "biteout," and horizontal dotted lines represent resonances at the same frequency on either side of a well-formed plasmapause. Magnetic conditions represented here are typical of poststorm recovery phase.

The low-latitude limit of FLRs is a function of geomagnetic activity, the incoming ULF wave power, the coupling of this power to FLRs, the ionospheric damping of the resonance and the resonance Q, and the efficacy of resonance detection techniques. As discussed in Section 6.6, a significant portion of low-latitude field lines is immersed in the ionosphere, and FLRs are not likely to be detected much below $L = 1.4$ (Menk, Waters, and Fraser, 2000). At low latitudes, the resonance width is typically 200–400 km, the resonance Q is in the range 1.5–3, and the normalized damping rate γ/ω_R is ~0.15–0.4 (Menk *et al.*, 1994; Menk, Waters, and Fraser, 2000; Ziesolleck *et al.*, 1993).

Figure 7.1 summarizes multistation observations of resonant frequencies within the inner magnetosphere. The peak in frequency near 38° latitude ($L = 1.6$) arises because plasma mass loading by ionospheric heavy ions decreases the frequency at lower latitudes, as discussed in Sections 4.4 and 6.7 (Hattingh and Sutcliffe, 1987;

Magnetoseismology: Ground-based remote sensing of Earth's magnetosphere, First Edition. Frederick W. Menk and Colin L. Waters.

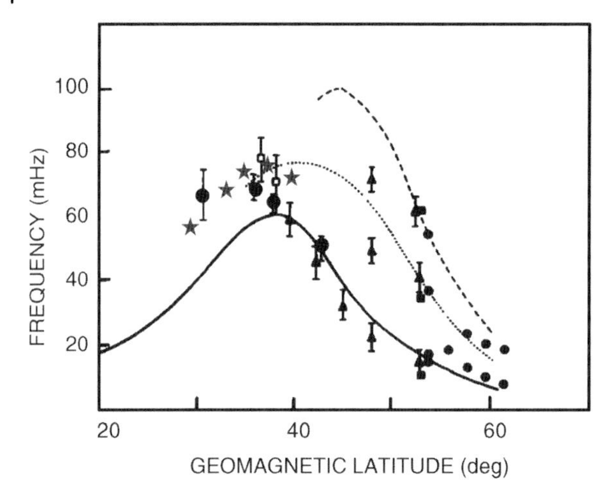

Figure 7.1 Summary of experimentally determined inner magnetosphere resonant frequencies. Symbols represent results from various studies: stars – Menk, Waters, and Fraser (2000); triangles with error bars – Waters, Menk, and Fraser (1994); solid circles at higher latitudes – Baransky *et al.* (1990); large solid circles with error bars – Hattingh and Sutcliffe (1987); solid squares at higher latitudes – Kurchashov *et al.* (1987); and open squares with error bars – Green *et al.* (1993). Solid, dotted, and dashed lines denote first three harmonics of the model toroidal mode eigenfrequency for spring equinox solar maximum and including $E \times B$ drifts. Adapted from Waters, Menk, and Fraser (1994).

Waters, Menk, and Fraser, 1994). This effect varies with local time. Also shown in Figure 7.1 are model toroidal mode eigenfrequencies for the first three harmonics at spring equinox (Poulter, Allan, and Bailey, 1988). Current modeling efforts outlined in Chapters 3 and 8 use 2.5D representations of wave propagation throughout the coupled magnetosphere–ionosphere system, including actual ionospheric conditions for the day in question.

At midlatitudes, the Alfvén velocity and density profile are affected by the plasmapause, which itself varies with magnetic activity. For example, Figure 7.2 shows the variation in plasma mass density between $L = 2.6$ and $L = 4.4$ during October 16–18, 1990, a magnetically quiet period following a $K_p = 4$ disturbance on October 15, 1990. Several features are evident: a sharp plasmapause near $L = 4.2$ on October 16 which largely decayed by the next day; diurnal and day-to-day variations in density; a strange shelf-like density feature around $L = 4$ on October 17 suggestive of heavy ion mass loading, and an offset between daytime and nighttime density values due to the longitudinal separation of the two magnetometer arrays. These effects are described in more detail in the following sections.

7.2 Plasmapause Dynamics

The plasmapause position and shape vary considerably with changing magnetic activity. Figure 7.3 shows ground array FLR measurements and the corresponding

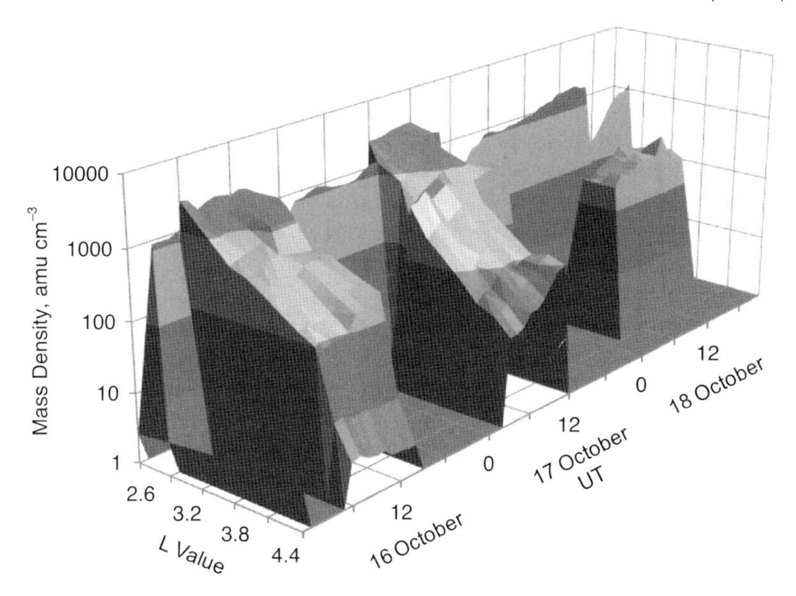

Figure 7.2 Plasma mass density map for October 16–18, 1990, based on observations from two magnetometer arrays separated by 10 h in local time. Adapted from Menk *et al.* (1999). (For a color version of this figure, please see the color plate at the beginning of this book.)

equatorial mass density estimates for January 1, 1998, following almost 3 weeks of magnetically quiet conditions (Menk *et al.*, 2004). In Figure 7.3a, filled circles and diamonds represent respectively cross-phase measurements obtained from closely spaced station pairs and H/D values from single stations. Vertical arrows indicate the plasmapause and magnetopause locations predicted by empirical models. No distinct plasmapause is evident, and the mass density profile resembles the Carpenter and Anderson (1992) model electron density values for a saturated plasmasphere.

Figure 7.4 illustrates how the noontime mass density profile changes with magnetic activity, represented by K_p in panel (b). The plots span March 7–19, 1998. Densities were estimated from FLR measurements with the SAMNET and IMAGE magnetometer arrays spanning the United Kingdom and Europe. Tick marks on the x-axis correspond to a step of $L = 1$, and each date label is at $L = 6$. Solid and dotted line profiles are for odd and even numbered days, respectively. These profiles depict the evolution of the plasmapause over a storm cycle. No plasmapause is evident on March 7, 1998, but following a storm on March 10 and 11, 1998 the plasmapause has eroded to $L \leq 3$. A well-defined plasmapause is present around $L = 3.4$, 4.2, and 4.2 on March 12–14, 1998, respectively. A steep plasmapause on March 15, 1998 decays to form two density gradients and by March 19, 1998, no distinct plasmapause remains.

Observations for an extremely disturbed interval are summarized in Figure 7.5 for the 2003 Halloween storm. Geomagnetic conditions during this interval are shown in the Figure 7.5a. D_{st} reached −383 nT at 2300 UT on October 30, 2003, in

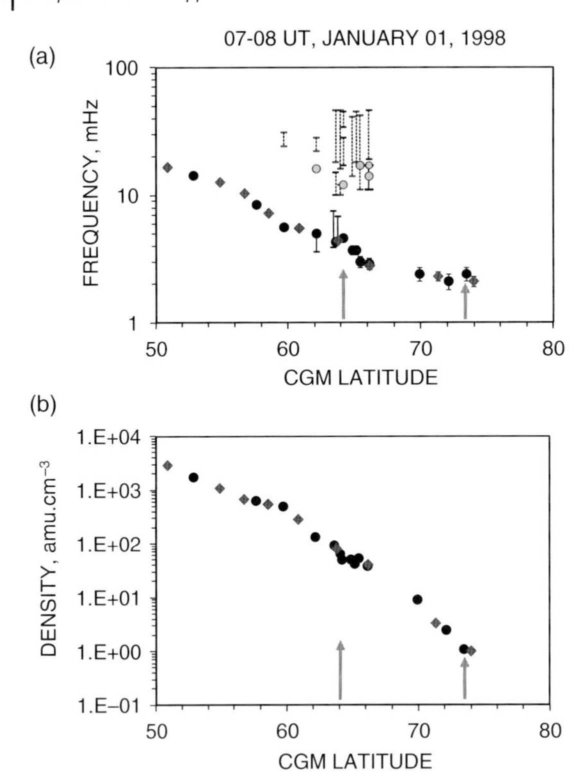

Figure 7.3 (a) Ground magnetometer-derived FLR measurements and (b) mass density estimates for January 1, 1998 under magnetically quiet conditions. Circles and diamonds represent respectively the cross-phase measurements obtained from closely spaced station pairs and *H*/*D* values from single stations. Vertical arrows indicate the plasmapause and magnetopause locations predicted by empirical models. FLR harmonics are seen near the expected plasmapause location, but no distinct plasmapause is evident. Adapted from Menk *et al.* (2004).

connection with solar wind speeds exceeding 1000 km s^{-1} and extreme IMF B_z conditions. Plasma mass densities estimated from the United Kingdom and European ground array cross-phase measurements and the T01 magnetic field model are presented in Figure 7.5b. Densities increased significantly immediately after the sudden commencements, possibly due to an increase in plasmaspheric O^+ concentration resulting from rapid ionospheric outflow. A similar density enhancement was found in the North American sector (Chi *et al.*, 2005) suggesting a globally enhanced heavy ion population. Recent analysis of toroidal mode FLRs measured by the CRRES spacecraft data (but with no direct He^+ measurements) has shown that an oxygen torus may readily occur in all local time zones near the plasmapause during storm recovery phases, most likely due to heating of the underlying ionosphere by plasmaspheric thermal electrons, resulting in an increased O^+ scale height (Nosé *et al.*, 2011). There is also likely to be an

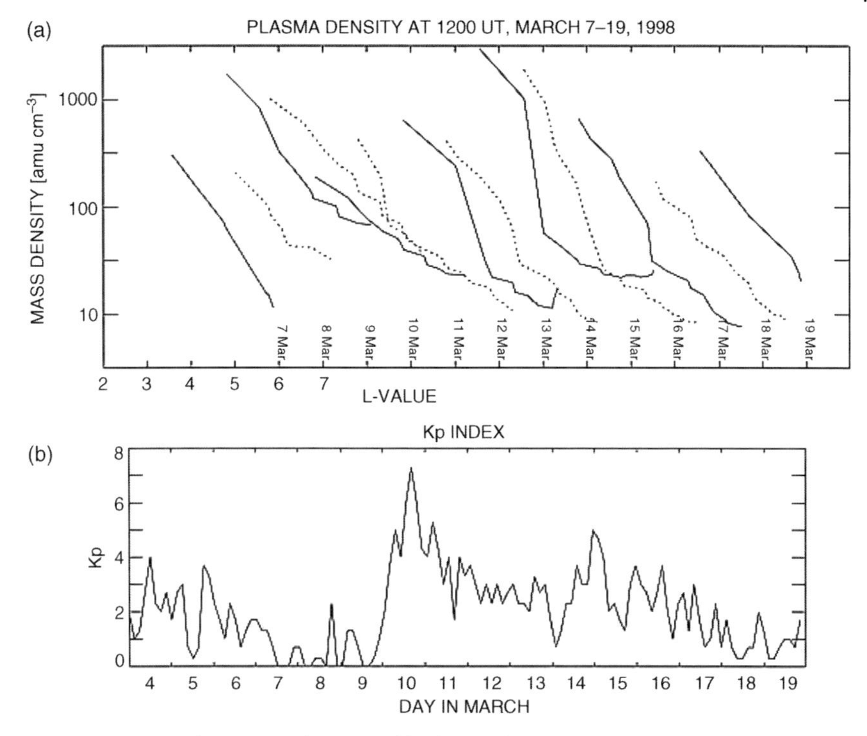

Figure 7.4 Noon plasma mass density profiles for March 7–19, 1998 compared with magnetic activity K_p. Each date label in panel (a) is at $L = 6$, and x-axis tick marks denote steps of 1.0 L. Figure 8 in Menk *et al.* (2004).

interactive process involving the oxygen torus, plasmasphere, and O^+ rich ring current.

FLR measurements provide a time resolution of order 20 min, and considerable variation can occur in radial density profiles over timescales of an hour. This is evident in Figure 7.6 that shows hourly mass density profiles between 0720 and 1220 UT on January 8, 1998, a mildly disturbed day ($\Sigma K_p = 14.7$) during which the signatures of large solar wind shocks were recorded by ground magnetometers at 0831 and 1017 UT (Menk *et al.*, 2004). The magnetopause is expected to have moved from $13.1 R_E$ at 0800 UT to $9.7 R_E$ at 1000 UT. The mass density profiles were determined using a realistic magnetic field model and show complicated changes in the position and shape of the plasmapause, which moved from $L = 5.1$ to 4.0 during this time. These are responses to rapid plasma redistribution caused by the penetration deep into the magnetosphere of enhanced convection electric fields that drive cross-L $E \times B$ drifts. Filled circles in the figure represent averages of several VLF-based electron density observations and are similar to the FLR-derived mass density estimates.

Rapid variations in plasmapause location are not adequately predicted by empirical models. Although FLR measurements from ground stations can be

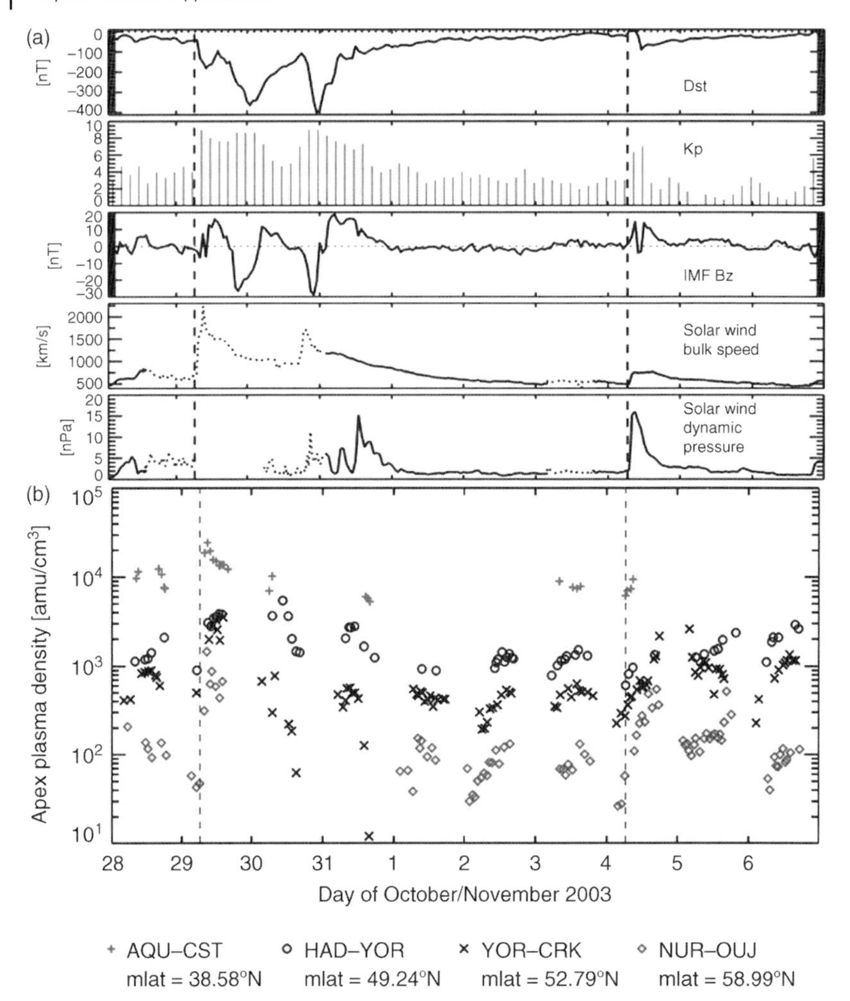

Figure 7.5 (a) D_{st}, K_p, IMF B_z, solar wind speed, and solar wind dynamic pressure during the 2003 Halloween storm. Spacecraft data have been time shifted to account for travel time to the magnetosphere. Vertical dashed lines identify sudden storm commencements. (b) Ground magnetometer-derived plasma mass densities for $L = 1.67$, 2.39, 2.79, and 3.84 (AQU-CST, HAD-YOR, YOR-CRK, and NUR-OUJ) for the same interval. Adapted from Kale *et al.* (2009).

used to determine the radial density profile, the plasmapause location may be identified by lack of a cross-phase signature in situations when it straddles two closely separated ground stations (Milling, Mann, and Menk, 2001). This is represented by the horizontal dotted lines in Figure 4.9 and provides the opportunity to track the plasmapause motion across suitable ground arrays. In fact, a variety of cross-phase signatures may be recorded, some of which are outlined in Section 7.3.

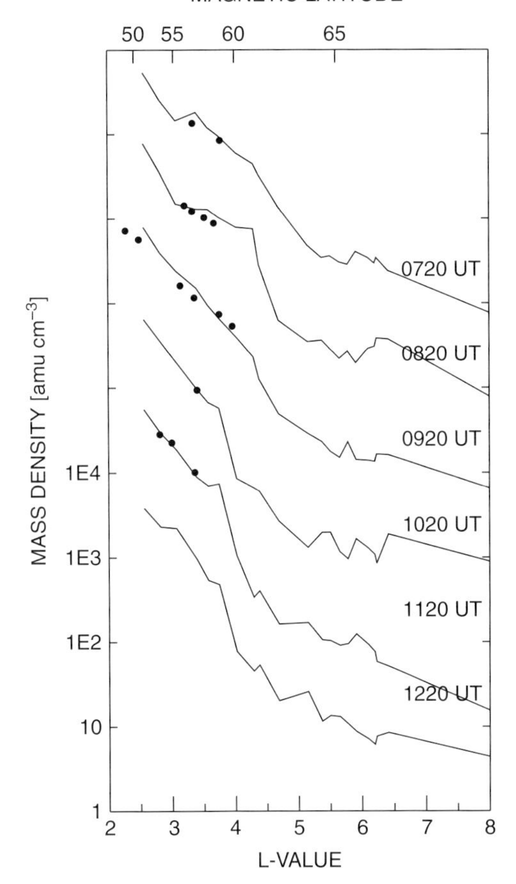

Figure 7.6 Hourly radial plasma mass density profiles, separated by an order of magnitude, spanning two solar wind pressure pulses at 0831 and 1017 UT. From Menk *et al.* (2004).

7.3 Density Notches, Plumes, and Related Features

The extreme ultraviolet (EUV) imager instrument on the IMAGE spacecraft provided dynamic measurements of the He^{+} intensity in the plasmasphere and yielded new insights into the spatial and temporal evolution of structures within the plasmasphere and at the plasmapause (Sandel *et al.*, 2003). These features include radial density depletions variously called notches, biteouts, cavities, or troughs. These may exist for as long as 60 h and have angular velocity typically at 85–90% of the corotation rate, thus slowly drifting westward across ground stations. Apart from being intrinsically interesting, biteouts are important features since they seem to be the source region of kilometric (100–800 kHz) radiation (Green *et al.*, 2002). Earlier observations of field-aligned plasmaspheric density cavities using the CRRES satellite

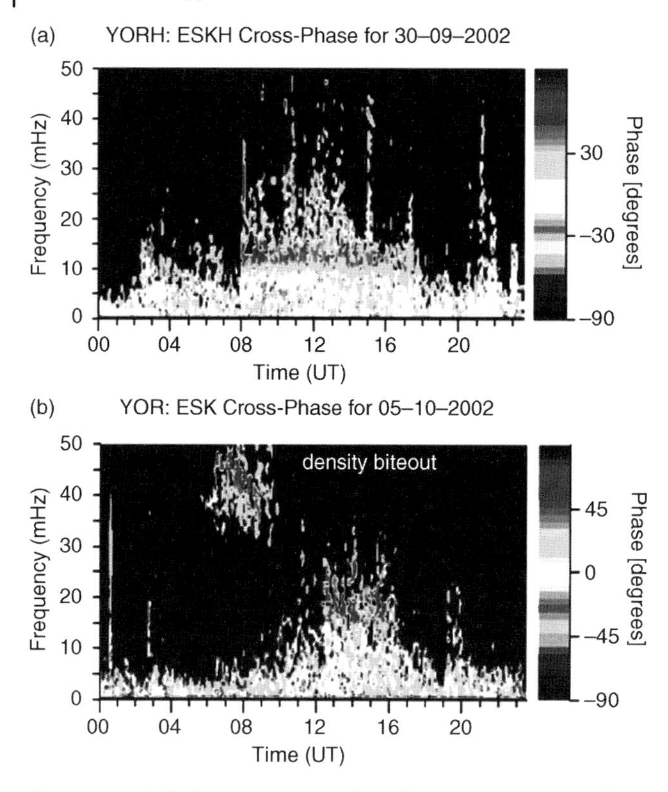

Figure 7.7 Whole-day $L = 2.67$ cross-phase frequency–time spectra for (a) September 30, 2002 and (b) October 5, 2002 when a density biteout occurred. (For a color version of this figure, please see the color plate at the beginning of this book.)

found that they occurred in about 13% of near-equatorial density profiles, particularly after plasmasphere erosion episodes, extending in L from $\Delta L = 0.5$–2 and $\geq 20^\circ$ wide in longitude (Carpenter *et al.*, 2000). Electron density in the cavities was found to be typically five times lower than that in the surrounding plasmasphere.

Density notches may be detected using ground magnetometers. This is illustrated in Figure 7.7, which compares whole-day cross-phase spectra from the YOR-ESK station pair in the United Kingdom (midpoint $L = 2.67$) on September 30, 2002 and October 5, 2002 about 3 days after a storm commencement. The first is a "normal" day and the band near 13 mHz identifies the resonant frequency. On October 5, this resonance band is shorter in duration and near 18 mHz, but the most striking feature is a resonance signature near 45 mHz in the local morning, representing a localized region of low density.

Properties of this notch feature were described in Grew *et al.* (2007). Figure 7.8 compares the FLR-derived mass density for 08–09 UT on October 5, 2002 (dotted line) with the same time on the previous day, and the tracer plot in top right shows the relative longitudes of the ground station array mapped to the equatorial plane at 1100 UT. The depletion on October 5, 2002 extends from $L < 2.4$ to $L \geq 4.5$ (well into

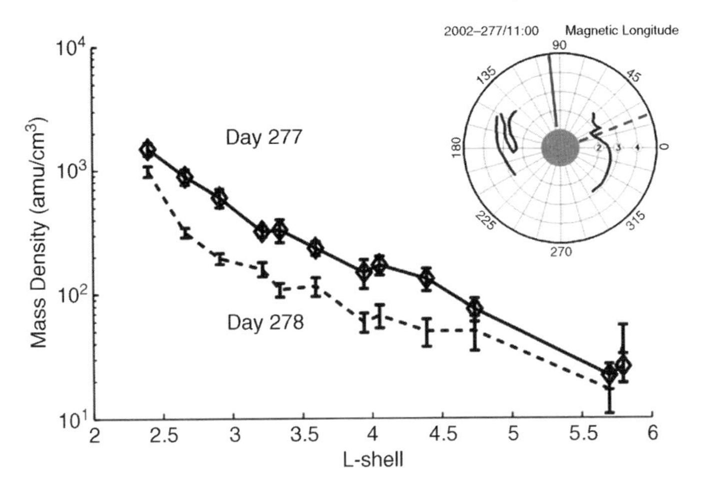

Figure 7.8 FLR-derived mass density profiles for 08–09 UT on October 4, 2002 (solid line) and October 5, 2002 (dotted line). Solid and dashed radial lines in the tracer plot in the top right represent the ground station longitudes projected to the equatorial plane at 1100 UT. From Grew *et al.* (2007).

the plasmatrough) with a maximum density decrease of order 60%. The plasmasphere corotation rate at this time was about 81%, so the duration of the notch signature represents a longitudinal width of $\leq$4 h. The obvious nature of the notch signature in the cross-phase spectrum accords with the schematic representation in Figure 4.9 and emphasizes the utility of routinely recorded ground FLR data.

Ground-based observations of VLF whistlers over four decades ago revealed the presence of detached regions of outlying plasmasphere-like density originating during storm recovery phases in the plasmapause bulge region, near dusk (Carpenter, 1970). These were reported as long-lived tail-like sunward surges of plasma attached to the main body of the plasmapause and drifting slowly azimuthally with a lag relative to the ground (Ho and Carpenter, 1976).

IMAGE EUV observations have shown that these plasma regions are drainage plumes that form in the dusk sector due to increased convection, 3–5 R_E in length and 0.5–1.0 R_E in width (Sandel *et al.*, 2003). Plumes are not only fundamental elements in magnetospheric plasma dynamics but also the likely site of wave–particle interactions resulting in the generation of ELF hiss and Pc1 and Pc2 EMIC waves (Yuan *et al.*, 2012). Plumes map to the ionosphere and are associated with total electron content (TEC) perturbations in GPS signals (Foster *et al.*, 2002; Yizengaw, Moldwin and Galvan, 2006). Ground magnetometer observations of enhanced mass density at $L > 4$ during the Halloween superstorm suggest that FLR measurements may be useful for detecting the formation of plasmaspheric drainage plumes under strong convection conditions (Chi *et al.*, 2005). In fact, a drainage plume was mapped by Grew *et al.* (2007) using mass density estimates from ground magnetometer stations spanning Europe. The results are summarized in Figure 7.9. A density enhancement associated with the plume near $L = 4.3$ is clearly evident in the mass density profile, in agreement with the EUV tracer plot if the plume feature is extrapolated to the ground array longitude.

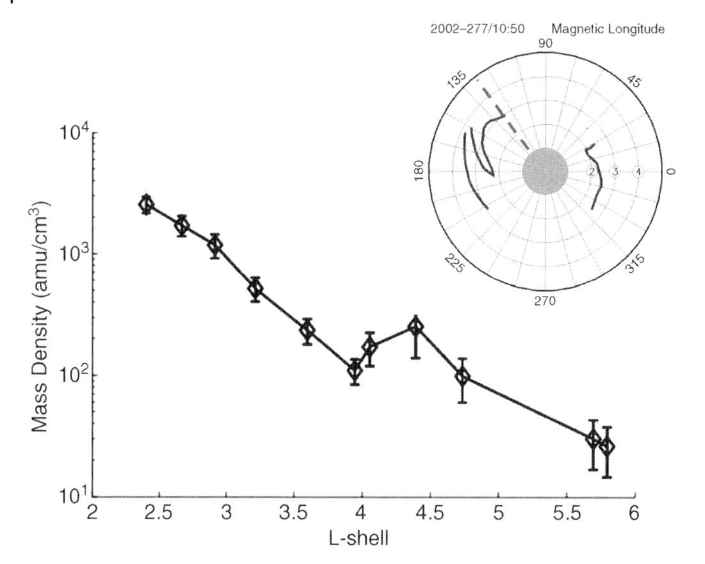

Figure 7.9 Radial mass density profile for 22–23 UT on October 3, 2002 showing evidence of a plasmaspheric drainage plume near $L = 4.3$. The ground station array longitude is represented by the radial line in the tracer plot at the top right. From Grew *et al.* (2007).

The mass density profile is similar to electron and He^+ density plume profiles obtained with the IMAGE EUV and RPI instruments (Garcia *et al.*, 2003).

Occasionally, a reversed cross-phase signature is seen between closely spaced ground magnetometer pairs. An example is shown in Figure 7.10, where arrows indicate sudden reversals in the sense of cross-phase. Such features identify a steep density gradient (Menk *et al.*, 2004). If the radial density variation is described by a power law of the form $r^{-\alpha}$, then in a dipole field geometry the toroidal mode eigenfrequency decreases with increasing L for $\alpha \leq 8$, is constant with L for $\alpha = 8$, and increases with increasing L for $\alpha > 8$. The latter situation corresponds to a cross-phase reversal (Kale *et al.*, 2007).

Cross-phase reversals may occur at plumes and other features as well as at a steep plasmapause. This is illustrated in Figures 7.11–7.13. Figure 7.11 shows dynamic cross-phase spectra for station pairs centered on $L = 3.9$ and $L = 3.2$ on June 11, 2001, a day after the moderate magnetic disturbance. In addition to a reversal in cross-phase near 0930 UT (UT$\sim$MLT) at $L = 3.9$, three different frequency structures occur in the $L = 3.2$ spectra, with a reversed cross-phase at the middle frequency. Electron density data are available from the IMAGE RPI experiment for this day and Figure 7.12a shows electron densities measured around 19–20 UT (07–08 MLT) within $\pm 30^\circ$ of the equatorial plane. The corresponding Alfvén speed profile derived from these electron density measurements and assuming a neutral proton-electron plasma is presented in Figure 7.12b. The inset shows an IMAGE EUV image for 1626 UT on this day, in which a drainage plume is evident around 09 LT. A detailed study of the formation of this plume was presented by Spasojević *et al.* (2003).

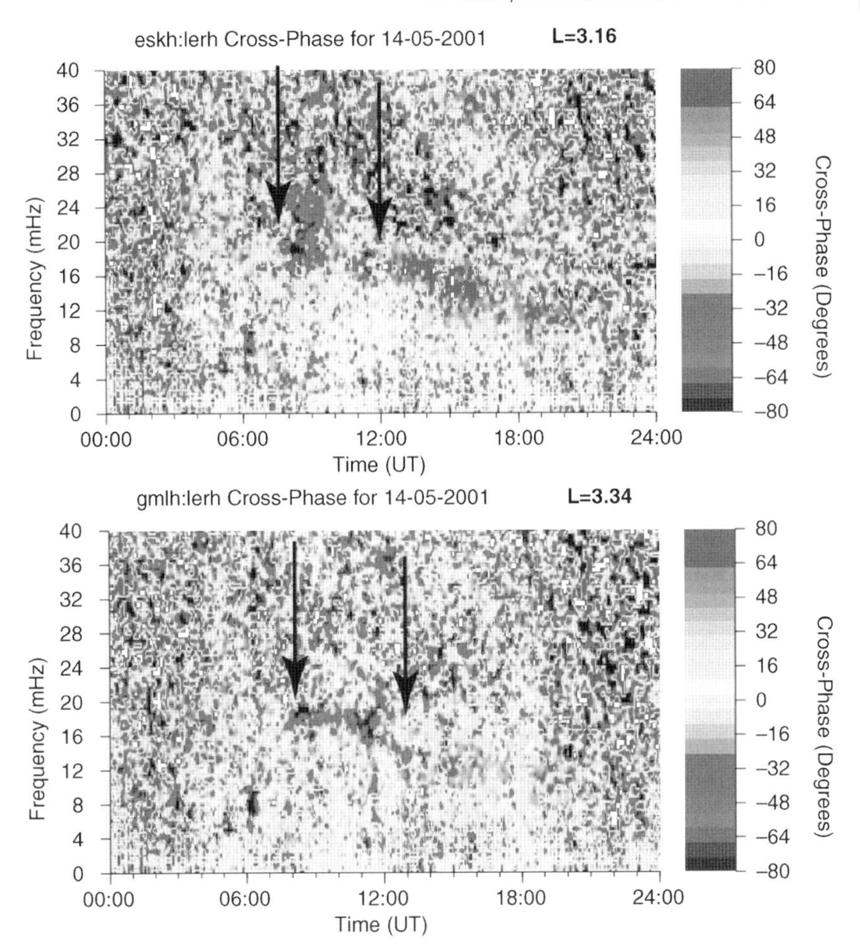

Figure 7.10 Dynamic cross-phase spectra for May 14, 2001, showing cross-phase polarity reversals, arrowed, at 0730 and 1200–1230 UT. From Kale *et al.* (2007). (For a color version of this figure, please see the color plate at the beginning of this book.)

The expected FLR frequencies and phases for this day have been determined by incorporating the Alfvén speed profile into a 2.5-dimensional numerical MHD model of the coupled magnetosphere–ionosphere, which itself used the IRI and MSIS models to represent the ionosphere for this time (Waters and Sciffer, 2008). The results are shown in Figure 7.13. The model magnetosphere was stimulated at the $L = 10$ R_E field line by a fast mode pulse with a Gaussian spatial distribution and uniform spectral content across the 1–100 mHz range. Azimuthal wave number was set to $m = 2$ and the ionospheric Pedersen and Hall conductances at 5 S. The input electron density and calculated resonant frequency profiles are given in panel (a), while panels (b) and (c) represent the expected ground-level power and cross-phase profiles. Reversals in expected resonance frequency and cross-phase spectra appear around $L = 3.7$ and $L = 4.4$.

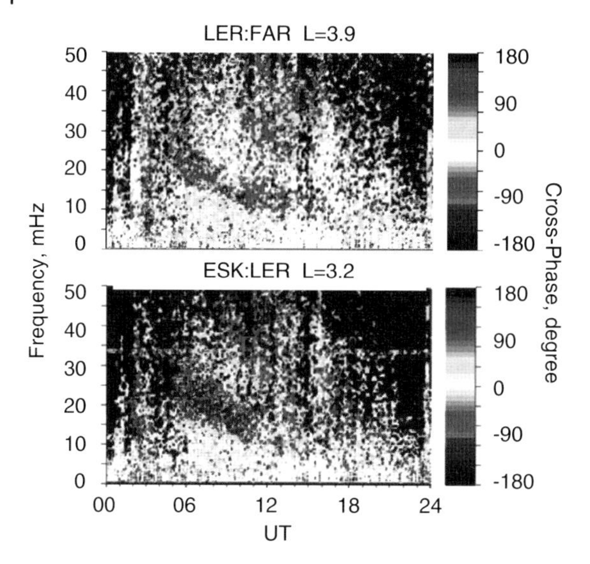

Figure 7.11 Dynamic cross-phase spectra for station pairs centered on $L = 3.9$ and $L = 3.2$ on June 11, 2001, a day after a $K_p = 6$ storm. A cross-phase reversal with time appears in the upper plot, and a reversal with frequency in the lower plot. (For a color version of this figure, please see the color plate at the beginning of this book.)

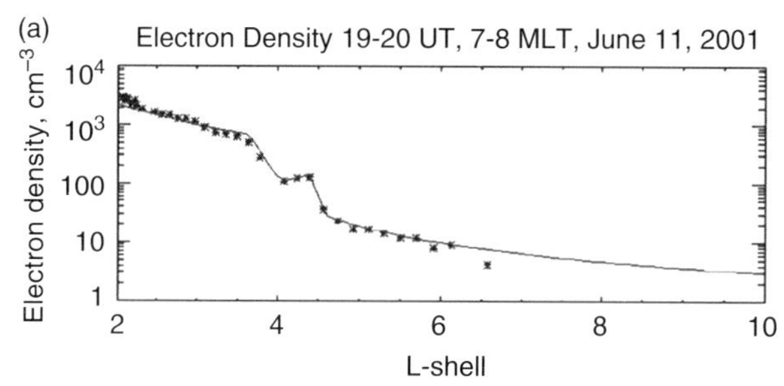

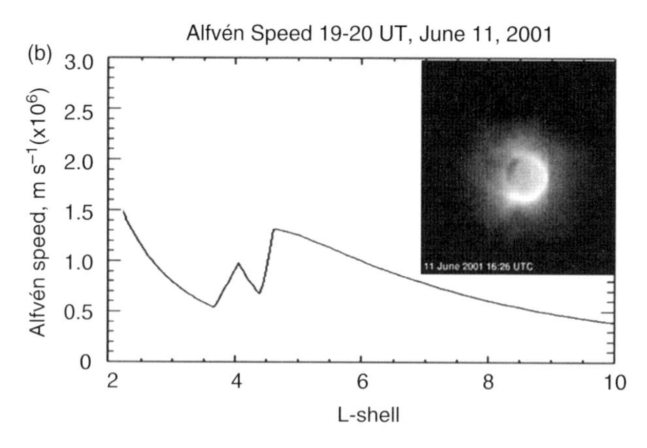

Figure 7.12 (a) Electron density profile for 19–20 UT on June 11, 2001 measured by the IMAGE RPI instrument. (b) Corresponding Alfvén velocity profile. IMAGE EUV image for 1626 UT on this day appears in the inset.

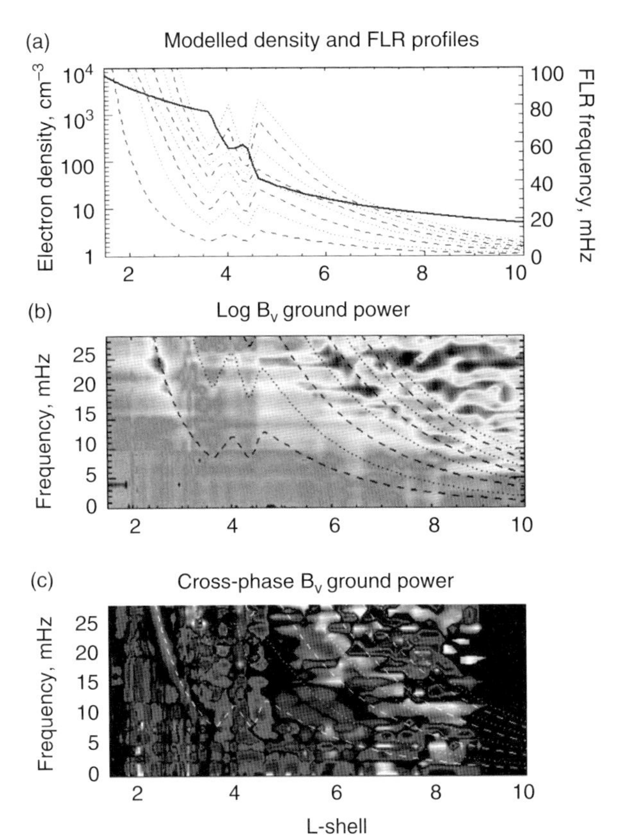

Figure 7.13 (a) Equatorial electron density (solid line) and resultant resonant frequency profiles for 19–20 UT on June 11, 2001 from a 2.5D numerical model. (b) Corresponding predicted power spectral density for the north–south ground-level magnetic perturbation. (c) Predicted ground cross-phase profile for an interstation spacing of 2°. (For a color version of this figure, please see the color plate at the beginning of this book.)

This work shows that (i) cross-phase reversals can identify the presence of plasmaspheric drainage plumes and other density features, (ii) such features may occur under moderately disturbed conditions, and (iii) numerical modeling tools permit realistic prediction of the frequency and phase profiles.

7.4 Refilling of the Plasmasphere

The plasmasphere is in dynamic equilibrium with the underlying ionosphere, and plasma flow up and down flux tubes is determined by the pressure gradient along the tube. Properties of the plasmasphere and ionosphere are therefore determined by diurnal effects and storm–time variations, when the plasmapause is eroded earthward and plasmaspheric plasma is convected via plumes into the outer magnetosphere. These processes reduce flux tube densities resulting in subsequent refilling from the ionosphere. The dynamic behavior of both the plasmasphere and ionosphere is therefore controlled by refilling processes and external drivers.

The refilling process has received considerable attention through observational and modeling studies (e.g. Lemaire and Gringauz, 1998; Reinisch *et al.*, 2004; Singh and Horwitz, 1992). Refilling of the plasmasphere following storm–time erosion takes some days. Measurement of refilling rates with spacecraft is difficult since the flux tubes corotate and may also expand and contract in volume and move in position in different local time sectors. Much early information on refilling rates was obtained from electron density estimates using VLF whistler measurements (Park, 1970, 1974), although this is difficult during local daytime and therefore VLF results often focus on summer data from high geographic latitudes. Refilling rates are latitude dependent and are specified in terms of the rate of density change in the equatorial plane per hour or per day, or the upward (or downward) fluxes typically across an area of $1\,\mathrm{cm}^2$ at 1000 km altitude. At midlatitudes, the upward electron flux (referenced to 1000 km altitude) is typically of order $3 \times 10^8\,\mathrm{el\,cm^{-2}\,s^{-1}}$. Downward fluxes at night are somewhat smaller (Obana, Menk, and Yoshikawa, 2010). There is some evidence that refilling is a two-step process, with smaller upward fluxes on the first day.

Measurements also come from incoherent scatter radars and satellites including the IMAGE RPI and EUV experiments (Denton *et al.*, 2012; Reinisch *et al.*, 2004, 2009; Sandel and Denton, 2007). These techniques mostly focus on measurements of rates of change of electron density, although transport rates and fluxes are different for different heavy ions, and the plasmaspheric density profile evolves differently for different heavy ion species (Singh and Horwitz, 1992). A notable exception is the use of the IMAGE EUV instrument to estimate He^+ fluxes (Sandel and Denton, 2007).

Section 7.2 demonstrated that ground-based measurements during and following magnetic storms show changes in resonant frequency due to rapid depletion and subsequent refilling of the plasmasphere. An example is illustrated in Figure 7.14 that shows the change in mass density during November 1–4, 1990 ($\Sigma K_\mathrm{p} = 17.0$, 18.0, 13.0, and 5.0) after a $K_\mathrm{p} = 5$ magnetic storm the previous day. Both diurnal and systematic refilling variations in density are evident. Horizontal dashed lines denote the expected quiet-time mass density from an empirical model (Berube *et al.*, 2005). Mass densities reached quiet-time levels at $L = 2.8$ within 2 days, taking 3 days at $L = 3.8$. It is relatively easy to estimate refilling rates and upward and downward ion fluxes from the density gradients. For example, in Figure 7.14, the refilling rate at $L = 3.7$ on November 1–2, 1990 is $\sim 20\,\mathrm{amu\,cm^{-3}\,h^{-1}}$.

Detailed studies of plasmaspheric refilling based on FLR measurements with ground magnetometer arrays and IMAGE satellite observations were described by Dent *et al.* (2006) and Obana, Menk, and Yoshikawa (2010). The latter study determined plasmaspheric refilling rates and upward fluxes in the recovery phase of three moderate magnetic storms, summarized in Figure 7.15. Refilling to prestorm levels took 2–3 days for $L = 2.3$ flux tubes, 3 days at $L = 2.6$, and over 4 days for $L > 3.3$. As seen in the top panel of the figure, refilling rates ranged from

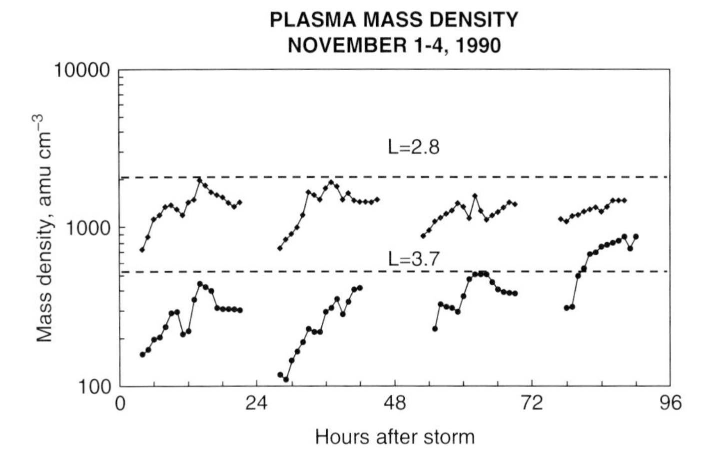

Figure 7.14 Plasmaspheric mass density at $L = 2.8$ (upper traces) and $L = 3.7$ during November 1–4, 1990, after a $K_p = 5$ magnetic storm on October 30, 1990. Dashed lines represent expected quiet time mass densities. Adapted from Menk *et al.* (1999).

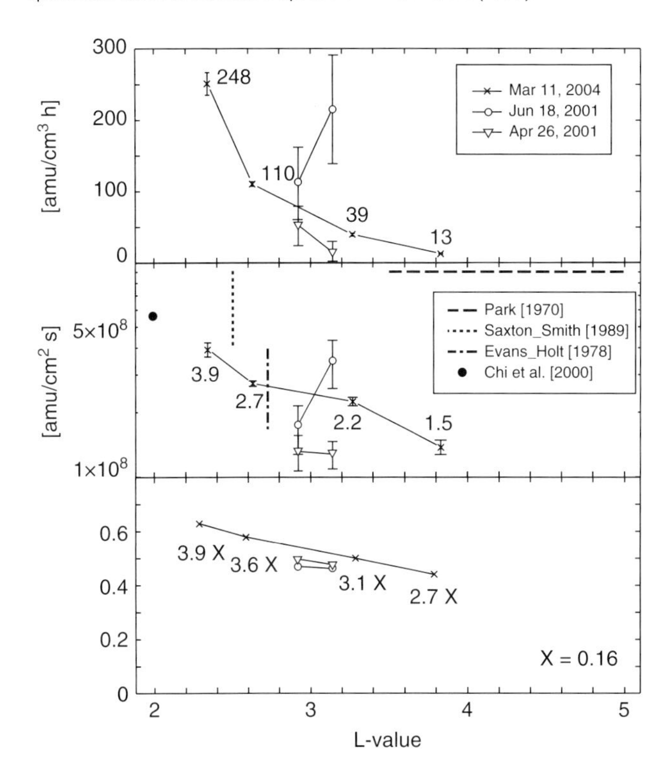

Figure 7.15 Summary of refilling rates at the equatorial plane (top) and upward fluxes at 1000 km altitude (middle) for three moderate magnetic storms. Bottom panel shows cosine of solar zenith angle averaged over the northern and southern hemispheres for these days. Symbols in middle panel refer to results from previous studies. From Obana, Menk, and Yoshikawa (2010).

$\sim$250 to 13 amu cm^{-3} h^{-1} from $L = 2.3$–3.8, superimposed upon diurnal variations. Refilling rates generally decrease with increasing latitude, although this trend was reversed on June 18, 2001 when both flux tubes were outside the plasmasphere and upward fluxes likely comprised a rich heavy ion component (Obana, Menk, and Yoshikawa, 2010). Upward plasma fluxes for the three storms are shown in the middle panel and are around 0.9–5.2 $\times$ 10^{8} amu cm^{-2} s^{-1}. These compare well with results from previous studies (except for Chi *et al.*, 2000) that are for electron fluxes. Nighttime loss rates, not shown in the figure, were typically 50 amu cm^{-3} h^{-1}. Daily averaged refilling rates, taking into account these downward nighttime fluxes, were about 420 amu cm^{-3} d^{-1} for $L = 2.9$–3.1.

Finally, the bottom panel of Figure 7.15 shows the cosine of the solar zenith angle averaged between the northern and southern hemispheres. This demonstrates that the change in solar illumination and hence ion production rate with latitude accounts for only about half the observed L-dependence of upward flux.

Such results point to the important contributions that ground-based measurements of ULF field line resonances may make to the understanding of magnetospheric refilling processes. Retrospective studies are possible using many years of magnetometer data available from various ground sites.

7.5 Longitudinal Variation in Density

The tilt and offset of the geomagnetic field with respect to Earth's geographic spin axis referred to in Section 2.1 result in a longitudinal asymmetry in solar illumination. This is illustrated in Figure 7.16a, where the footprint of the $L = 2.5$ shell is mapped to a geographic Mercator projection. The large offset of the dipole field near $-70°$E longitude (e.g., the eastern United States) means that the $L = 2.5$ field line has its footprint near 65° geographic latitude in the southern hemisphere but near 42° latitude in the northern hemisphere. During austral summer, the southern end of this flux tube is immersed in an almost continuously sunlit ionosphere; during austral winter, the local ionosphere is almost continuously dark. This has profound effects on ionospheric conductance and the formation of half- and quarter-wave field line resonances, discussed in Section 8.7. Another consequence is a substantial longitudinal and annual variation in plasmaspheric density due to the variation in electron and ion supply to the flux tube, discussed here.

It has long been known that electron density of the ionospheric F region is unusually high in December months compared with that in June months (Rishbeth and Müller-Wodarg, 2006). A related annual variation in plasmaspheric electron density appears in VLF whistler data. This is illustrated in Figure 7.16b, which summarizes 8 years of whistler data from Stanford, CA ($L \approx 2.5$; $-122°$E). The ordinate axis represents whistler dispersion $D = t\,(f\,)^{1/2}$, where t is the travel time at frequency $f = 5$ kHz. The quantity D is proportional to equatorial electron density $N_{eq}^{1/2}$ along the whistler flux tube. The figure shows that electron densities in December exceed those in June by a factor of $\sim$1.5. This trend has been verified with *in situ* observations and also causes an annual variation in the spectrum of

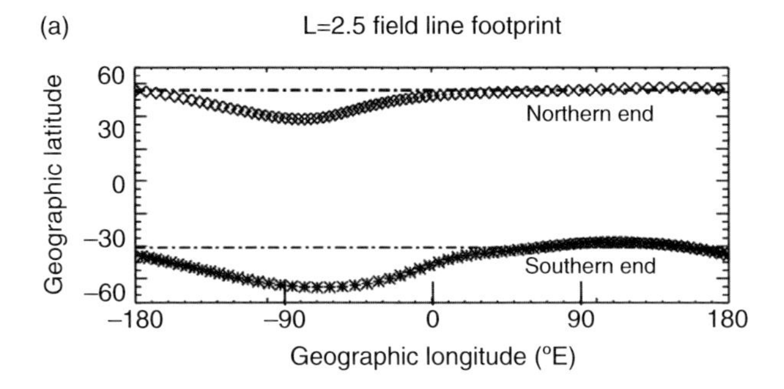

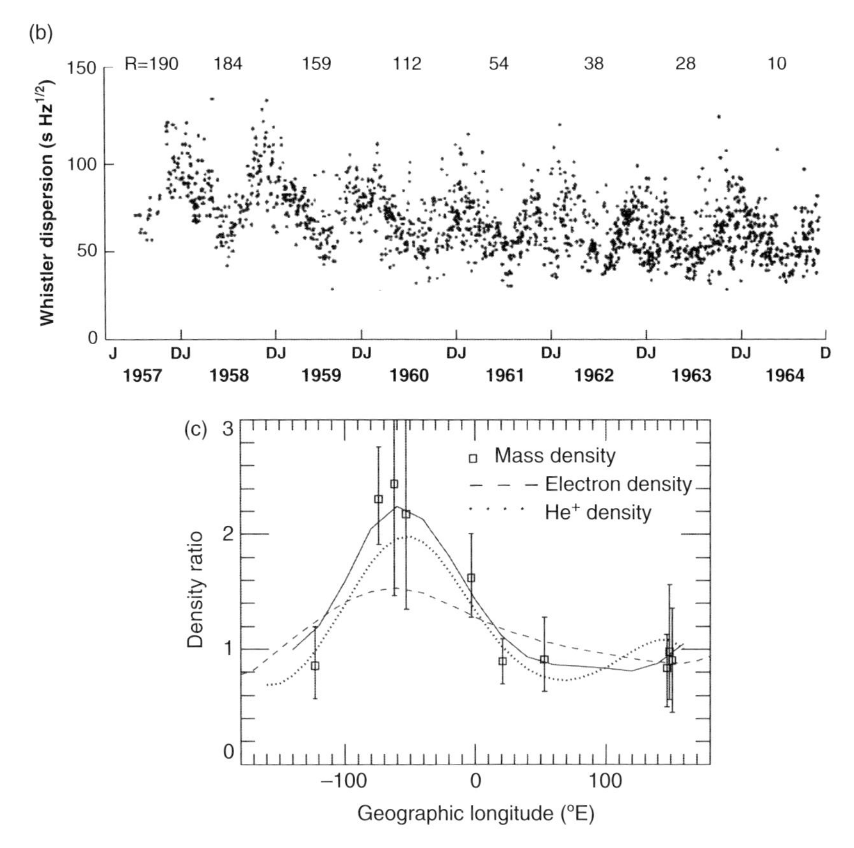

Figure 7.16 (a) Footprint of $L = 2.5$ field lines mapped to the ground. (b) Annual variation in VLF whistler dispersion (proportional to $\sqrt{}$electron density) over 8 years at California. (c) Annual and longitudinal variation in ratio of mass density, He^+ density, and electron density at $L = 2.5$. Adapted from Clilverd *et al.* (2007), Park, Carpenter, and Wiggin (1978), and Menk *et al.* (2012).

pitch-angle scattered radiation belt electrons precipitating into the atmosphere (Clilverd *et al.*, 2007).

A longitudinal and annual variation has also been found in plasmaspheric ion density. Figure 7.16c compares the ratio of annual maximum/minimum mass density (from ground magnetometer FLR measurements), He^+ density (IMAGE EUV), and electron density (IMAGE RPI) at $L = 2.5$ at solar maximum (Menk *et al.*, 2012). Maximum asymmetry clearly occurs near −60°E longitude in each case. This needs to be accounted for in empirical density models. Furthermore, the seasonal variation in the relative ion concentrations can be estimated as outlined in Section 6.5. In this case, it was found that at both −3°E and −74°E longitudes, the He^+ concentration was about 5% by number, but O^+ concentrations appear to be substantially higher at −3°E than at −74°E. The reason for this is not clear, but may be related to the proximity to the South Atlantic anomaly. It has previously been established that FLR frequencies at low latitudes are higher near the anomaly, presumably due to changes in O^+ production or loss rates (Sutcliffe, Hattingh, and Boshoff, 1987).

7.6 Solar Cycle Variations in Density

Comparison of whistler dispersion data in Figure 7.16b with yearly averaged sunspot numbers listed above the plot shows that plasmaspheric electron densities decreased from solar maximum to minimum by a factor of ~1.5 at $L \approx 2.5$, although this trend was not obvious for $L \geq 3$ (Park, Carpenter, and Wiggin, 1978). The period and occurrence frequency of Pc3 field line resonances recorded at $L = 1.9$ during 1957–1973 show well-defined variations with solar cycle (Figure 7.17a). This is consistent with the observed variation in resonant frequency of Pc3 ULF waves at $L = 1.6$, indicating an increase in plasmasphere density by a factor of ~2 from solar minimum to maximum (Figure 7.17b) (Bencze and Lemperger, 2011; Vellante *et al.*, 1996). However, this effect may be masked by an increase in the frequency of the driving upstream waves with increasing solar activity, while at such low latitudes, mass loading from ionospheric O^+ strongly affects the resonant frequency. The variation in F2 region electron density also shown in Figure 7.17 is somewhat larger than that in mass density (Bencze and Lemperger, 2011). Enhanced ionospheric density results in increased plasmaspheric density, increasing damping of FLRs (Verö and Menk, 1986) and decreasing occurrence frequency.

A strong solar cycle effect also appears in mass density data from near-geosynchronous satellites ($L = 6.8$), illustrated in Figure 7.18 (Takahashi, Denton, and Singer, 2010b). Panel (a) shows that the third harmonic of the resonant frequency, f_{T3}, is in antiphase with the $F_{10.7}$ solar radio flux variation over a solar cycle, while panel (b) demonstrates that the plasmatrough density and $F_{10.7}$ are closely related. Regression analysis yields a linear relationship for 27 day average quantities ($R = 0.937$):

$$\log(\rho_{eq\_27d}) = 0.421 + 0.00390 F_{10.7\_27d}, \tag{7.1}$$

where ρ_{eq_27d} is in units of amu cm^{-3}.

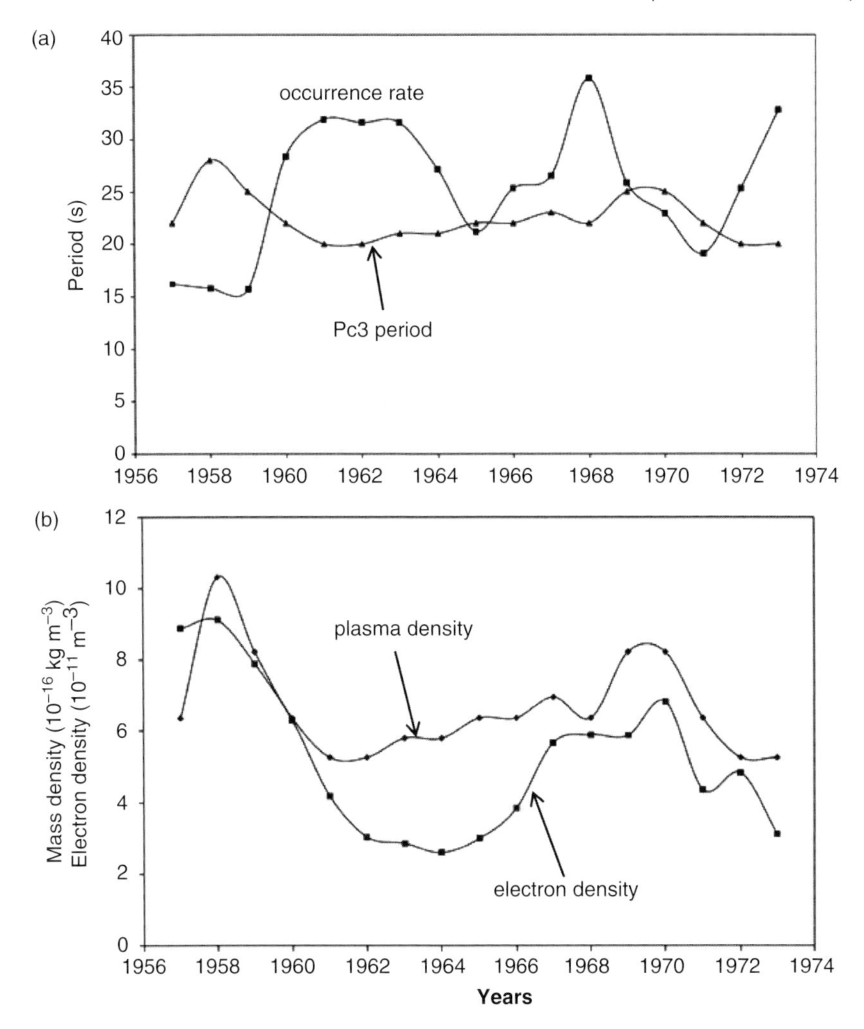

Figure 7.17 (a) Solar cycle variation in period and occurrence rate of Pc3 field line resonances at $L = 1.9$. (b) Corresponding variation in FLR-derived equatorial mass density and ionospheric electron density. From Bencze and Lemperger (2011).

The observed solar cycle variation in resonant frequency at $L = 6.8$ is similar to that at $L = 1.6$. This implies that the $F_{10.7}$ parameter controls the solar cycle mass density variation in both the plasmatrough and plasmasphere in the same quantitative manner, through the dependence of ionospheric conditions on solar UV/EUV radiation. These results suggest that routine measurements of field line resonant frequency can provide a proxy for solar flux or, conversely, that routine observations of solar output can be used to statistically predict mass density inside both the plasmasphere and the plasmatrough. The solar cycle variation in mass density will also affect magnetospheric phenomena that depend on density or resonant

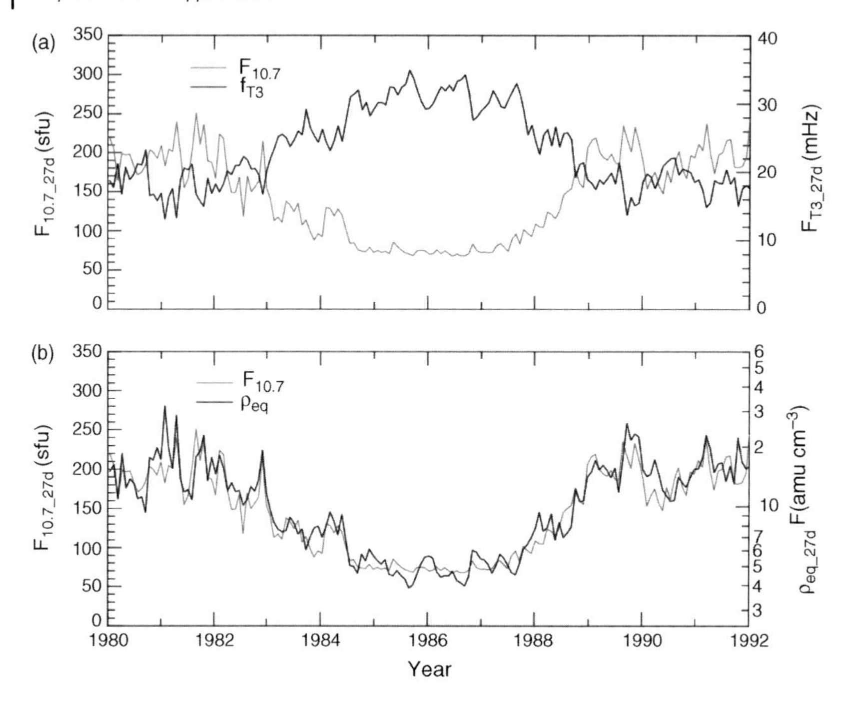

Figure 7.18 (a) Variation of the third harmonic of the quiet time 27 day average resonant frequency f_{T3} at $L = 6.8$, and solar radio flux $F_{10.7}$ during 1980–1992. (b) Corresponding variation in equatorial mass density and $F_{10.7}$. From Takahashi, Denton, and Singer (2010b).

frequency. For example, a solar cycle variation in the efficiency of resonance-driven radial transport is likely (Takahashi, Denton, and Singer, 2010b).

7.7
Determining the Open/Closed Field Line Boundary

At high latitudes, FLR measurements provide an indication of the location of the last closed field line (open–closed boundary (OCB)) and the occurrence of reconnection at the dayside magnetopause. This is illustrated in Figure 7.19. Under average IMF conditions, cusp region ground stations pass just equatorward of the OCB, and cross-phase measurements show azimuthal phase delays consistent with wave propagation away from a Kelvin–Helmholtz instability source region near noon. Figure 7.19a and b depicts the cross-phase between two sites both at −74.5°CGM latitude but separated by 100 km in longitude, across the 2.7–4.7 mHz band (Φ5), as a function of IMF B_z (a) before and (b) after local noon (Ables and Fraser, 2005). The local time and IMF-dependent differences are due to the variation in cusp latitude and hence proximity of the OCB with IMF B_z. The two boxed sectors herein indicate outlier regions at large negative B_z. Two case

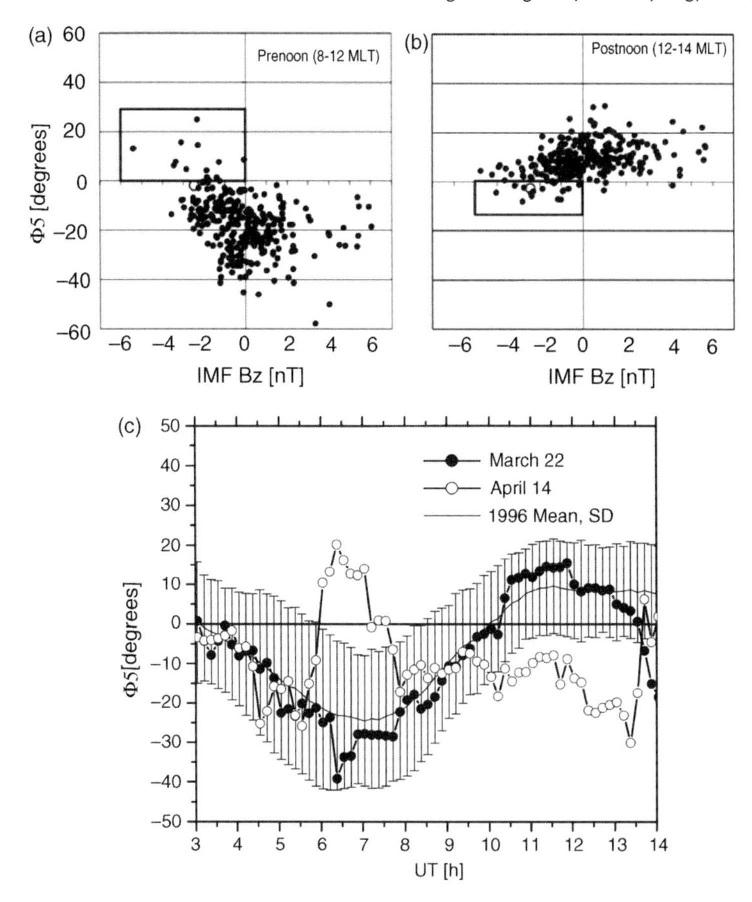

Figure 7.19 (a and b) Variation in 2.7–4.7 mHz cross-phase (Φ5) between longitudinally separated Antarctic ground stations with IMF B_z (a) before and (b) after local noon, over all of 1996. (c) Diurnal variation in Φ5 for two selected days, and the mean and standard deviation in Φ5 for all of 1996. Local noon is near 1030 UT. From Figures 1 and 2 in Ables and Fraser (2005).

studies are summarized in Figure 7.19c. On March 22, 1996, the diurnal variation in Φ5 is typical for closed field line conditions, with propagation away from noon. On April 14, the pattern is reversed. This occurs on about 5% of days, when IMF $B_z < -2$ nT and the azimuthal component of propagation is toward noon, indicative of newly opened flux tubes unbending under the influence of IMF B_y.

7.8 Determining the Magnetospheric Topology at High Latitudes

A rich variety of plasma waves in the 0.1–5 Hz (Pc1–Pc2) range are evident in frequency–time records from ground magnetometers in the polar regions.

Historically, the names given to these wave types described their appearance in such data sets and similarities with the appearance of VLF whistlers: hydromagnetic chorus, hm whistler, periodic hm emission, hm emission burst, pearls, narrowband Pc2, unstructured Pc1 noise, 4 s band, intervals of pulsations of diminishing period (IPDP) or rising period (IPRP), and so on (e.g. Arnoldy *et al.*, 1988; Fukunishi *et al.*, 1981; Nagata *et al.*, 1980). It was generally supposed that these emissions mostly resulted from ion cyclotron waves generated under a variety of conditions in the outer magnetosphere.

In order to identify the wave source regions, Menk *et al.* (1992, 1993) mapped the occurrence of different types of Pc1–Pc2 waves using observations from ground stations distributed between the plasmapause and the polar cap. They used particle data from the DMSP low-Earth orbit satellites, ionograms, riometer data, and Pc5 magnetic pulsations to identify different magnetospheric regions and show that specific types of Pc1–Pc2 waves are associated with the outer magnetosphere, boundary layer, and cusp and plasma mantle regions, and therefore that ground-based observations of Pc1–Pc2 pulsations may be used to identify and monitor the high-latitude magnetosphere topology. This is because large regions of the outer magnetosphere characterized by distinct particle populations map to relatively small regions on the ground at high latitudes.

The situation is represented schematically in Figure 7.20 and has been confirmed by detailed triangulation measurements from a closely spaced station array (Neudegg *et al.*, 1995). Intense unstructured broadband (~0.15–0.45 Hz) diffuse

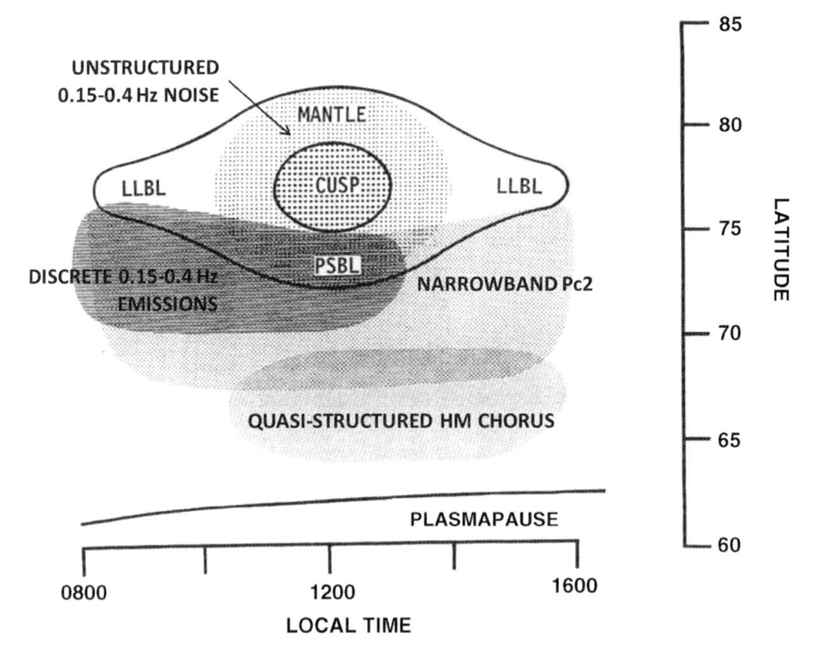

Figure 7.20 Schematic representation of Pc1–Pc2 source regions in high-latitude ground magnetometer data. From Menk *et al.* (1992).

noise-like emissions occur within 2° poleward of the poleward edge of the cusp, and most likely originate in the plasma mantle (Dyrud *et al.*, 1997). Narrowband unstructured emissions typically around 0.2 Hz occur under and a few degrees equatorward of the plasma sheet boundary layer in the noon sector and are mostly likely due to ion cyclotron waves generated in the equatorial plasmatrough (Hansen *et al.*, 1991). A variety of discrete Pc1 emissions comprising short packets or bursts with frequencies <0.4 Hz are probably due to ion cyclotron waves generated by resonance with 1–5 keV ions in the equatorial boundary layer region (Hansen *et al.*, 1992). Reconnection events and solar wind compressions may increase the temperature anisotropy of trapped ions in this region to promote ion cyclotron wave growth.

Finally, accumulating evidence suggests that structured Pc1–Pc2 emissions that occur during local afternoon in the plasmatrough in the recovery phase of magnetic storms are generated at the strong density and temperature gradients occurring at the edge of and inside plasma drainage plumes (Engebretson *et al.*, 2008; Morley *et al.*, 2009).

In summary, ground magnetometer measurements of Pc1–Pc2 pulsations can provide information on the location of the magnetospheric mantle, cusp, low-latitude boundary layer, and plasma drainage plumes. Although the waves may propagate away from their source regions in the ionospheric duct, the waveguide attenuation is relatively high and the direction of propagation points to the source (Neudegg *et al.*, 1995).

7.9 Wave–Particle Interactions

The criterion for resonance interaction between a ULF wave of frequency ω and azimuthal wave number m and the trapped particles bouncing between conjugate points with frequency ω_b and drifting around Earth at frequency ω_d was discussed in Section 4.6 and is given by

$$\omega - m\omega_d = N\omega_b. \tag{7.2}$$

Such interactions may be important in energizing radiation belt particles (Ozeke and Mann, 2008). The ULF waves involved are guided poloidal mode waves with high azimuthal wave number and may be generated by unstable energetic particle distributions in the ring current (Hughes *et al.*, 1978b). Due to spatial integration effects, such waves are attenuated on the ground but are readily detected in the ionosphere with radio sounders (Ponomarenko *et al.*, 2001; Wright and Yeoman, 1999). This aspect is described further in Section 8.5. Here, we present one representative example.

On January 6, 1998, the Finland and Iceland East CUTLASS SuperDARN HF radars were operating in a special high time-resolution mode optimized for ULF wave detection. A sequence of westward-moving wave-like features was recorded in the returned echo Doppler shift from the overlapping radar fields of view for over 4 h. This is illustrated in Figure 7.21 that shows a detrended Doppler

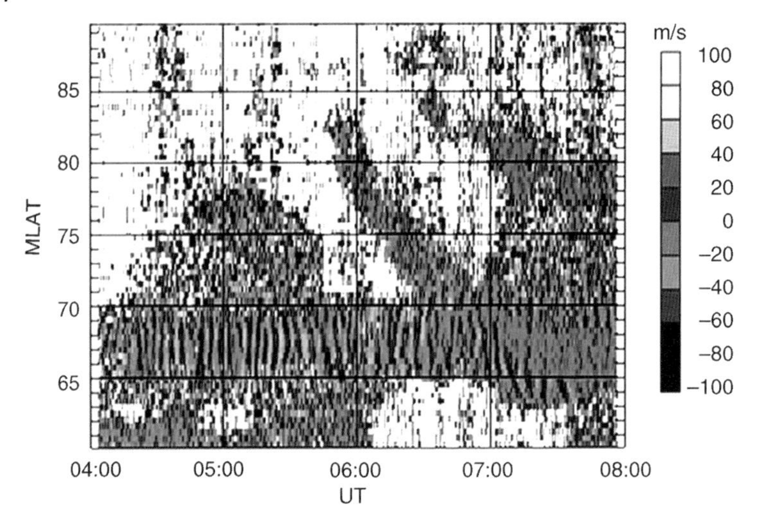

Figure 7.21 Doppler velocity oscillations in beam 5 of the Finland (Hankasalmi) HF radar from 0400–0800 UT on January 6, 1998. (For a color version of this figure, please see the color plate at the beginning of this book.)

velocity–latitude–time plot for beam 5 of the Finland (Hankasalmi) radar located at 62.3°N, 26.6°E geographic, over 0400–0800 UT. Further details on the radars, nearby ground magnetometers and data analysis techniques, appear in Menk *et al.* (2003). Alternating vertical bands between 65° and 70° magnetic latitude represent motion of F region plasma toward and away from the radar. These echoes are from a region between Scandinavia and Greenland identified by a shaded patch in the map presented in Figure 7.25. Representative power spectra along this beam are shown in Figure 7.22. The velocity oscillations are in the 3–4 mHz frequency range, and are not obvious in ground magnetometer records (not shown here).

Observations from overlapping beams of the Finland and Iceland radars allow the frequency, wave number, and phase speed of the oscillations to be determined. These are presented in Figure 7.23 and show that the oscillation frequency varied between 3 and 4 mHz, with azimuthal wave number $m \approx -70$ and westward phase speed of about 800 m s^{-1}. At the same time, FLRs were identified using cross-phase measurements from the IMAGE array of ground magnetometers spanning Scandinavia and used to determine equatorial mass densities. Near the latitude of the radar oscillations (66°–70°), the resonant frequencies were around 4 mHz, but moving to lower L the resonant frequencies increased stepwise at 66° MLAT to 10–14 mHz. This change in frequency relates to a strange density enhancement, seen in Figure 7.24. This resembles the mass density signature of a plasma plume, as shown in Figure 7.9, although there are no further observations to support this. During this time, the POLAR satellite was passing overhead, as shown by the track mapped to the ground in Figure 7.25a. The ion distribution function was

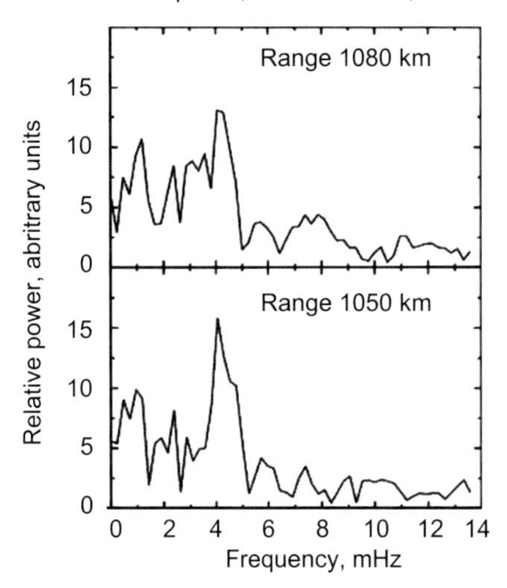

Figure 7.22 Representative power spectra of the oscillations in beam 5 at two range gates.

determined using the TIMAS and CAMMICE instruments and is shown in Figure 7.25b. A bump in the distribution near 20 keV indicates an unstable particle population.

Waves can grow at the expense of particle energy if there is a non-Maxwellian distribution of energetic particles. For example, ring current protons injected at a substorm undergo longitudinal drift, but these drift paths may open at certain L-shells due to changes in magnetic activity, violating the third adiabatic invariant. Assuming that waves have frequency $f = 3$ mHz and $m = -70$, then from Equation 7.2 $\omega_d = 2.7 \times 10^{-4}$ rad s^{-1} if we assume only drift resonance, that is, $N = 0$. These particles take 6.5 h to drift around Earth. A possible substorm injection occurred near 1600 UT on the previous day, so the particles would have drifted almost twice around Earth to reach the observation zone.

Magnetospheric protons undergo gradient, curvature, and $E \times B$ drifts due to the convection and corotation electric fields. In a dipole field, the bounce-averaged azimuthal drift can be described by (Chisham, 1996; Ozeke and Mann, 2008)

$$\omega_d = -\frac{6WLP(\alpha)}{B_S R_E^2} + \frac{2\psi_0 L^3 \sin\phi}{B_S R_E^2} + \Omega_E, \tag{7.3}$$

where W = proton energy, $P(\alpha) \approx 0.35 + 0.15 \sin\alpha$ is a function of the particle's equatorial pitch angle, ϕ is the azimuth angle, ψ_0 is the convection electric potential and can be expressed empirically as a function of K_p, and Ω_E is Earth's rotation rate. Using measured parameters and putting $W = 21$ keV, $\alpha = 45°$, and $L = 6.2$ gives $\omega_d = 2.7 \times 10^{-4}$ rad s^{-1}, in agreement with observations.

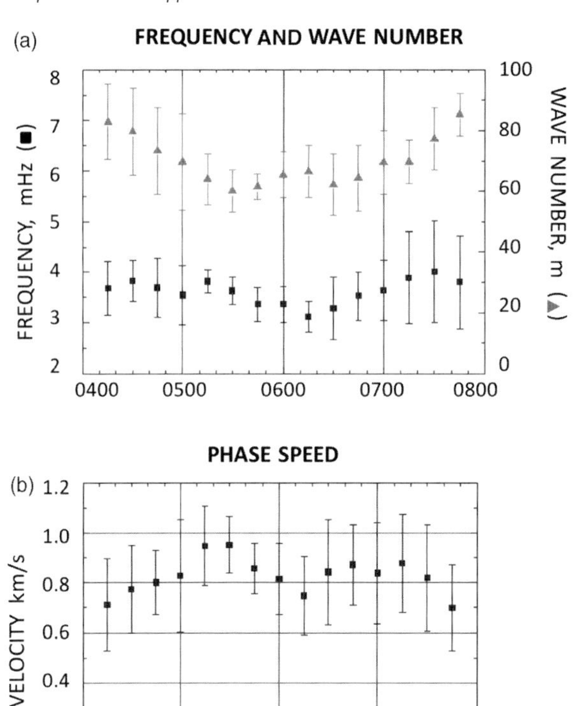

Figure 7.23 (a) Time variation of frequency (squares) and azimuthal wave number (triangles) for Doppler oscillations detected by the Finland and Iceland East radars. (b) Equivalent phase speed of the westward propagating wave.

The $N = 0$ drift resonance requires the wave to be a fundamental poloidal mode standing wave. Magnetic conditions on both January 5 and early January 6, 1998 were quiet, suitable for drift resonance with injected particles (Chisham, 1996). The event is similar to a giant pulsation event recorded by multiple satellites and ground stations outside the plasmasphere (Takahashi *et al.*, 2011). In our case, the high azimuthal wave number inhibits ground observations, although toroidal mode FLRs were detected at a similar frequency using cross-phase measurements. It is possible these waves are generated by a drift wave instability, which requires a plasma density or temperature gradient (Green, 1985) such as may be expected near a drainage plume. On the other hand, the good agreement between observed particle and wave properties points to a drift resonance. The peak in FLR-derived mass density is intriguing. While similar to the signature of a drainage plume, it may also indicate the presence of a heavy ion population, although magnetic conditions were quiet.

In summary, this example has shown that (i) ionospheric sounders can readily detect high-wave number ULF waves that are difficult to detect with ground

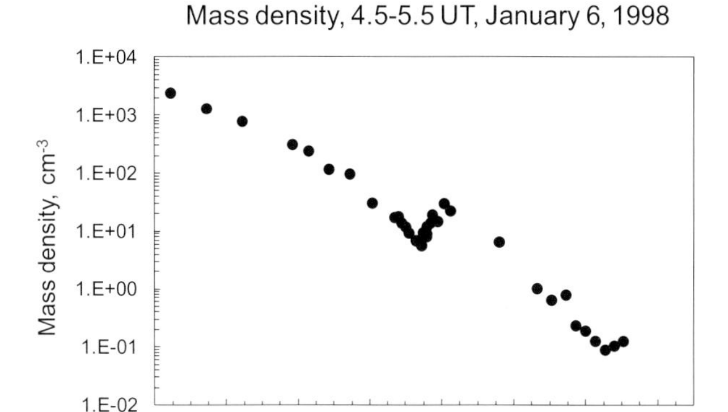

Figure 7.24 Radial mass density profile determined from cross-phase measurements across the IMAGE magnetometers spanning Scandinavia. A density peak appears near 66° magnetic latitude.

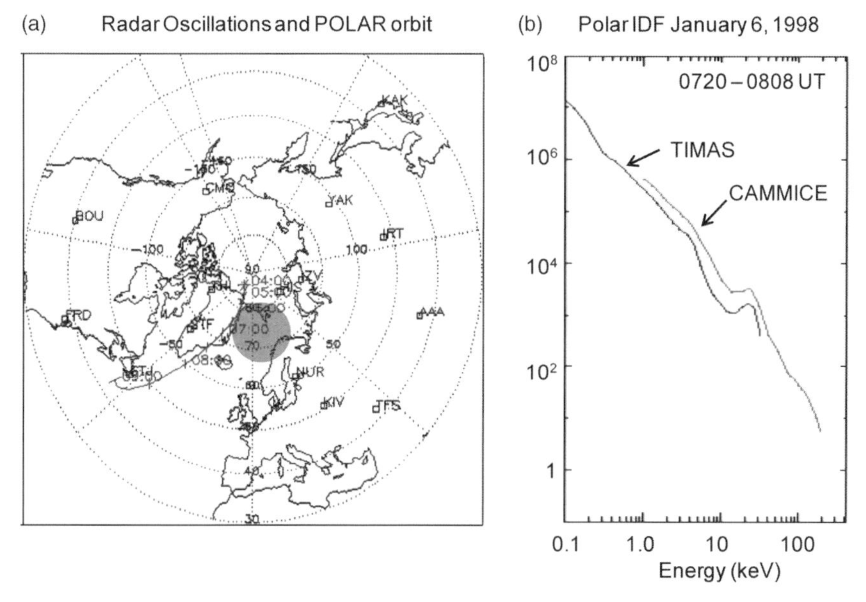

Figure 7.25 (a) Map showing location of ionospheric Doppler oscillations (shaded circle) and projection of POLAR satellite orbit. (b) Ion distribution functions recorded by POLAR when passing over the wave observation region. Courtesy of L. Baddeley.

magnetometers alone, (ii) the waves are associated with FLRs and most likely guided poloidal mode standing oscillations, and (iii) the waves are most likely generated through resonance interaction with westward drifting ring current particles outside the magnetosphere, possibly associated with a drainage plume or significant heavy ion population.

7.10 Radial Motions of Flux Tubes

Compressional mode ULF plasma waves propagating through the magnetosphere may couple to shear Alfvén field line oscillations, as described in Section 3.5. However, the azimuthal wave electric field may also drive $E \times B$ cross-L plasma drifts producing radial oscillations of flux tubes, resulting in Doppler shifts of ducted whistler mode VLF signals (Andrews, 1977). Fundamental toroidal mode standing field line oscillations exhibit an electric field antinode in the equatorial plane, and any VLF Doppler oscillations due to such azimuthal flux tube motions are at double the frequency of the ULF wave. However, poloidal mode standing field line oscillations have an electric field node at the equator, and the Doppler oscillations are at the same frequency for the fundamental mode but at double the ULF frequency for the second harmonic. This allows the oscillation mode to be determined and enables comparison of the magnetic disturbance in the magnetosphere with that seen on the ground. In this way, (i) the 90° rotation of Pc4–Pc5 wave fields by the ionosphere was observationally verified, (ii) the azimuthal component of the wave electric field in the equatorial plane was measured, and (iii) the strength of the north–south wave magnetic field at the base of the magnetosphere was estimated, by Andrews (1977).

Occasionally, VLF Doppler oscillations are excited simultaneously on more than one adjacent flux tube. In this case, the relative phasing between the Doppler oscillations may be used to estimate the phase velocity and wave normal direction of the fast mode ULF wave in the magnetosphere. In one such study of 13 mHz Doppler oscillations at $L \approx 2.4$ (well below the local FLR frequency), the ULF wave phase velocity in the equatorial plane was estimated at $0.92 \times 10^6\,\mathrm{m\,s^{-1}}$ with a wave normal direction of 52° anticlockwise from the Sun–Earth line, that is, from the region of the afternoon magnetopause (Yearby and Clilverd, 1996). This phase velocity was comparable to but a bit less than the Alfvén speed.

A study of about 100 ULF Doppler oscillation events recorded over 1 year using colocated magnetometers and VLF receivers at $L = 2.5$ was reported by Menk *et al.* (2006). It was shown that: (i) ULF waves and VLF Doppler oscillations occurred at frequencies corresponding to ion cyclotron generation in the upstream solar wind; (ii) the Doppler oscillations are due to radial flux tube motions of ~3–5 km driven by fast mode waves with azimuthal electric field in the equatorial plane of order $1\,\mathrm{mV\,m^{-1}}$ and northward magnetic field ~0.8–1.2 nT; (iii) when the wave frequency and the local field line eigenfrequency matched, the waves coupled to standing toroidal mode FLRs detected with cross-phase ground magnetometer measurements; (iv) localized poloidal mode standing oscillations were also produced and detected with cross-phase measurements in the magnetometer D components on the ground; and (v) scale size of whistler flux tubes in the equatorial plane is much smaller than that for FLRs, so VLF Doppler oscillations can provide more precise spatial information on ULF wave fields in the magnetosphere than can be obtained from ground-based measurements.

These studies demonstrate that VLF receivers can directly monitor ULF wave fields in the magnetosphere, and in combination with ground magnetometer observations can provide information on wave sources and properties.

8
ULF Waves in the Ionosphere

8.1
Introduction

ULF waves are temporal variations in the geomagnetic field that imply some nearby temporal change in electric current. In near-Earth space there are no metallic conductors, so for a given electric field, larger currents flow where the electrical conductivity is largest. In the magnetosphere, effects from the Lorentz force favor field-aligned movement of charge until the upper ionosphere is encountered. The shear Alfvén ULF wave mode carries a field-aligned current (FAC). As the neutral particle density increases with decreasing altitude, the field-aligned current is diverted more horizontally. The geomagnetic field and collisions of neutral and ionized particles in the ionosphere produce the Hall current, which together with the field-aligned current and associated "direct" conductivity and the usual current parallel to the electric field (Pedersen conductivity) comprises the components of the anisotropic ionosphere conductivity. These ionospheric currents generate the magnetic signature of ULF waves detected by ground-based magnetometer networks.

Coordinated studies of ULF waves and their properties began in earnest during the 1950s, and the importance of the ionosphere and its effects on ground magnetic signatures of ULF waves was readily appreciated. Energy deposition into the ionosphere for the higher frequency ULF waves ($\sim$1 s period) was estimated by Dessler (1959), and by the early 1960s it was well known that the ionosphere acted "as a sort of filter" to ULF waves, particularly the amplitude attenuation features (Akasofu, 1965). Results from a computer numerical solution for kilohertz waves using magnetoionic theory were described by Pitteway (1965) and studies by Hughes (1974) and Hughes and Southwood (1976) formed the basis for the comprehensive analysis of ULF wave propagation through the ionosphere to the ground.

The interaction of ULF waves with the ionosphere involves many parameters. These include the geomagnetic field $\boldsymbol{B}_0$ and dip angle I, the direct, Pedersen and Hall conductances and their variation with altitude, and wave properties such as frequency, phase speed, and spatial scale dimensions. Analytic treatments assume time and spatial variations of the ULF fields of the form $\exp[i(\boldsymbol{k}\cdot\boldsymbol{r} - \omega t)]$

Magnetoseismology: Ground-based remote sensing of Earth's magnetosphere, First Edition. Frederick W. Menk and Colin L. Waters.

and a thin current sheet ionosphere described by height integrating the conductivities to give the direct, Pedersen and Hall conductances. Most treatments have also assumed a vertical B_0. This is because for ideal MHD in the magnetosphere and vertical B_0 in the ionosphere, the shear and fast mode waves can be described in terms of divergence-free and curl-free expressions of the ULF electric field, with a simplifying orthogonality property. These are the two solutions of Kato and Tamao (1956), Dungey (1963), and Hughes (1974), employed more recently by Yoshikawa and Itonaga (2000). Developments since then have explored the important role of the ionosphere Hall current in the modification of ULF wave properties as they interact with the magnetosphere–ionosphere–atmosphere ground system.

8.2 Electrostatic and Inductive Ionospheres

The development of mathematical models to describe the interaction of ULF waves with the magnetosphere–ionosphere–atmosphere ground system has necessarily invoked a number of simplifying assumptions. The analysis requires equations that can seamlessly model the transition from the combined electromagnetic and fluid perturbations in the magnetosphere (magnetohydrodynamics) to an ionosphere with anisotropic conductivity, and emerge as an electromagnetic wave in the neutral atmosphere, all at subhertz frequencies. This has similarities with multilayer electromagnetic wave reflection and transmission problems encountered in undergraduate electromagnetic theory and may be formulated using reflection and transmission coefficient matrices (e.g. Alperovich and Federov, 2007). The development in terms of the wave fields, while requiring more mathematical steps, provides insight into the details of the process and assumptions imposed and is followed here.

Any electromagnetic perturbation can be described by the Ampère and Faraday equations:

$$\nabla \times \boldsymbol{H} = \boldsymbol{J} + \frac{\partial \boldsymbol{D}}{\partial t}, \tag{8.1}$$

$$\nabla \times \boldsymbol{E} = -\frac{\partial \boldsymbol{B}}{\partial t}, \tag{8.2}$$

and the simplified, point form of Ohm's law, $\boldsymbol{J} = \sigma \boldsymbol{E}$. The most common simplifications are to (i) treat the ionosphere as "electrostatic," (ii) confine the ionosphere currents to a vertically thin and horizontally uniform sheet, and (iii) assume a vertical geomagnetic field $\boldsymbol{B}_0$.

An electrostatic ionosphere has $\nabla \times \mathbf{E} \rightarrow 0$, which is usually envisioned as a very slow time rate of change in the magnetic field, typically less than $\sim$10 mHz. However, from the left-hand side of Equation 8.2, this condition may also be achieved for waves with large spatial scales or, equivalently, small wave numbers. An impedance analysis for the electrostatic ionosphere was expressed in terms of the conductance in Scholer (1970), where it was shown that the reflection coefficient AA for shear Alfvén waves is

$$AA_{\text{static}} = \frac{\Sigma_a - \Sigma_P}{\Sigma_a + \Sigma_P}, \tag{8.3}$$

where Σ_P is the Pedersen conductance (height-integrated conductivity) and $\Sigma_a = 1/(\mu_0 V_A)$ is the Alfvén wave conductance.

The vertical $\boldsymbol{B}_0$ condition discussed in the literature assumes the reader has in mind a horizontal ionosphere. A more precise description is that $\boldsymbol{B}_0$ is perpendicular to the thin, horizontally uniform ionospheric current sheet. This configuration is a reasonable approximation for the high-latitude regions. In this case, the divergence-free and curl-free (toroidal and poloidal) wave fields are the two ULF modes (shear Alfvén and fast modes) on the magnetosphere side, as shown by Dungey (1963) and developed further by Hughes (1974). The two conditions of perpendicular $\boldsymbol{B}_0$ and horizontally uniform thin current sheets were also used to simplify the physics of wave reflection and mode conversion from an inductive ionosphere (nonelectrostatic condition), since the field of the toroidal part is orthogonal to the field of the poloidal part (Yoshikawa and Itonaga, 1996).

At lower latitudes, the field line resonant frequencies increase and both the vertical $\boldsymbol{B}_0$ and the low-frequency approximations are invalid. The electrostatic approximation was removed in Yoshikawa and Itonaga (1996) and in subsequent work by examining the consequences of the back-emf induced by the ionosphere Hall current, which essentially puts back the right-hand side of Equation 8.2. A concise description of the process is shown in Figure 8.1 (Yoshikawa and Itonaga, 2000). The process begins by assuming that a shear Alfvén wave mode is incident onto the ionosphere from an ideal MHD region. In practice, both the shear Alfvén and fast modes may be incident from the magnetosphere onto the ionosphere. The incident shear Alfvén mode is represented in Figure 8.1 by a field-aligned current directed vertically downward. In the horizontal ionospheric current sheet, the $\nabla \cdot \mathbf{E}$ currents are produced. Both the Pedersen and Hall currents are present for the electrostatic ionosphere condition.

The Hall current complicates magnetosphere–ionosphere coupling and is often omitted. This can lead to inconsistent descriptions. For example, ULF waves incident from the magnetosphere (b_A^i) onto the ionosphere and attenuated on the ground (b_g) have often been modeled using (Nishida, 1964)

$$b_g \sim 2\frac{\Sigma_H}{\Sigma_P}\exp(-|k_\perp|d)b_A^i, \tag{8.4}$$

which includes the Hall conductance Σ_H, but no inductive terms. This might be followed by a relationship between the electric and magnetic wave fields (Allan and Knox, 1979):

$$b = \mu_0 \Sigma_P E, \tag{8.5}$$

where Σ_H is omitted. It is important to have a consistent description when, for example, mapping the ground magnetic signatures of ULF waves to the fields in the ionosphere.

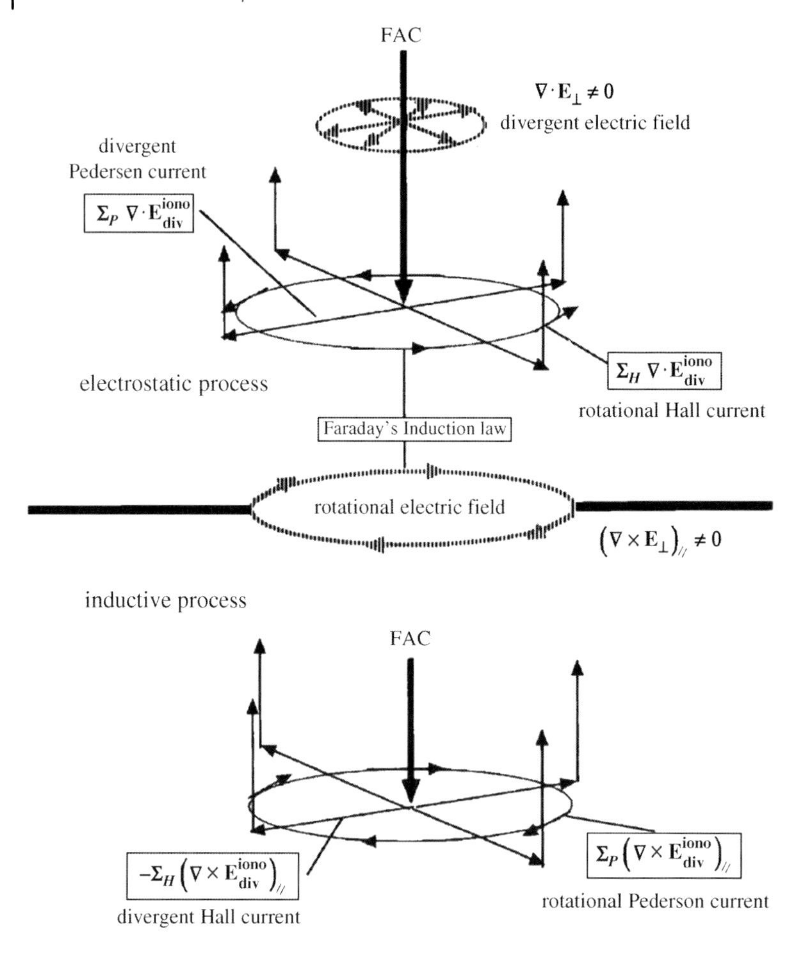

Figure 8.1 Illustration of the current density in the ionosphere for an incident shear Alfvén mode with a vertical geomagnetic field. *Top:* The divergent Pedersen and rotational Hall currents are driven by the divergent ULF wave electric field. *Bottom:* The divergent Hall and rotational Pedersen currents are also excited by the inductive ULF electric field. From Yoshikawa and Itonaga (2000).

An appreciation of the role of the ionosphere Hall currents appears to have coincided with the recent emphasis on inductive effects of the ionosphere. However, it is important to realize that the Hall current term must be included even for an electrostatic ionosphere approximation. This is the mechanism that produces the ground magnetometer signal for this particular geometry.

The inductive feedback process is illustrated in the lower portion of Figure 8.1. One might view the process as a concept similar to the back-emf in electric motors. Equation 8.4 describes an exponential decrease in the ULF magnetic field from the ionosphere to the ground. The divergent Hall current in the inductive process opposes the rotational Hall current, which is an additional physical mechanism that reduces the ground magnetometer signal as the ULF frequency increases. This and other effects of the ionosphere conductances are discussed in the following sections.

8.3
ULF Wave Solution for a Thin Sheet Ionosphere

Removing the condition for vertical $\mathbf{B}_0$, ULF wave interaction with a thin sheet ionosphere was described by Sciffer and Waters (2002). The geometry for the southern hemisphere has X pointing north, Y the west and Z is upward. The background magnetic field lies in the XZ plane at an angle, I from the X axis so that

$$\mathbf{B}_0 = |\mathbf{B}_0|[\cos(I), 0, \sin(I)] \tag{8.6}$$

The magnetic field consists of a static (geomagnetic) field, $\mathbf{B}_0$ with a superimposed ULF time variation, **b**. There is no static electric field and the ULF electric field is **E**. Assuming a temporal dependence of the ULF wave fields in Equations 8.1 and 8.2 as $e^{-i\omega t}$ gives

$$\nabla \times \mathbf{b} = -\frac{i\omega}{c^2}\varepsilon\mathbf{E} \tag{8.7}$$

$$\nabla \times \mathbf{E} = i\omega\mathbf{b} \tag{8.8}$$

where ε is the permittivity tensor. The one-dimensional (1D) solution is obtained by assuming a horizontal spatial dependence of $\exp[i(k_x x + k_y y)]$ and solving for the wave fields as a function of height. We begin in the magnetosphere where ideal MHD conditions require the field aligned ULF wave electric field, $E_{//}$ to be zero, that is,

$$E_{//} = E_x\cos(I) + E_z\sin(I) = 0 \tag{8.9}$$

The two ULF wave modes in the magnetosphere are the fast mode and the shear Alfv'en mode. The fast mode has

$$\nabla \bullet \mathbf{E} = 0 \quad (\nabla \times \mathbf{E})_{//} \neq 0 \tag{8.10}$$

with dispersion relation

$$\frac{\omega^2}{V_A^2} = k_x^2 + k_y^2 + k_{z,f}^2 \tag{8.11}$$

Therefore, the wave electric field of the fast mode is

$$\mathbf{E}_f = \beta[-k_y\sin(I), k_x\sin(I) - k_{z,f}\cos(I), k_y\cos(I)] = \beta\mathbf{P}_f \tag{8.12}$$

where β is a (complex) constant to be determined. The shear Alfv'en wave mode has

$$\nabla \bullet \mathbf{E} \neq 0 \quad (\nabla \times \mathbf{E})_{//} = 0 \tag{8.13}$$

and dispersion relation

$$\frac{\omega^2}{V_A^2} = [k_x\cos(I) + k_{z,a}\sin(I)]^2 \tag{8.14}$$

Equations (8.11) and (8.14) give both the incident and reflected solutions for $k_{z,f}$ and $k_{z,a}$. The wave electric field of the shear Alfv'en mode is

$$\mathbf{E}_a = \alpha[\{k_x \sin(I) - k_{z,a} \cos(I)\}\sin(I), k_y, -\{k_x\sin(I) - k_{z,a} \cos(I)\}\cos(I)] = \alpha\mathbf{P}_a \tag{8.15}$$

for α a (complex) constant.

The total ULF wave fields in the magnetosphere region are

$$\begin{bmatrix} E_x^{mag} \\ E_y^{mag} \\ E_z^{mag} \end{bmatrix} = \alpha^r \hat{\mathbf{P}}_a^r + \alpha^i \hat{\mathbf{P}}_a^i + \beta^r \hat{\mathbf{P}}_f^r + \beta^i \hat{\mathbf{P}}_f^i \tag{8.16}$$

where **P** are unit vectors that contain the wave polarization information.

The horizontally uniform, thin ionosphere current sheet is located at Z=0 and the conductance tensor is

$$\begin{aligned} \vec{\Sigma} &= \begin{bmatrix} \Sigma_d \cos^2(I) + \Sigma_P \sin^2(I) & \Sigma_H \sin(I) & (\Sigma_d - \Sigma_P)\sin(I)\cos(I) \\ -\Sigma_H \sin(I) & \Sigma_P & \Sigma_H \cos(I) \\ (\Sigma_d - \Sigma_P)\sin(I)\cos(I) & \Sigma_H \cos(I) & \Sigma_d sin^2(I) + \Sigma_P \cos^2(I) \end{bmatrix} \\ &= \begin{bmatrix} \Sigma_{11} & \Sigma_{12} & \Sigma_{13} \\ \Sigma_{21} & \Sigma_{22} & \Sigma_{23} \\ \Sigma_{31} & \Sigma_{32} & \Sigma_{33} \end{bmatrix} \end{aligned} \tag{8.17}$$

where Σ_d, Σ_P and Σ_H are the direct, Pedersen and Hall conductances. The relative permittivity tensor relates the conductance tensor, vacuum permittivity and wave frequency

$$\varepsilon = 1 + \frac{i\vec{\Sigma}}{\varepsilon_0 \omega} = \begin{bmatrix} \varepsilon_{11} & \varepsilon_{12} & \varepsilon_{13} \\ \varepsilon_{21} & \varepsilon_{22} & \varepsilon_{23} \\ \varepsilon_{31} & \varepsilon_{32} & \varepsilon_{33} \end{bmatrix} \tag{8.18}$$

The ionosphere currents alter the horizontal magnetic flux components across the sheet according to the standard magnetic field boundary conditions

$$\hat{\mathbf{a}}_n \times \Delta\mathbf{b} = (-\Delta b_y, \Delta b_x, 0) = \mu_0 (j_x^{ion}, j_y^{ion}, 0) \tag{8.19}$$

From (8.19), (8.1) and $\mathbf{j}=\Sigma\ \mathbf{E}$ using (8.17), the electric fields in the magnetosphere, ionosphere and the atmosphere are related by

$$\begin{aligned} &\mu_0\left(\Sigma_{11} - \frac{\Sigma_{31}\Sigma_{13}}{\Sigma_{33}}\right) E_x^{ion} + \mu_0\left(\Sigma_{12} - \frac{\Sigma_{32}\Sigma_{13}}{\Sigma_{33}}\right) E_y^{ion} \\ &\quad - \frac{i}{\omega}\left(\lim_{z\to 0^+}\left[\frac{dE_x^{mag}}{dz} - \frac{dE_z^{mag}}{dx}\right] - \lim_{z\to 0^-}\left[\frac{dE_x^{atm}}{dz} - \frac{dE_z^{atm}}{dx}\right]\right) = 0 \end{aligned} \tag{8.20}$$

$$\begin{aligned} &\mu_0\left(\Sigma_{21} - \frac{\Sigma_{31}\Sigma_{23}}{\Sigma_{33}}\right) E_x^{ion} + \mu_0\left(\Sigma_{22} - \frac{\Sigma_{32}\Sigma_{23}}{\Sigma_{33}}\right) E_y^{ion} \\ &\quad - \frac{i}{\omega}\left(\lim_{z\to 0^+}\left[\frac{dE_z^{mag}}{dy} - \frac{dE_y^{mag}}{dz}\right] - \lim_{z\to 0^-}\left[\frac{dE_z^{atm}}{dy} - \frac{dE_y^{atm}}{dz}\right]\right) = 0 \end{aligned} \tag{8.21}$$

There are no currents in the atmosphere and from the dispersion relation for electromagnetic waves in free space

$$k_z^{atm} = \sqrt{\frac{\omega^2}{c^2} - k_x^2 - k_y^2} \tag{8.22}$$

The horizontal components of the electric field are continuous across the ionosphere-atmosphere interface. The ground is located at Z=−d where we set the electric fields to zero. Since the frequencies are in the mHz range, the atmosphere wave number in Equation (8.22) is usually imaginary so the electric fields in the atmosphere have the following form

$$E_{x,y}^{atm} \propto \frac{\sinh[ik_z^{atm}(z+d)]}{\sinh[ik_z^{atm} d]} \exp i(k_x x + k_y y) \tag{8.23}$$

$$E_z^{atm} \propto \frac{\cosh[ik_z^{atm}(z+d)]}{\sinh[ik_z^{atm} d]} \exp i(k_x x + k_y y) \tag{8.24}$$

Furthermore, the proportionality constants for Equation (8.23) are the x and y components of Equation (8.16). We now have the appropriate functional forms of the equations for the magnetosphere, ionosphere sheet and atmosphere. Substituting Equations (8.23, 8.24) and (8.16) for Z=0 (at the ionosphere current sheet) into Equations (8.20) and (8.21) gives one 2×2 element matrix, θ^i, for the incident signal and a similar matrix, θ^r, for the reflected signal. The 2×2 element reflection and mode conversion matrix is obtained by multiplying the incident matrix, θ^i by the inverse of θ^r.

The IDL code (*rcoeff_ionos.pro*) available from Appendix 1 shows how the reflection and mode conversion coefficient matrix is computed. The input variables are the ULF frequency, ionosphere current sheet height, Alfvén speed for the magnetosphere, horizontal ULF wave numbers, and the ionosphere conductance. The upward and downward vertical wave numbers for the shear Alfvén and fast modes are calculated, followed by the electric field unit vectors **P** from Equations 8.12 and 8.15. The conductivity tensor elements of Equation 8.17 are then used to compute the coefficients of the electric fields in Equations 8.20 and 8.21. The 2×2 θ^i and θ^r matrices are then calculated and combined to give the final result. The code plots the four coefficients as a function of $\boldsymbol{B}_0$ dip angle, as shown in Figure 8.2. For the parameters chosen, the atmosphere wave number is imaginary, so Equations 8.23 and 8.24 are valid.

Figure 8.2 shows the magnitudes and phases of the reflection and mode conversion coefficients as a function of geomagnetic field dip angle. The A and F refer to the shear Alfvén and fast modes, respectively. The AA and the FF panels show wave reflection values for the shear and fast modes, respectively. These are reflection coefficients from the ionosphere–atmosphere ground system.

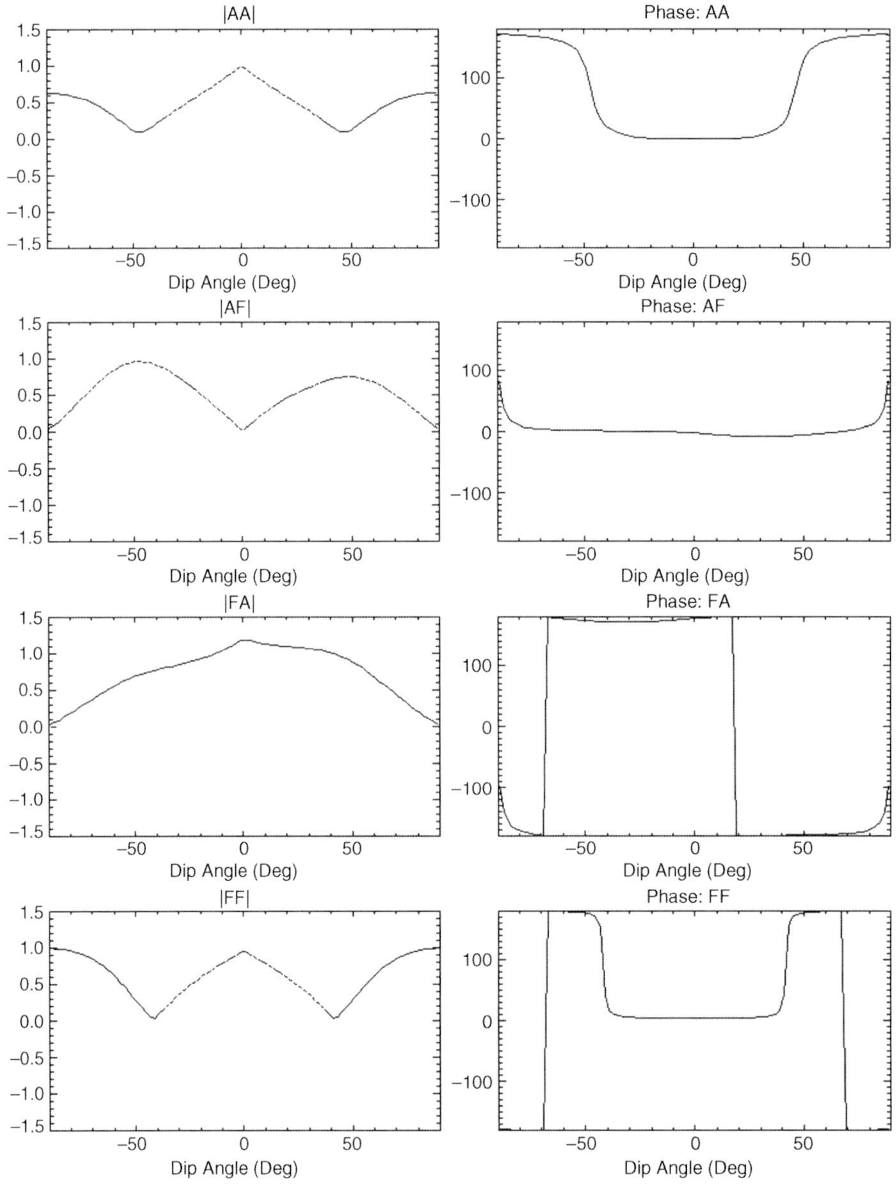

Figure 8.2 Reflection and mode conversion coefficients for $k_x = k_y = 8 \times 10^{-8}\,\mathrm{m}^{-1}$, ionosphere height, $d = 125$ km, Alfvén speed $V_A = 2\pi d$, direct height-integrated conductivity $\Sigma_d = 1 \times 10^6$ S, Pedersen height-integrated conductivity $\Sigma_P = 4$ S, Hall height-integrated conductivity $\Sigma_H = 8$ S, and ULF frequency $f = 20$ mHz.

The AF and FA panels show the mode conversion values. For example, the AF panel shows the conversion from shear Alfvén to fast mode produced by the system that peaks for midlatitudes, while the conversion from any incident fast mode into the shear Alfvén mode is maximum near the equator and low latitudes, for the given parameters. While the coefficients plotted with dip angle appear symmetrical about the equator, this is not necessarily the case. Using the code, we can reduce the conductances to values representative of a nighttime ionosphere. The AF and FA coefficients become hemisphere dependent, while the AA coefficient reduces for most latitudes. A large value for AA produces high-quality (longer ring time) field line resonances.

The combined ULF wave fields and the various upward and downward components of the shear Alfvén and fast modes may also be calculated. The electric fields in the magnetosphere may be obtained from Equation 8.16 and the associated magnetic wave fields are then calculated using Equation 8.8. In the atmosphere, the ULF wave electric fields are obtained from Equations 8.23 and 8.24. The IDL program *ulf_updown_fields.pro* (see Appendix 1) shows the details of the calculations and how the continuity of the tangential electric field components are used to match the magnetosphere and atmosphere solutions. Most of this code is identical to *rcoeff_ionos.pro* to obtain the reflection and wave mode conversion matrix. Thereafter, the appropriate amplitudes of the various solutions in the different regions are found and stitched together to form the final altitude solution. In addition to providing the solution for any $\boldsymbol{B}_0$ dip angle, the user may also alter the mix of incident fast to shear Alfvén mode and the phase difference between the two modes. The solutions are valid provided the atmosphere wave number in Equation 8.22 is imaginary. An example of the code output is shown in Figure 8.3. All the horizontal electric field components in Figure 8.3a are zero on the ground, which is assumed to be a perfect conductor. The reader might set the code parameters for 100% input shear mode (mix = 1.0) and zero k_y. The downward fast mode wave fields would then be zero and the switch from b_y in the magnetosphere to b_x in the atmosphere and on the ground for the total field components should be evident. This demonstrates the well-known change in ULF wave polarization through the ionosphere, which is seen in the bottom left panel of Figure 8.3b even for the 80% shear Alfvén mix default setting in the code.

8.4
ULF Wave Solution for a Realistic Ionosphere

A number of instruments are available to probe ULF wave properties in the ionosphere. These include HF radars and Doppler sounders, which will be discussed in Section 8.6. Therefore, a height-dependent model for ULF wave passage through the ionosphere is required. The thin ionosphere current sheet approximation can be removed and replaced with an altitude-dependent conductivity tensor. The ULF wave solution for this case must be solved numerically. Performing the curl operations in Equations 8.7 and 8.8 for our one-dimensional approximation yields four first-order differential equations for the horizontal ULF wave field components:

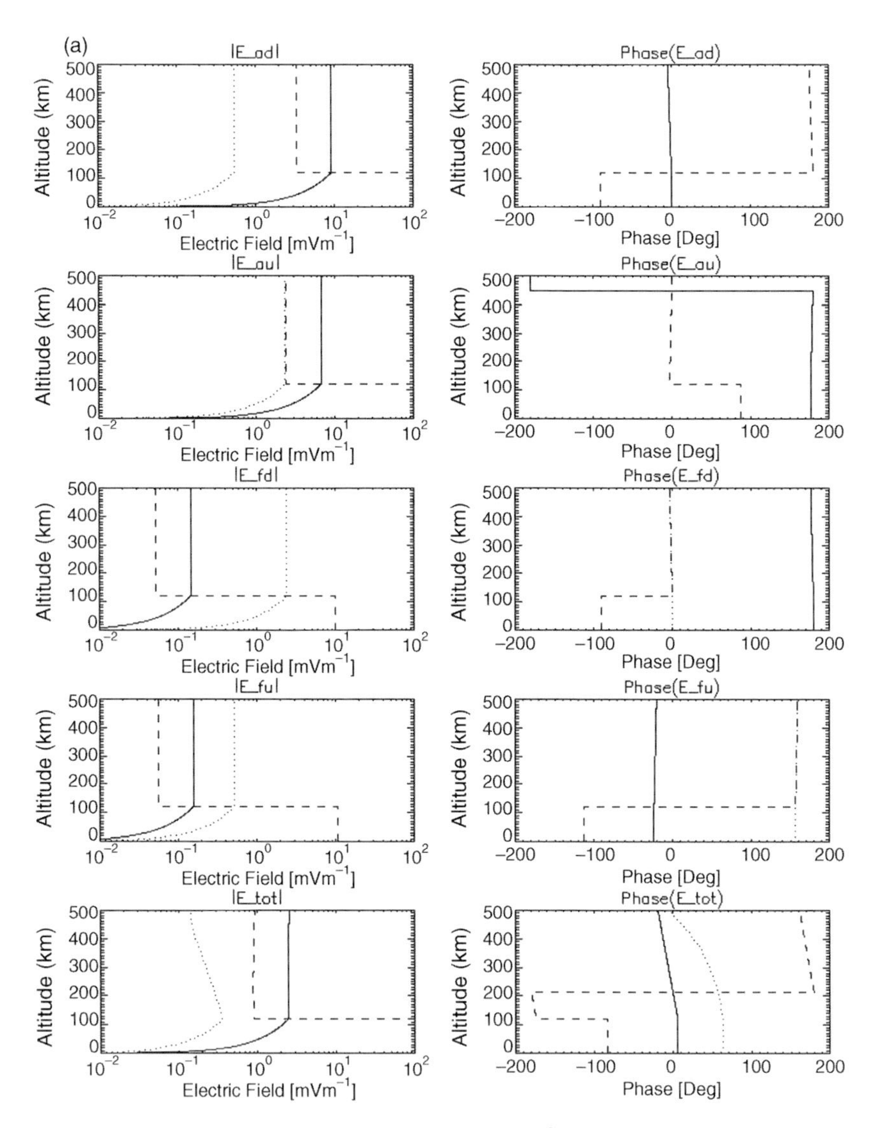

Figure 8.3 The upward and downward components of ULF wave fields for a geomagnetic field dip angle 70°, $k_x = 8 \times 10^{-8}\,\mathrm{m}^{-1}$, $k_y = 8 \times 10^{-9}\,\mathrm{m}^{-1}$ ionosphere height, $d = 120$ km, Alfvén speed $V_A = 2\pi d$, direct height-integrated conductivity $\Sigma_d = 1 \times 10^6$ S, Pedersen height-integrated conductivity $\Sigma_P = 6.2$ S, Hall height-integrated conductivity $\Sigma_H = 7.6$ S, and ULF frequency $f = 20$ mHz. (a) ULF electric field. (b) ULF magnetic fields.

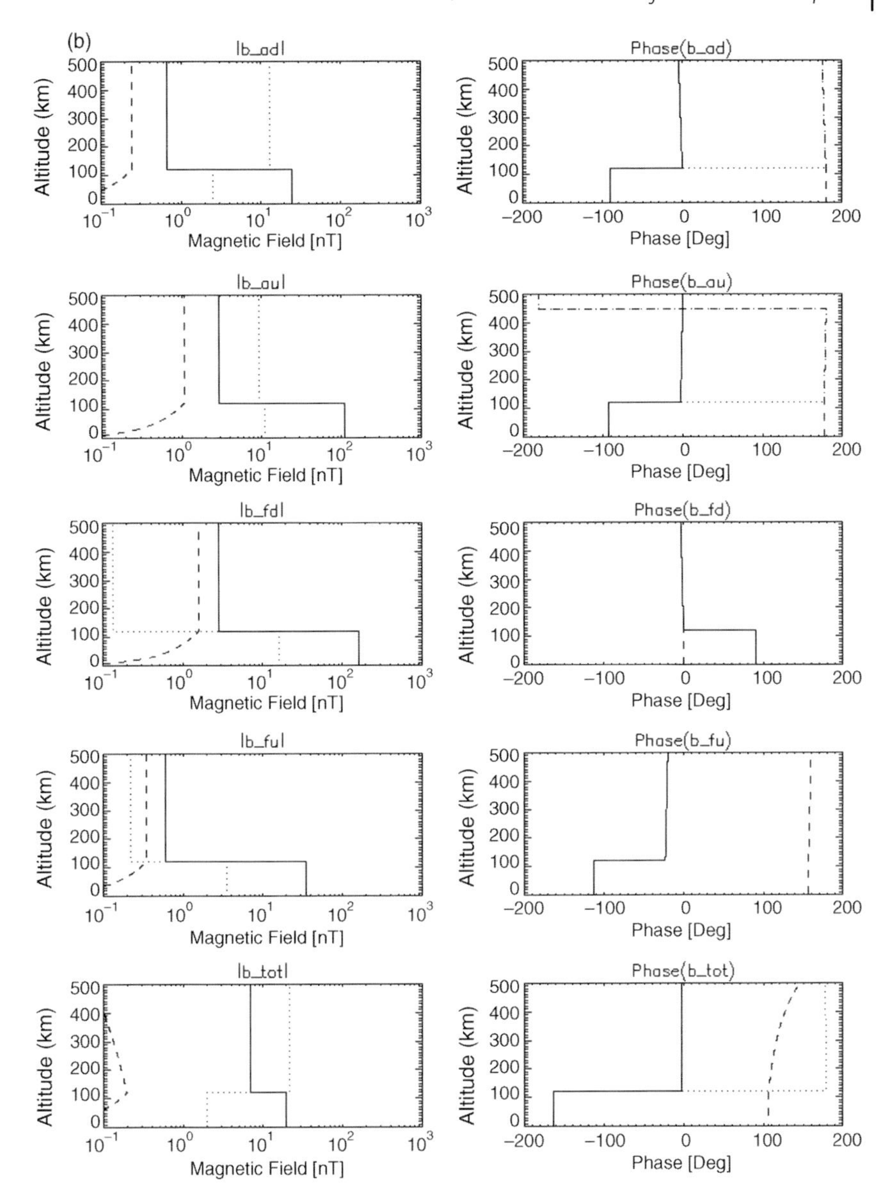

Figure 8.3 (*Continued*).

$$
i\left[\frac{k_y^2}{\omega}-\frac{\omega}{c^2}\left(\varepsilon_{11}-\frac{\varepsilon_{13}\varepsilon_{31}}{\varepsilon_{33}}\right)\right]E_x-i\left[\frac{k_xk_y}{\omega}+\frac{\omega}{c^2}\left(\varepsilon_{12}-\frac{\varepsilon_{13}\varepsilon_{32}}{\varepsilon_{33}}\right)\right]E_y-ik_y\frac{\varepsilon_{13}}{\varepsilon_{33}}b_x
+\frac{\mathrm{d}b_y}{\mathrm{d}z}+\frac{ik_x\varepsilon_{13}}{\varepsilon_{33}}b_y=0, \tag{8.25}
$$

$$i\left[\frac{k_x k_y}{\omega} - \frac{\omega}{c^2}\left(\varepsilon_{21} - \frac{\varepsilon_{23}\varepsilon_{31}}{\varepsilon_{33}}\right)\right]E_x - i\left[\frac{k_x^2}{\omega} - \frac{\omega}{c^2}\left(\varepsilon_{22} - \frac{\varepsilon_{23}\varepsilon_{32}}{\varepsilon_{33}}\right)\right]E_y + ik_y\frac{\varepsilon_{23}}{\varepsilon_{33}}b_x + \frac{db_x}{dz} - \frac{ik_x\varepsilon_{23}}{\varepsilon_{33}}b_y = 0, \tag{8.26}$$

$$ik_y\frac{\varepsilon_{31}}{\varepsilon_{33}}E_x + \frac{dE_y}{dz} + ik_yE_y + i\left(\omega - \frac{c^2k_y^2}{\omega\varepsilon_{33}}\right)b_x + ik_xk_y\frac{c^2}{\omega\varepsilon_{33}}b_y = 0, \tag{8.27}$$

$$\frac{dE_x}{dz} + ik_x\frac{\varepsilon_{31}}{\varepsilon_{33}}E_x + ik_x\frac{\varepsilon_{32}}{\varepsilon_{33}}E_y - ik_xk_y\frac{c^2}{\omega\varepsilon_{33}}b_x - i\left(\omega - \frac{c^2k_x^2}{\omega\varepsilon_{33}}\right)b_y = 0, \tag{8.28}$$

and two analytic equations for the vertical field components:

$$E_z = -\frac{\varepsilon_{31}}{\varepsilon_{33}}E_x - \frac{\varepsilon_{32}}{\varepsilon_{33}}E_y - k_y\frac{c^2}{\omega\varepsilon_{33}}b_x + k_x\frac{c^2}{\omega\varepsilon_{33}}b_y, \tag{8.29}$$

$$b_z = -\frac{k_y}{\omega}E_x + \frac{k_x}{\omega}E_y. \tag{8.30}$$

Equations 8.25–8.28 may be solved as a boundary value problem (Sciffer, Waters, and Menk, 2005) and the reflection and mode conversion matrix calculation is no longer required. However, these numerical solutions do not provide details of the various ULF wave mode up/down components that are available from the analytic solutions.

Experimenting with typical values for the ULF wave spatial scale, frequency, dip angle, and ionosphere conductance shows that a number of physical processes alter the properties of ULF waves as they interact with the ionosphere. The best-known effect is a rotation of the ULF wave polarization azimuth in the horizontal plane as seen in the b_x and b_y components of Figure 8.3b. A simplified explanation for this effect begins with Ampère's law, where the curl of the ULF wave magnetic field changes from a field-aligned current associated with the shear Alfvén mode to zero current in the atmosphere. Therefore, the components of the magnetic field must rotate to set the curl to zero. However, this does not include effects from the Hall current. If we begin with an incident shear Alfvén mode, then depending on the various parameters (dip angle, spatial scale, frequency, and ionosphere conductance), the shear Alfvén to fast mode conversion coefficient AF will be nonzero, resulting in fast mode in the topside ionosphere. Therefore, in general, the lower altitudes of the magnetosphere, near the ionosphere, will have some mix of the shear Alfvén and fast modes, even if we approximate the ULF properties by assuming a pure shear Alfvén incident mode. The fast mode generated in the anisotropic ionosphere is evanescent upward. Therefore, the horizontal plane, polarization azimuth change from the ground to the magnetosphere depends on all ULF/ionosphere interaction parameters and the altitude of the measurement. The polarization properties are further altered if the ionosphere conductivity contains horizontal gradients. These aspects are illustrated in Figure 8.4.

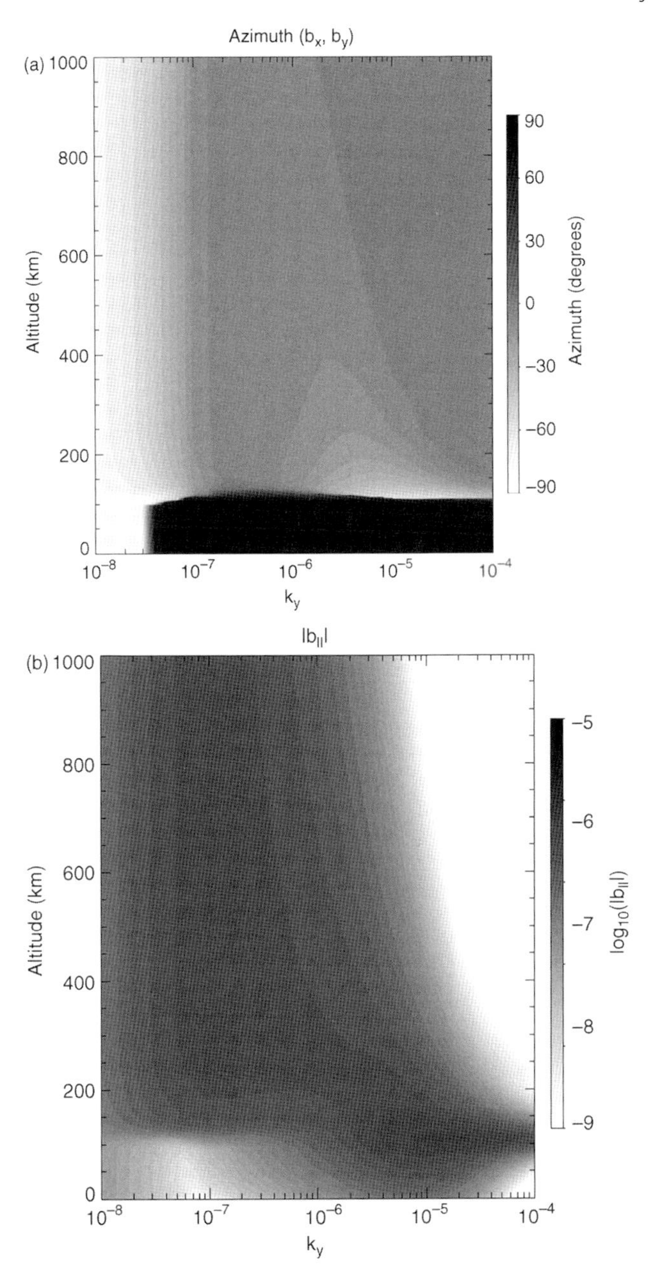

Figure 8.4 (a) The polarization azimuth computed from ULF b_x and b_y for a frequency of 16 mHz and dip angle of $I = 70^\circ$ at solar maximum ionosphere conditions. The wave numbers are $k_x = 10^{-10}\,\mathrm{m}^{-1}$ and k_y varying between 10^{-8} and $10^{-4}\,\mathrm{m}^{-1}$. (b) The amplitude of the field-aligned (compressional) component of the ULF wave magnetic field for the parameters used in panel (a). From Sciffer, Waters, and Menk (2005). (For a color version of this figure, please see the color plate at the beginning of this book.)

The attenuation in amplitude of ULF waves propagating from space, through the ionosphere to the ground, has been known for quite a long time in the form of Equation 8.4 or some similar expression (e.g. Nishida, 1964). Essentially, the MHD disturbance encounters a boundary (atmosphere) where the propagation constant normal to the boundary switches from real to imaginary, an evanescent mode. While Equation 8.4 does involve the Hall conductance, it is an "electrostatic ionosphere" approximation, which does not necessarily mean small frequencies. For ULF wave and ionosphere parameters that give an inductive ionosphere, the ULF amplitude on the ground is also attenuated by the back-emf currents of the inductive effect. This is illustrated in Figure 8.5 where the incident ULF magnetic field has been given unit magnitude. The fields expected on the ground for an electrostatic ionosphere are shown with $*$ symbols. The rollover in amplitude is more pronounced as the ULF frequency increases, due to the inductive feedback onto the ionosphere Hall current. The finite ionosphere conductance also means that the ULF wave electric field is not zero here, as is often assumed in the calculation of FLRs.

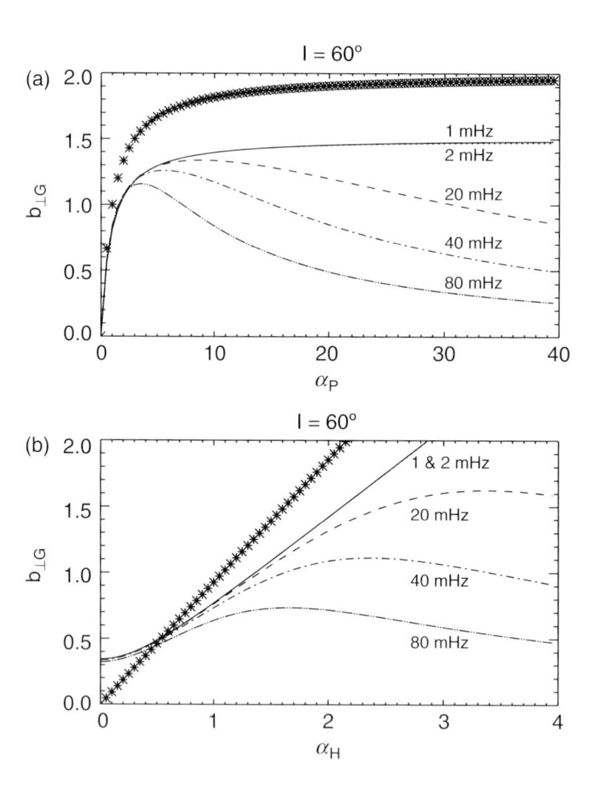

Figure 8.5 The normalized, horizontal ULF magnetic field on the ground for a dip angle of 60°, $V_A = 1 \times 10^6$ m s^{-1}, $k_x = 0$, $k_y = 1/d$, and $d = 100$ km. (a) Variation with $\alpha_P = \Sigma_P/\Sigma_A$ for $\Sigma_H = 2\Sigma_P$ and Σ_A is the Alfven conductance. (b) Variation with $\alpha_H = \Sigma_H/\Sigma_P$ for $\Sigma_P = 10$ S. The $*$ symbols show the $\Sigma_H/(\Sigma_P + \Sigma_A)$ curve. Adapted from Sciffer, Waters, Menk (2004).

8.5 FLRs and the Ionosphere

As discussed in Chapters 5 and 6, remote sensing the plasma mass density in the magnetosphere using FLRs requires a determination of the FLR frequencies at given locations. Although the ionosphere is often represented as the resonance nodal points for FLR electric fields, Section 8.4 showed nonzero electric fields there. Furthermore, the ionosphere introduces neutral particle collisions into the resonance with associated damping. From considerations of simple harmonic motion, damping is known to alter the resonant frequency. If the ionosphere introduces shifts in frequency for the FLRs or affects them in any way that influences plasma mass estimation, then we need to include these effects in magnetoseismology algorithms. In this section, the effects of the ionosphere on the identification and selection of FLR frequencies are discussed.

The formation of FLRs depends on suitable energization in the magnetosphere and reflection coefficients at the ionosphere. Magnetoseismology depends on being able to detect FLRs over a range of latitudes, so models that are limited to assuming a vertical $\boldsymbol{B}_0$ at the ionosphere are inadequate. The FLR frequencies increase to the Pc3 band when probing the inner plasmasphere, so we need to include the inductive ionosphere features. Furthermore, FLRs have an amplitude profile with latitude that is inconsistent with an $\exp(ik_x x)$ spatial model, so a one-dimensional description is inadequate.

A ULF wave model for the magnetosphere, ionosphere, and atmosphere requires solutions to elliptic, parabolic, and hyperbolic differential equations. For example, the combined fluid and electromagnetic, ideal MHD equations of Section 3.5 are fine for the magnetosphere, but not for the atmosphere. A ULF wave model that solves the MHD equations in the magnetosphere, tracks ULF wave fields from the magnetosphere through the ionosphere to the ground, including oblique $\boldsymbol{B}_0$, Pedersen and Hall currents, and solves in a meridian plane was described by Waters and Sciffer (2008). In this setting, the FLR spatial structure and frequencies are solved, so that latitude-dependent ionosphere and magnetosphere (Alfvén speed) parameters are included with shear Alfvén and fast mode coupling described in both the magnetosphere and the ionosphere.

The ULF waves are described in the most general way as an electromagnetic disturbance, using the effective permittivity discussed in Section 3.6. The coordinate system must be compatible with the dipole-like magnetosphere and the spherical ionosphere and atmosphere. In order to simplify the ionosphere boundary conditions yet model ULF waves in a dipole geometry, a distorted dipole coordinate system was developed (Lysak, 2004). We define a coordinate set (u^1, u^2, u^3) that is a function of the standard spherical coordinates (R, θ, φ). Then, we define two basis vector sets, one tangential to the geomagnetic field and the other set orthogonal to the ionosphere current sheet. For R the position vector in spherical coordinates, the tangential basis vectors v_i are defined by

$$\mathbf{v}_i = \frac{\partial \mathbf{R}}{\partial u^i} = \partial_i \mathbf{R}. \tag{8.31}$$

The cotangent (reciprocal, dual, or gradient) basis vectors $\mathbf{v}^i$ are

$$\mathbf{v}^i = \nabla u^i. \tag{8.32}$$

These basis vectors are related such that $\mathbf{v}^i \bullet \mathbf{v}_j = \delta_j^i$. The Jacobian of the transformation from (R, θ, φ) to $(u^1; u^2; u^3)$ is

$$J = \frac{\partial(u^1, u^2, u^3)}{\partial(R, \theta, \varphi)} = \mathbf{v}^1 \cdot (\mathbf{v}^2 \times \mathbf{v}^3).$$

The coordinate system that morphs from spherical into dipolar has (Lysak, 2004)

$$u^1 = -\frac{R_\mathrm{I}}{R}\sin^2\theta, \quad u^2 = \varphi, \quad u^3 = -\frac{R_\mathrm{I}^2 \cos\theta}{R^2 \cos\theta_0}, \tag{8.33}$$

where R_I is the radius of the ionosphere, R is the radial distance from the center of Earth to the particular point in space, and θ_0 is the colatitude on the ground. The model assumes an azimuthal dependence of all electric and magnetic fields of the form $e^{im\varphi}$ where m is the azimuthal wave number.

The Maxwell equations may be reformulated using the covariant and contravariant basis vectors (D'haeseleer *et al.*, 1991). In the magnetosphere, the basis vector component v_3 is tangential to the geomagnetic field, so the ULF electric field component $E_3 = 0$. The equations to be solved are

$$\frac{\partial E^1}{\partial t} = \frac{V^2}{J}(imb_3 - \partial_3 b_2), \tag{8.34}$$

$$\frac{\partial E^2}{\partial t} = \frac{V^2}{J}(\partial_3 b_1 - \partial_1 b_3), \tag{8.35}$$

$$\frac{\partial b^1}{\partial t} = \frac{1}{J}\partial_3 E_2, \tag{8.36}$$

$$\frac{\partial b^2}{\partial t} = -\frac{1}{J}\partial_3 E_1, \tag{8.37}$$

$$\frac{\partial b^3}{\partial t} = \frac{1}{J}(imE_1 - \partial_1 E_2), \tag{8.38}$$

for $V^2 = 1/(\mu_0 \varepsilon)$, which uses the permittivity defined in Equation 3.67. The popular finite-difference time-dependent techniques are used to solve Equations 8.34–8.38, given a spatial distribution of the Alfvén speed. Energy for the FLRs is provided by driving a high-latitude field line with a compressional signal that initiates fast mode propagation.

The ionosphere current density is described by $j = \Sigma \cdot E$ and the standard electromagnetism field boundary conditions apply. The horizontal electric fields E_1 and E_2 and the radial magnetic field b^3 are continuous across the ionosphere current sheet.

The ionospheric boundary equation relating the electric and magnetic fields is

$$\mu_0 \Sigma \cdot E = \hat{\boldsymbol{r}} \times \Delta \boldsymbol{b}, \tag{8.39}$$

where Σ is the conductance tensor and Δb is the discontinuity in the ULF magnetic field across the current sheet. The atmosphere is described by $\nabla \cdot b = \nabla \times b = 0$, so the ULF magnetic field may be expressed in terms of a scalar potential ψ where $b = -\nabla\psi$ and $\nabla^2\psi = 0$. This Laplace equation in the atmosphere is solved using spherical harmonic basis functions to obtain the magnetic field below the current sheet. For ULF waves, the ground is a good conductor, so the radial component of the wave magnetic field on the ground is set to zero. At the ionosphere current sheet, continuity in the radial magnetic field component is imposed.

A snapshot of the ULF wave fields from the model is shown in Figure 8.6. Here the excitation was a band-limited fast mode at the outer field line with a center frequency of 5 mHz. The top left panel shows the compressional component of the magnetic field perturbations b_μ, while the top right panel is the electric field component E_ν in the plane of the page, perpendicular to the dipole field. This is the radial direction

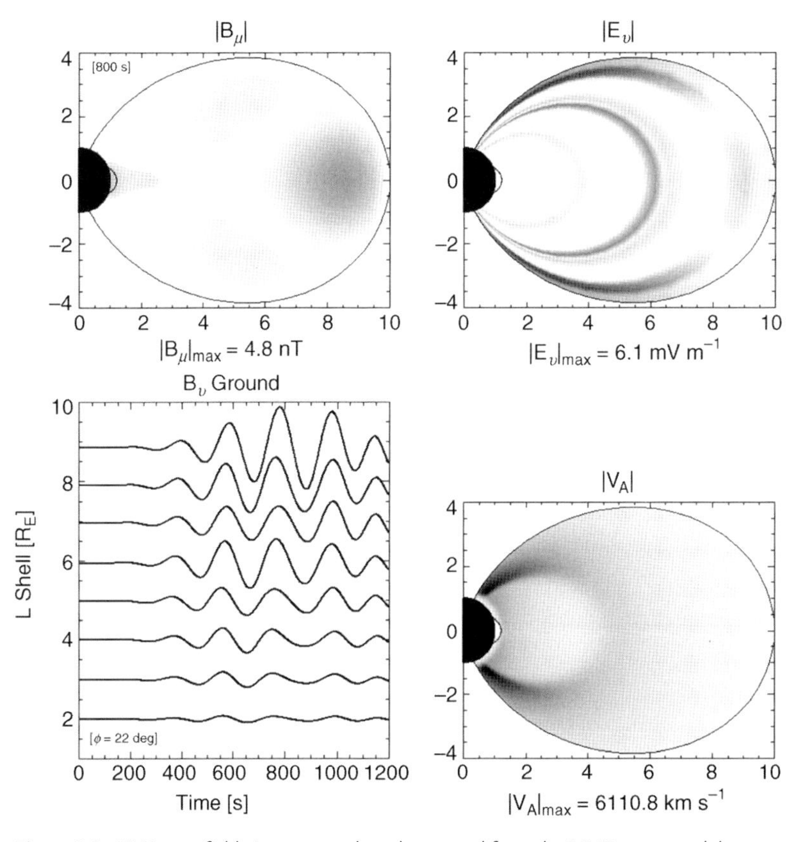

Figure 8.6 ULF wave fields in space and on the ground from the MHD wave model.

when in the equatorial plane. The E_ν component shows the FLR structure, with a fundamental harmonic along the field near $L \approx 6$ and a third harmonic for $L \approx 9$. The driver was set to excite odd mode harmonics. The lower left panel shows the resultant time series of the north–south magnetic field on the ground for various latitudes (L-shell). Such simulation of ULF fields from the magnetosphere to the ground allows the investigation of remote sensing applications.

The model was used to assess the effect of the ionosphere on FLR frequency and location (Waters and Sciffer, 2008). The main result is shown in Figure 8.7. A broad, but band-limited fast mode was used to excite a fundamental FLR in the plasmasphere at 20 mHz. The Pedersen and Hall conductances were varied to see if the 20 mHz resonance moved in latitude. Figure 8.7a shows a slight shift of the peak in power with latitude as the conductance increases. This shows features very similar to that of a simple harmonic oscillator with damping. As discussed in Chapter 6, the FLRs are detected using the cross-phase of data obtained from latitudinally spaced sites. The cross-phase computed from the model output data at fixed sites near 40° latitude selects the 20 mHz resonance as shown in Figure 8.7b. While the maximum of the cross-phase decreases with decreased conductance (increased damping), the frequency selected by the cross-phase maximum is consistently at the same FLR frequency. A similar result is obtained for the amplitude difference method, as seen in Figure 8.7c. Therefore, over the range of realistic ionosphere conductances, detection of the FLR frequency using the cross-phase, within experimental uncertainty, is independent of the ionosphere conductance. This is consistent with results presented in Chapter 6.

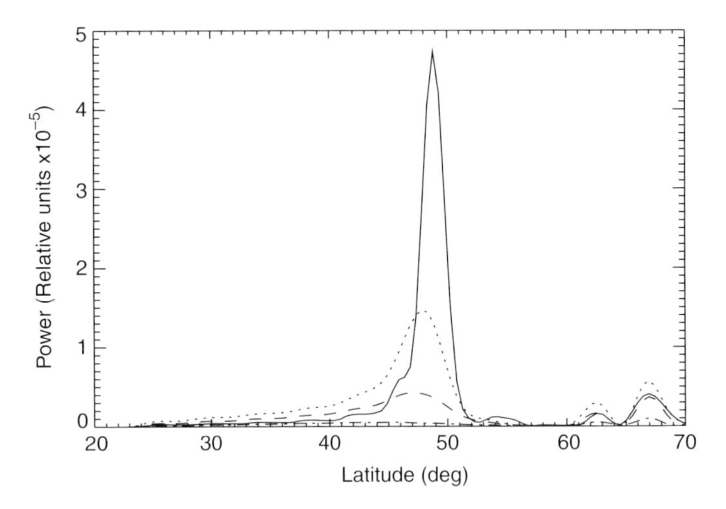

Figure 8.7 (a) Power at 20 mHz in the north–south magnetic field perturbation versus latitude for grid points located on the ground for different values of ionosphere Pedersen conductance (solid: 10 S; dotted: 2 S; dash: 1 S; dash-dotted: 0.4 S). (b) Cross-phase (top) and (bottom) amplitude difference spectra for the 20 mHz fundamental FLR for different values of ionosphere Pedersen conductance (dotted: 10 S; dashed: 2 S; dash-dotted: 1 S; dash-dot-dotted: 0.4 S). From Waters and Sciffer (2008).

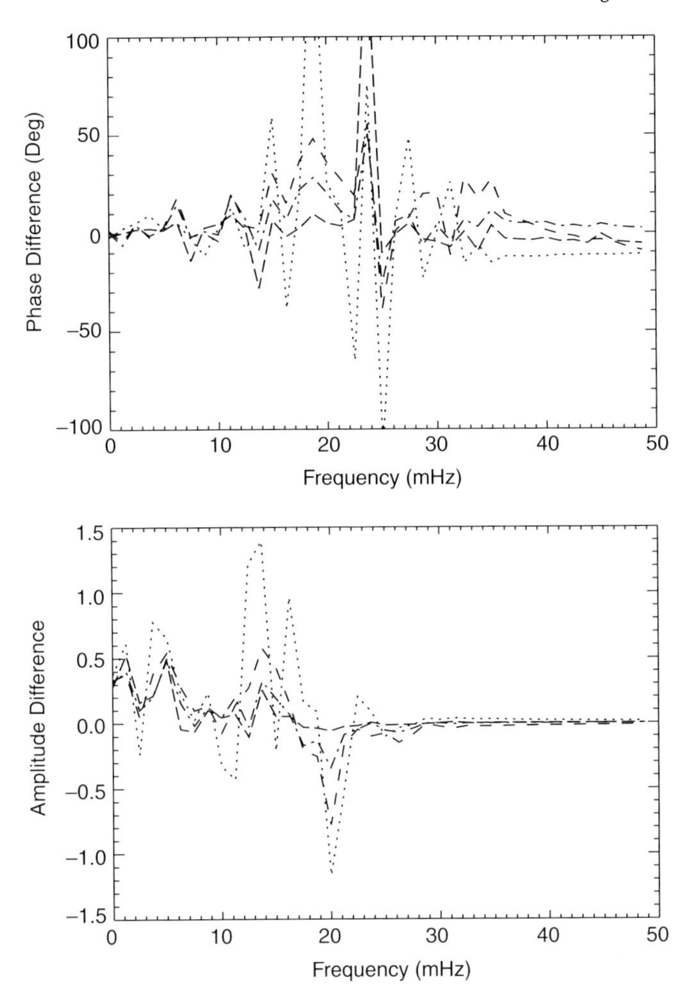

Figure 8.7 *(Continued)*.

8.6 Remote Sensing ULF Electric Fields in Space

ULF wave reflection properties at the ionosphere are critical parameters for magnetoseismology. ULF field line resonances exhibit enhanced amplitudes in the magnetosphere and for odd harmonics, the electric field amplitudes are largest in the equatorial plane. Electric fields in space accelerate charged particles. ULF waves, particularly in the lower Pc5 frequency range, are likely to play an important role in MeV electron dynamics in the radiation belts as discussed in Section 7.9. However, there are few *in situ* measurements of ULF electric fields in space, and it is difficult to obtain multi-spaced measurements. Ground magnetometer arrays provide a promising data source for developing a magnetosphere electric field remote sensing algorithm (Ozeke, Mann, and Rae, 2009). Given the number of parameters that influence the ULF wave transition from space to the ground, it may seem impossible to obtain reasonable estimates of the wave electric field magnitudes in the magnetosphere using ground data. However, this can be achieved provided the ionosphere Hall current and associated effects are included.

Assume a fundamental FLR frequency has been detected at a particular latitude on the ground. This provides the north–south magnetometer time series b_ν^g. The idea is to map this amplitude into the east–west magnetic field component (assuming 90° rotation) in the ionosphere, b_φ^i followed by a solution for the resonance wave function along the field to estimate the equatorial electric field E_ν^{eq}. The various ratios of the electric and magnetic fields from these three locations provide a convenient way to organize the results of an investigation of the process. Using the model described in Section 8.5, Sciffer and Waters (2011) showed that the tricky part is to map the ground fields up into the ionosphere. Using a 5 mHz fast mode driver for the model, a fundamental resonance appeared at $L = 6.3$. The various electric and magnetic field ratios were determined for model runs with azimuthal variation of $m = 2$ and $m = 0$ and various ionosphere conductance levels, and the results are summarized in Table 8.1. The resonant widths are also included.

Table 8.1 shows that the ratio of the equatorial plane electric field to the ionosphere magnetic field (second last column) is relatively constant. The process to take b_φ^i from the ionosphere to b_ν^g at the ground depends on conductance and m number (wave spatial scale). In order to illustrate what is happening, the wave fields for the

Table 8.1 Conductances, resonance widths, and wave amplitude ratios for the shear mode in the ionosphere and magnetosphere and on the ground for $m = 1$ and $m = 2$.

Σ_P (S)	Σ_H (S)	$\Delta\theta^i$ (°)	$\Delta\theta^g$ (°)	E_ν^{eq}/b_ν^g	E_ν^{eq}/b_φ^i	b_φ^i/b_ν^g
$m = 0$						
5	1	2.5	5.0	0.02	0.06	0.41
5	5	1.5	3.0	0.04	0.06	0.72
5	10	1.5	3.0	0.03	0.06	0.50
$m = 2$						
1	0	5.0	—	—	0.05	
1	0.2	6.0	12.0	0.16	0.05	3.26
1	1	6.0	12.0	0.03	0.05	0.68
$m = 2$						
5	0	2.5	—		0.05	—
5	1	3.0	6.0	0.20	0.05	3.95
5	5	2.5	5.0	0.04	0.05	0.81
5	10	2.0	4.0	0.02	0.05	0.46
$m = 2$						
10	0	1.5	—	—	0.05	—
10	2	2.0	4.0	0.22	0.05	4.37
10	10	2.0	4.0	0.05	0.05	0.95
10	20	2.0	4.0	0.03	0.05	0.56

From Sciffer and Waters (2011).

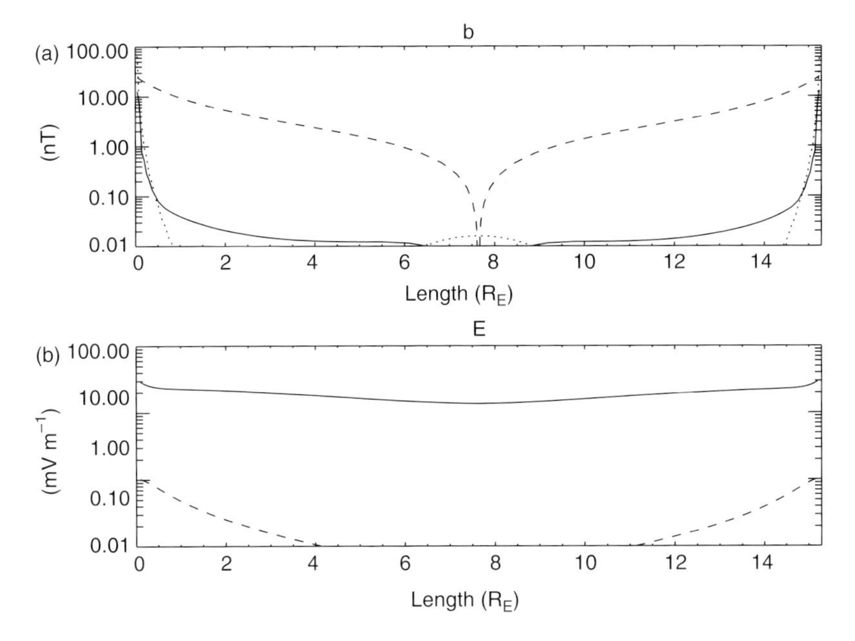

Figure 8.8 Amplitude of the equatorial ULF (a) magnetic field and (b) electric field for $m = 2$ at the resonant L-shell ($6.3 R_E$) as a function of distance along the field line. The conductance values were $\Sigma_P = 5$ S and $\Sigma_H = 10$ S. The solid lines show the ν components, the dashed lines show the φ components, and the dotted line shows the μ (field-aligned) component. From Sciffer and Waters (2011).

$L = 6.3$ resonance are shown in Figure 8.8. The decay in amplitude of the fast mode (b_μ and b_ν) with height is the key feature, and this is generated by the ionosphere rotational current system, resulting in an upward evanescent fast mode. These fields influence the ground magnetic field magnitudes. A common perception is that ULF wave amplitudes measured on the ground are less than those at the ionosphere, due to the exponential decrease of Equation 8.4. Ratios less than unity in the last column of Table 8.1 and the results in Sciffer, Waters, and Menk (2004) and Yoshikawa and Itonaga (2000) show that ULF magnetic field observed on the ground may be larger than those above the ionosphere. The reason is the production of fast mode in the ionosphere. This ratio of magnetic field in the ionosphere to that on the ground is the trickiest step when trying to determine E_ν^{eq}/b_ν^{g}. More research is required in order to develop a simple analytic relationship. In the meantime, E_ν^{eq} can be estimated from ground data using the 2D model of Waters and Sciffer (2008).

8.7 Quarter-Wave Modes

Magnetoseismology using FLRs depends on the identification of the correct harmonic. If the ionosphere has large conductance values in both hemispheres for a particular field line, then the ULF wave electric fields will be a minimum there. Therefore, the usual choice is an odd harmonic, typically the fundamental. However, the offset of the solar-driven properties of the ionosphere, which are ordered by geographic coordinates, compared with magnetospheric processes such as FLRs which are described using geomagnetic coordinates, can result in quarter-wave resonant modes when only one hemisphere supports a ULF wave electric field node (Allan and Knox, 1979b; Obana *et al.*, 2008). Owing to the relative geometries of the geographic and geomagnetic coordinate systems, quarter-wave modes are more likely to occur in the American longitude sector and around dawn and dusk in winter or summer, as shown for a representative day in Figure 8.9.

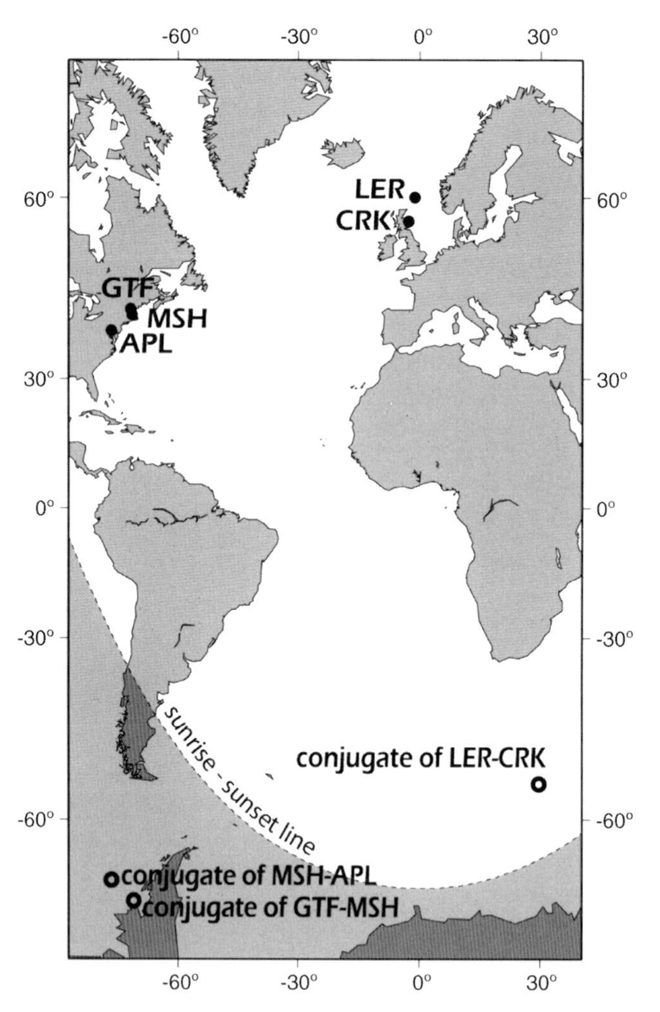

Figure 8.9 Map showing the terminator location (dashed line) and sunlit and dark hemispheres at 1200 UT on June 27, 2001, and various magnetometer station pairs (filled circles) and their conjugate midpoints (open circles). From Obana *et al.* (2008).

Consider the predawn situation when the ionosphere electron density is relatively depleted. For a south (north) hemisphere midwinter (midsummer), a magnetic field line in the summer ionosphere will encounter sunrise before the opposite winter hemisphere. This produces quite different FLR ionosphere boundary conditions with the possibility of a reasonable "node" in the summer ionosphere and an "antinode" at the winter one. As the sunrise progresses into morning, the winter ionosphere also increases in conductance, switching the FLR to the usual half-wave resonance.

Figure 8.10 shows the temporal sequence for the detection of FLRs using the cross-phase technique when the ionosphere conductances in the southern hemisphere are quite different from those in the northern hemisphere. Due to solar EUV around dawn, the summer north hemisphere allows a ULF electric field quasi-node, while the south winter hemisphere lags in electron density replenishment and the resonance at this end is like an open pipe condition with an electric field quasi-antinode, forming a quarter-wave mode. The FLR frequency of the quarter-wave mode is about half the value of the more common half-wave mode resonance. The

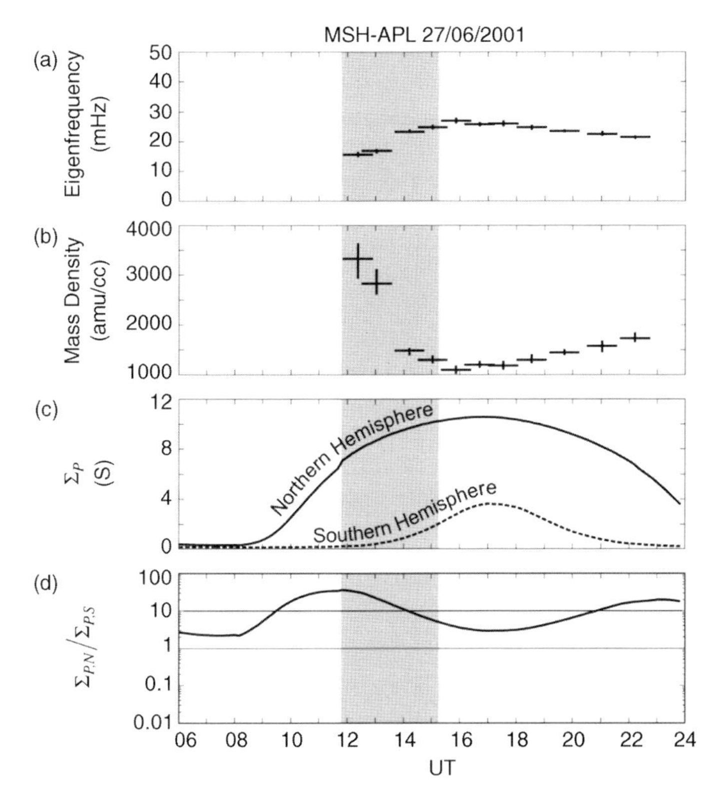

Figure 8.10 Time series of (a) eigenfrequency and (b) inferred mass density for conditions represented in Figure 8.9. (c) Height-integrated Pedersen conductivities at northern (solid line) and southern (dashed line) ends of the field line. (d) Interhemispherical (north–south) ratio of Pedersen conductance. From Obana *et al.* (2008).

quarter-wave frequencies appear in the 12–13 UT section of Figure 8.10. As the winter hemisphere ionosphere conductance increases, the resonance transforms into the half-wave mode. The consequences of assuming half-wave modes for plasma mass density estimation are shown in Figure 8.10b. Therefore, when estimating plasma mass density from FLRs around dawn and dusk we must consider the possibility of quarter-wave modes, especially at American longitudes.

8.8 Detection of ULF Waves in the Ionosphere

Many ionosphere and ULF wave parameters can alter ULF wave properties in transition from space to the ground, and it is important to obtain ULF wave data at ionosphere heights in order to understand the process. Radio sounders operating in the HF (3–30 MHz) band provide data from the various ionosphere altitudes, since the electron density N_e and plasma frequency are related by $f_p \approx 9\sqrt{N_e}$ Hz. Variations in the frequency of HF signals that correlated well with geomagnetic field variations in the ULF range were first reported in the 1960s (Chan, Kanellakos, and Villard, 1962; Duffus and Boyd, 1968) for oblique radio propagation paths. More detailed studies using vertical incidence found that the HF frequency shift due to ULF waves was proportional to the HF frequency (Marshall and Menk, 1999; Menk, Cole, and Devlin, 1983) and to the magnitude of the geomagnetic field variation (Menk, 1992; Watermann, 1987). This provides a mechanism for obtaining ULF wave information in the ionosphere using HF signals.

The path s of HF electromagnetic signals propagating through the ionosphere depends on the frequency f_{HF} and the refractive index μ that is given by the Appleton–Hartree equation. If the refractive index changes with time, then the associated change in f_{HF} is (Bennett, 1967)

$$\Delta f = -\frac{f_{HF}}{c}\int_s \frac{\partial \mu}{\partial t}\cos\varphi \, ds, \tag{8.40}$$

where $c = 3 \times 10^8\ \mathrm{m\,s^{-1}}$ and φ is the angle between the direction of energy transport and the wave normal. For vertical incidence, the Doppler frequency shift is often written in terms of the Doppler velocity V^* (Poole, Sutcliffe, and Walker, 1988):

$$\Delta f = 2f_{HF}\frac{V^*}{c}, \tag{8.41}$$

where V^* is the time rate of change of the phase height. Terms such as "Doppler velocity" arose from radar research where the velocity of a target gives an associated Doppler frequency, although Δf does not necessarily require movement of the reflection layer (Poole, Sutcliffe, and Walker, 1988). The time rate of change of electron number density is

$$\frac{\partial N_e}{\partial t} = q - l - \mathbf{v}\cdot\nabla N_e - N_e\nabla\cdot\mathbf{v}. \tag{8.42}$$

If charged particle production and loss are equal (so $q - l = 0$), then the Doppler velocity arises from magnetic field and electron density effects on the refractive index, expressed as

$$V^* = \int_0^{z_r} \left[\frac{\partial \mu}{\partial B_L} \frac{\partial B_L}{\partial t} + \frac{\partial \mu}{\partial B_T} \frac{\partial B_T}{\partial t} + \frac{\partial \mu}{\partial N_e} \frac{\partial N_e}{\partial t} \right] dz, \tag{8.43}$$

where B_L and B_T are the parallel and transverse components of the geomagnetic field and z^r is the real reflection height. ULF waves alter the refractive index and hence the frequency of HF signals by perturbing the ionosphere electron density and the geomagnetic field.

The continuous wave (CW) Doppler sounder transmits a very stable frequency upward into the ionosphere and uses a phase-locked loop in the receiver circuit to detect Δf (Menk *et al.*, 1995). An example of Doppler sounder and magnetometer time series from low latitudes is shown in Figure 2 in Marshall and Menk (1999). The examination of such data in the spectral domain reveals a number of harmonics and how they vary with latitude. We transform each data sequence into the frequency domain using an FFT. An example showing the ratio of the Doppler to the magnetometer spectral amplitude across a range of latitudes is shown in Figure 8.11. The shear mode at resonance enhances the Doppler signal giving peaks in the ratio at the resonant frequencies, as seen in Figure 8.11a. This reveals the decrease in the fundamental frequency with increasing latitude, and the presence of harmonics up to 120 mHz. The cross-phase between the ionosphere Doppler and ground magnetometer signal (Figure 8.11b) shows sharp 180° phase changes at the FLR frequencies. Field line resonance signatures may be identified in the ionosphere at higher harmonics than on the ground. The observational results provide input to models, which then provide information on the resonance width and wave mode mix at the ionosphere (Menk, Waters, and Dunlop, 2007).

Ionosondes and over-the-horizon radars use pulsed signals that also detect ULF wave signatures in the ionosphere. The international HF radar research community operates over 25 installations of the Super Dual Auroral Radar Network (SuperDARN) design. A comprehensive description of the system and research topics can be found in Chisham *et al.* (2007). A technique to extract the ULF signals from the Doppler velocities obtained by the SuperDARN instrumentation was described in Ponomarenko, Menk, and Waters (2003). The line-of-sight Doppler velocities are obtained along the radar beam at various ranges (e.g. Figure 7.21). Each of the 16 possible beams can be sampled at 3 (or 6) s intervals. An example from the radar on Bruny Island, Tasmania where Doppler velocities up to 500 m s^{-1} are seen is shown in Figure 8.12. The SuperDARN Doppler data include the convection electric field-driven velocities described in Chapter 2. In order to extract the ULF signatures, the larger convection-driven velocities are excluded using an autoregressive smoothing filter, leaving the ULF oscillations shown in Figure 8.12c.

The SuperDARN instruments are a crucial component of space physics research, particularly in the southern hemisphere where oceans dominate the surface. The radars provide a spatial resolution down to plus/minus a range gate (~45 km), measurements over sea where ground magnetometers cannot be located, remote configurable scan sequences, and ULF data at ionospheric heights (~300 km) over a

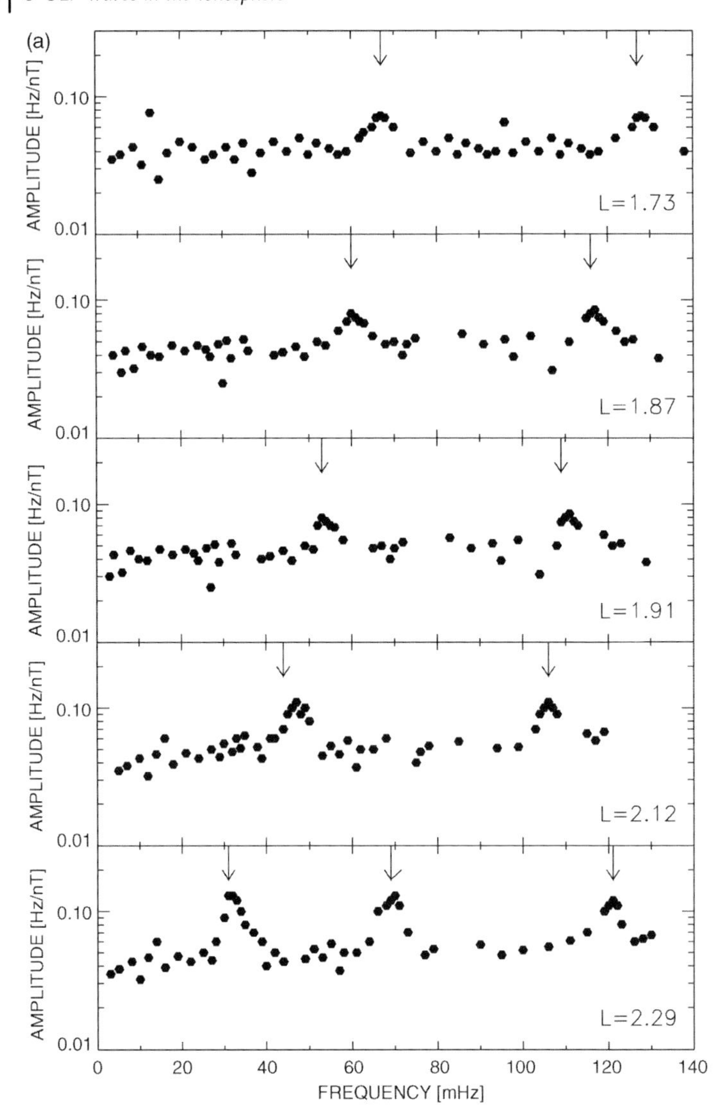

Figure 8.11 (a) Normalized amplitude of ionospheric ULF Doppler oscillations, correlated with magnetic pulsations on the ground, on January 12, 1994 as a function of frequency and L-value. (b) Ionosphere–ground phase difference for ULF Doppler oscillations on this day as a function of frequency and L-value. Arrows indicate local field line resonance frequencies. From Menk, Waters, and Dunlop (2007).

large spatial area. Up to 1000 magnetometer installations would be required in order to match the equivalent location data stream from one radar. Many of the clearest latitude profiles of FLRs have come from HF radar data (e.g. Figure 5.4). They provide experimental data on high-m ULF wave events, which are often screened from detection on the ground due to their localized spatial structure.

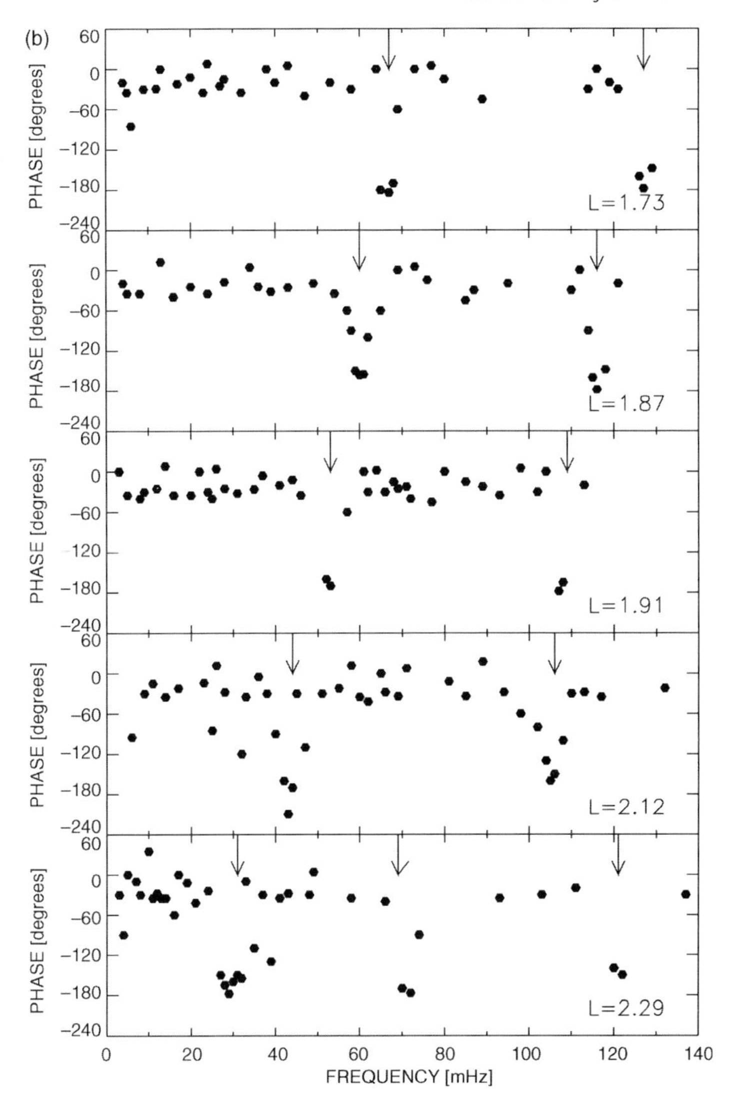

Figure 8.11 (*Continued*).

The combination of radar and ground magnetometer data provides information on the transfer function between the ionosphere and the ground. The radar configuration of the two Australian SuperDARN radars and the location of a ground magnetometer on the only available land in the field of view (Macquarie Island) are shown in Figure 8.13. Around 10 UT (20 MLT) on October 19, 2006, a substorm was preceded by two Pi2 signals. The mean-subtracted time series from the Macquarie Island fluxgate magnetometer is shown in Figure 8.14. The first Pi2 is seen ~0930 UT followed by a second ~1000 UT, which is most clear on the

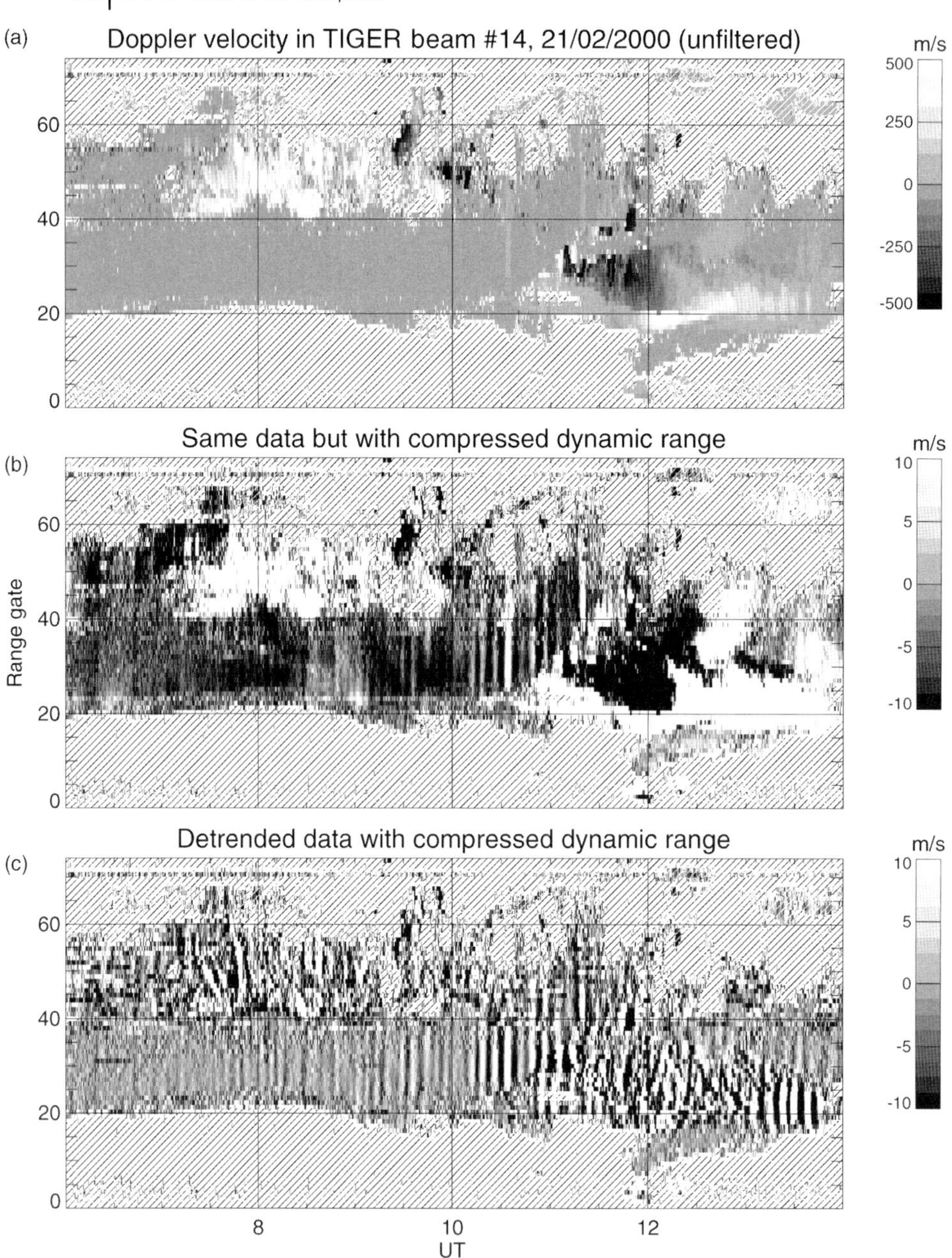

Figure 8.12 Doppler velocity variations in beam 14 of the Bruny Island (Tasmania) SuperDARN HF radar, for 0600–1400 UT on February 21, 2000. Range–time cells with no valid data have diagonal shading. Panel (a) shows unmodified data obtained via FITACF procedure. Panel (b) shows the same data but with an artificially saturated velocity amplitude scale ($\pm 10\,\mathrm{m\,s^{-1}}$). Panel (c) results from removing an autoregressive smoothing trend (window size 600 s) from the data and using the saturated amplitude scale. Bands of stripes in the traces represent periodic vertical oscillations of the ionospheric reflection point driven by the ULF wave fields. From Ponomarenko, Menk, and Waters (2003).

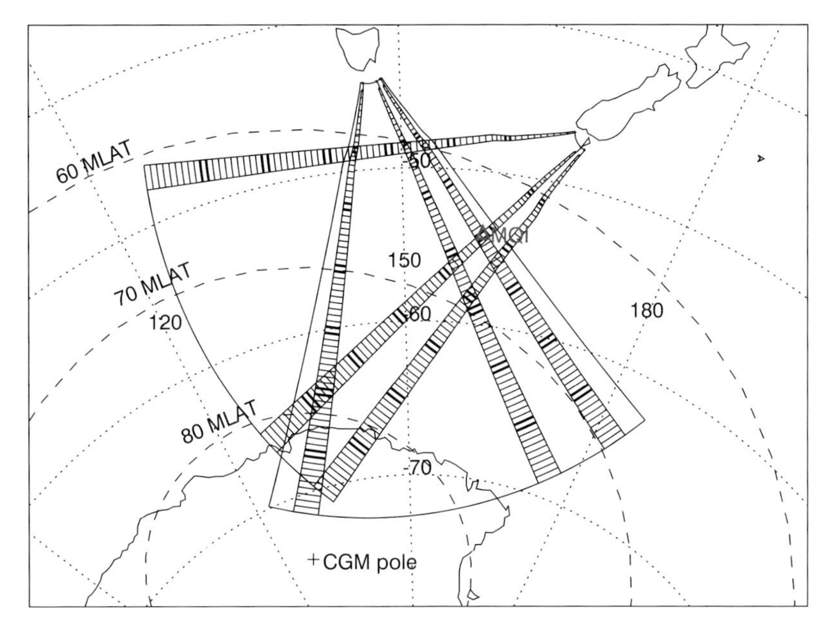

Figure 8.13 Bruny Island (Tasmania) and Invercargill (New Zealand) HF radar fields of view. The highlighted beams are used in a special operating mode to detect ULF wave signatures.

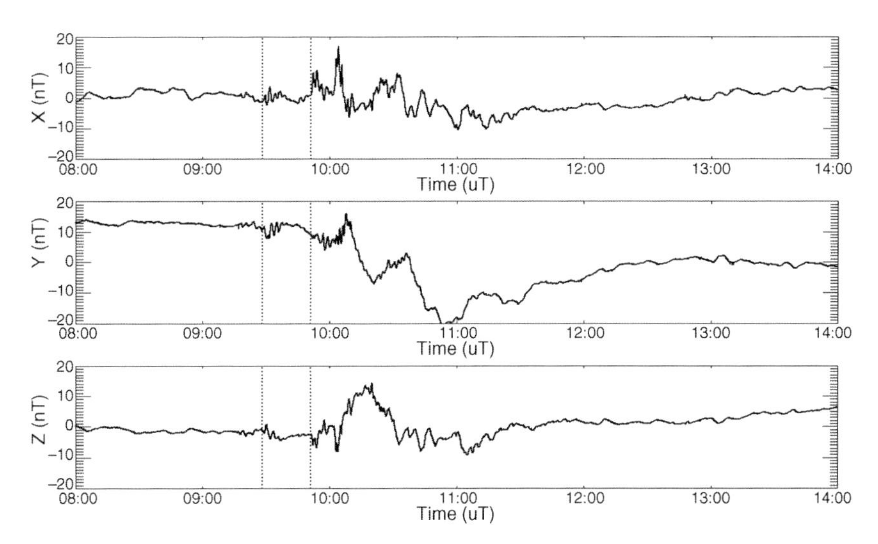

Figure 8.14 Mean removed, fluxgate magnetometer time series from Macquarie Island, 08–14 UT, October 19, 2006. Two Pi2 signals are indicated by the vertical dashed lines.

Y-component. The crossed beam configuration of the radars over Macquarie Island provides the horizontal ionosphere electric fields of the Pi2. Focusing on the first Pi2 event, the magnetometer and radar time series, projected onto orthogonal north–south and east–west basis vectors, are shown in Figure 8.15. A wave polarization

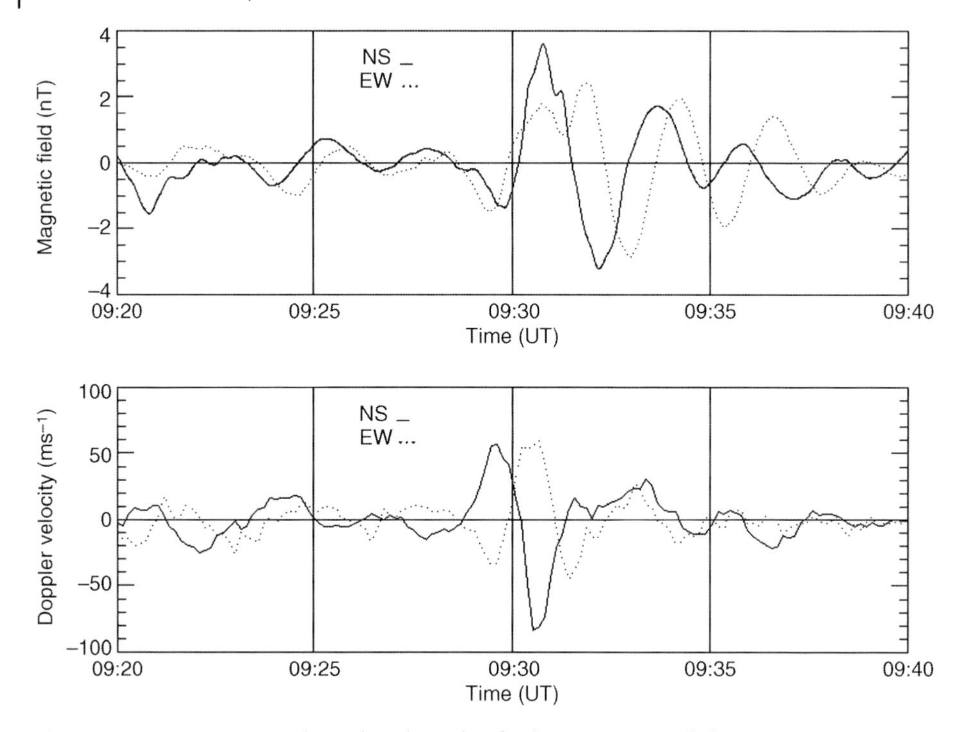

Figure 8.15 Magnetometer and Doppler velocity data for the Pi2 event recorded at 0930 UT on October 19, 2006.

analysis shows that the polarization azimuth in the ionosphere, compared with the ground, differs by 90°, consistent with a shear Alfvén mode. The ionosphere transfer ratio of this Pi2 was 0.025 nT/m s^{-1}.

8.9 Consequences for Radio Astronomy

In this final section, we highlight some aspects of studies of ULF waves in the ionosphere that impact the astronomy community. Experimental evidence for ULF-induced Doppler shifts superposed on HF signals in the ionosphere has been available for over three decades. Of course, these are not the only nor the largest causes of ionospheric perturbations. Traveling ionosphere disturbances (TIDs), atmospheric gravity waves, and large changes in TEC during magnetic storms impact HF properties in the ionosphere. However, the focus here is on ULF waves, and in this section, on those ULF waves that are seen in both ground magnetometer and ionosphere data. This excludes some high-*m* ULF waves, which are not seen on the ground.

An investigation of ULF wave effects on interferometer-based radio astronomy data was described in Waters and Cox (2009). Using the formulation outlined in Section 8.4, the ULF fields through the ionosphere were computed, followed by estimates of the variations in electron number density using Equation 8.42, which

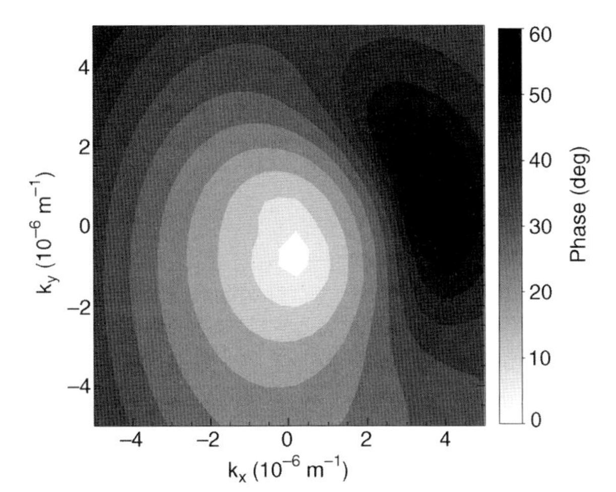

Figure 8.16 The variation in differential phase for a 70 MHz signal due to changes in TEC from a 15 mHz ULF wave with ULF wave mix of 80% shear Alfvén mode at 1000 km altitude, as a function of the ULF wave spatial scale size. Conditions were for local noon using the divergence term (last term in Equation 8.42) only. From Waters and Cox (2009). (For a color version of this figure, please see the color plate at the beginning of this book.)

are related to variations in the total electron content (TEC). These are similar to the calculations required to model Doppler shifts from HF signals that "reflect" from the ionosphere, such as those discussed in Section 8.8. The main difference for the higher frequencies that pass through the ionosphere is the treatment of the ∇N_e term. If there are no horizontal gradients, then this term is zero, since N_e is zero at both ends of the integration path, that is, the ground and space. The calculations discussed here were for the LOw Frequency ARray (LOFAR) radio telescope in The Netherlands ($\sim 53°$ latitude).

The phase differences in a 70 MHz signal passing through the midday ionosphere (no horizontal gradients) at separations equal to the ULF quarter wavelength are shown in Figure 8.16. The ULF wave horizontal spatial scales are shown on the vertical and horizontal axes. These are quite significant phase changes, particularly if an interferometer is to operate over large ground distances and given the time variability we see in the ULF wave activity. If horizontal gradients in the ionosphere electron density (around dawn and dusk) are included, then the phase differences can reach 360° and beyond. Clearly, multidimensional ULF wave models and the combined ionosphere and ground measurements of ULF waves are very important for developing remote sensing applications involving ULF waves.

9
Magnetoseismology at Other Planets and Stars

9.1
Magnetoseismology at Other Planets

It seems reasonable to imagine that standing eigenoscillations of magnetic field lines may exist at any planet possessing a magnetosphere, such as the gas giants. It has long been known that auroral activity, decametric and kilometric radio emissions, plasma wave emissions (e.g., at the upper hybrid resonance frequency), and lightning-triggered field line-guided whistlers are produced at these planets (Clarke *et al.*, 1980; Kurth *et al.*, 1985). This allowed Jupiter's radial electron density profile to be determined over three decades ago (Gurnett *et al.*, 1981).

Figure 9.1 shows a representative Jovian radial electron density profile. Several interesting features are evident: a magnetopause typically near $60R_J$, very low density just inside the magnetopause, the presence of the Io plasma torus at $6R_J$ (discussed further below), and the radial decrease in electron density $N_e \propto (1/R)^{\alpha}$ with a power law index varying from $\alpha = 7.4$ near the Io torus to $\alpha = 4.8$ in the outer magnetosphere. It is likely that the entire plasma distribution in the outer magnetosphere originates from plasma escaping radially outward from the Io torus. The plasma scale height of the torus determined from whistler dispersion measurements ranges from 1.5 to $2.5R_J$. Detailed analysis of Voyager spacecraft data has revealed the presence of a variety of complex plasma wave modes and density features in Jupiter's magnetosphere (Barnhart *et al.*, 2009). Furthermore, observations of Jovian synchrotron radiation during the Cassini fly-by indicate the presence of ultrarelativistic electrons with energies up to 50 MeV, which may be repeatedly accelerated through an interaction with plasma waves (Bolton *et al.*, 2002).

Jupiter's magnetosphere is dominated by the outflow of gas from the intensely volcanic moon Io. This becomes ionized to form a plasma torus comprising mostly electrons, sulfur, and oxygen ions, initially corotating with Jupiter's 10 h period, compared with the 42 h orbital period of the moon and the neutral gas ring. The coupling of Jovian magnetospheric field lines with torus plasma produces large electric currents resulting in large-scale Alfvén and slow mode magnetic field and plasma perturbations. An auroral spot occurs at the footprint of the Io flux tube. There is accumulating evidence that Jupiter's substantial auroral and hydromagnetic wave activities are controlled by the Io torus interaction, including the breakdown of

Magnetoseismology: Ground-based remote sensing of Earth's magnetosphere, First Edition. Frederick W. Menk and Colin L. Waters.

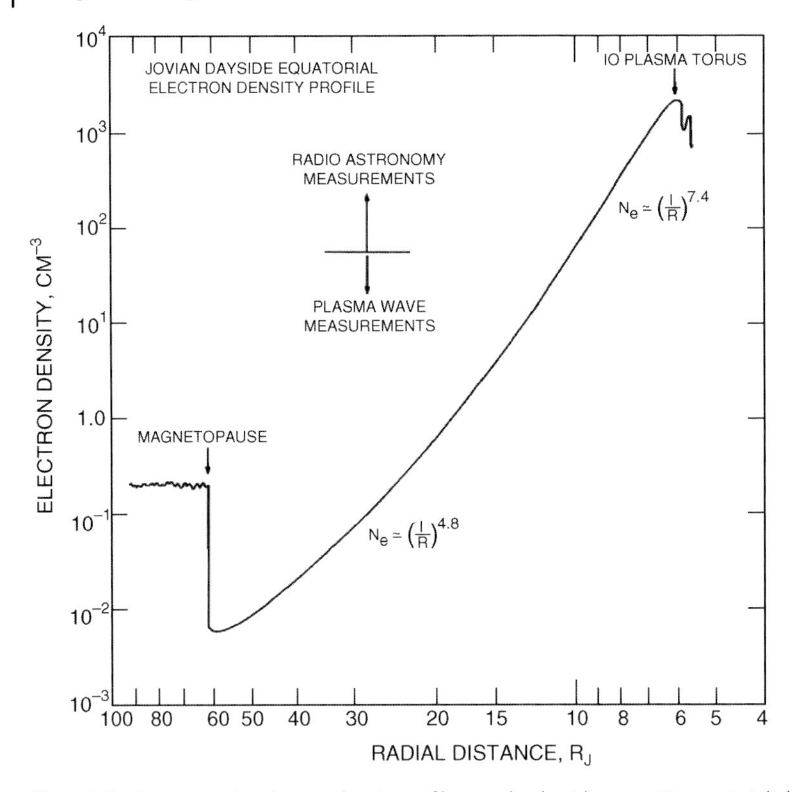

Figure 9.1 Representative electron density profile near the dayside magnetic equatorial plane for Jupiter. Radial distance is in units of Jovian radii (R_J). From Gurnett *et al.* (1981).

corotation as Io flux tubes become progressively more stretched closer to Jupiter (Bonfond *et al.*, 2012).

Standing Alfvén waves have been detected at the Io plasma torus by Voyager 1 (Acuña, Neubauer, and Ness, 1981; Glassmeier *et al.*, 1989) and Pioneer 10 (Walker and Kivelson, 1981), and in Jupiter's middle magnetosphere (Khurana and Kivelson, 1989). Strictly speaking, global standing waves are not expected since the field line eigenperiod (between conjugate ionospheres) is very long relative to the Alfvén speed, and comparable to the planet's rotation period. Field line resonances are also not expected in the Jovian magnetosphere. However, in regions of enhanced mass density and hence low Alfvén speed, standing waves may be locally excited. The Io torus is such a region. Decoupled axisymmteric toroidal and poloidal mode eigenoscillations of the entire torus, with periods of about 1200 and 800 s, respectively, have been observed (Glassmeier *et al.*, 1989). A mass density of order 10^4 amu cm^{-3} has thus been inferred at the Io torus, representing a mass loading of 7–9 compared with the local electron density. The Alfvén wave power dissipated by Io plasma interactions is $\sim 10^{12}$ W, mostly into the Jovian ionosphere, inner magnetosphere, and torus region. However, the generation mechanism of the MHD waves observed in the inner magnetosphere is not clear. Observations by the Ulysses spacecraft of 6800 s (0.15 mHz) oscillations in particle fluxes, plasma density, and

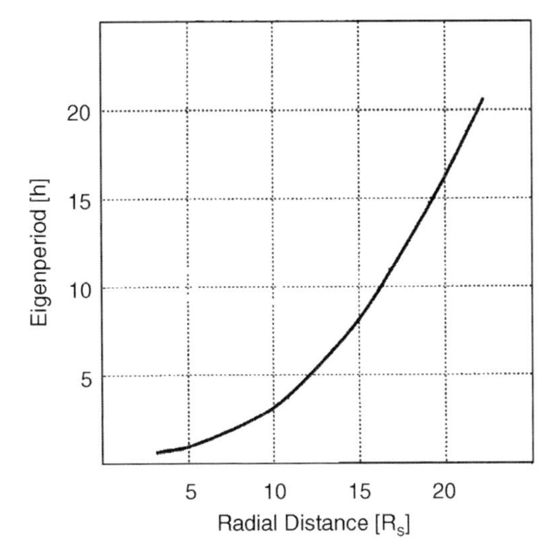

Figure 9.2 Estimated eigenperiods for dipolar field lines in Saturn's magnetosphere assuming the electron density at the magnetopause is $N_e = 1 \times 10^5\,\text{m}^{-3}$ and an r^{-4} radial dependence of mass density. From Cramm *et al.* (1998).

magnetic field intensity have been interpreted as second-harmonic poloidal mode flapping of the magnetic–equatorial surface near $21 R_J$ (Schulz *et al.*, 1993). Solving the familiar MHD equations for poloidal eigenmodes and assuming a power law field-aligned density distribution yield equatorial density estimates of about 24 amu cm^{-3}, suggesting significant heavy ion mass loading.

Due to the low solar wind pressure, Saturn's magnetosphere is very large. Plasma in the outer magnetosphere corotates at about 80% of the rigid-body corotation rate and is mainly concentrated around the equatorial plane. The magnetosphere comprises mostly H^+ and O^+ and by assuming a radial dependence of mass density, the Alfvén speed and expected field line eigenperiod can be readily determined, as shown in Figure 9.2. The predicted eigenperiod is several hours, comparable to the planetary rotation period of 10 h 39 min. Thus, standing waves along field lines are also unlikely at Saturn, since magnetic field and plasma conditions are unlikely to be invariant during an eigenperiod.

At Earth, resonant coupling occurs when the frequency of an incoming driving fast mode matches the local eigenfrequency for standing oscillations with suitable ionospheric boundary conditions. At Saturn, the ionosphere does not influence the resonant coupling of propagating waves and resonant mode coupling occurs if the field-aligned component of the phase velocity of the driving fast mode wave equals the local Alfvén velocity at the resonant point. For example, KHI-driven magnetopause surface waves that produce fast mode perturbations may resonantly couple to propagation-guided Alfvén mode field line oscillations at the local resonant frequency, which is not necessarily an eigenfrequency of the field line (Cramm *et al.*, 1998).

In fact, the observation in Cassini magnetometer data of a time-stable modulation of Saturn's magnetic field at a period of 10 h 47 min is regarded as a proxy measurement of the planet's rotation rate, although the cause of this modulation is not clear (Giampieri *et al.*, 2006). Transverse Alfvén oscillations with similar periods are superimposed on the main field modulation and appear to originate in separate magnetospheric regions independent of local time, latitude, or radial distance (Kleindienst *et al.*, 2009). The cause of such features has not been established, but may involve a corotating magnetic field anomaly caused by a current system or some internal asymmetry. Detailed analysis of Cassini magnetospheric magnetic field data shows that magnetic field modulation rates on open field lines in the northern and southern hemispheres are different (10.6 and 10.8 h) and are superimposed on the rotating quasi-uniform inner field on closed field lines, suggestive of field-aligned current systems that rotate with different periods (Andrews *et al.*, 2010). This is an exciting and topical area of work.

There is a subtle but important aspect to such investigations of magnetic field perturbations at other planets. A variety of phenomena may be modulated by the magnetic perturbations. For example, Cassini studies show that the modulation of kilometric radiation fluxes is synchronous with the magnetic field perturbations (Andrews *et al.*, 2010). This means that phenomena that could be more conveniently studied remotely, for example, from Earth, may provide information on magnetic field variations, which in turn relate to properties such as the configuration and mass loading of flux tubes and hence the structure of the magnetosphere, current systems, and core field. This expands opportunities for remote sensing of other planets and in fact other stars, as will be seen in Section 9.4.

Mercury possesses a weak magnetic field and minimagnetosphere, but no significant ionosphere. Nevertheless, transverse 2 s ULF waves with narrowband appearance similar to field line resonances were detected at $1.3 R_M$ by Mariner 10 in 1974, when the magnetopause was at $1.72 R_M$ (Russell, 1989). Assuming a cool electron–proton plasma of density $3\,\text{cm}^{-3}$, the expected field line eigenperiod is 8 s; so the observed waves could be a third to fifth harmonic, although other harmonics were not detected.

Observations from the MESSENGER spacecraft that reached Mercury in January 2008 show frequent occurrence of Kelvin–Helmholtz waves at the dayside magnetopause and extensive ULF wave activity in a magnetosphere that is dominated by Na^+ ions (Boardsen *et al.*, 2009; Slavin *et al.*, 2008). Narrowband, mostly right-hand polarized 2–6 s period waves have been seen at frequencies between the He^+ and H^+ cyclotron frequencies, with up to four harmonics. The generation mechanism of these waves is not yet clear and Mercury and its wave environment will prove to be a strange and interesting world.

9.2 Magnetoseismology of the Solar Corona

Alfvén predicted the existence of MHD waves within the Sun (Alfvén, 1942), and the use of such waves as a remote plasma diagnostic technique was proposed some decades ago (Roberts, Edwin, and Benz, 1984; Uchida, 1970). The development of modern imaging techniques from ground- and space-based platforms, including the

SOHO (Solar and Heliospheric Observatory) and TRACE (Transition Region and Coronal Explorer) satellites, has revealed that MHD waves and oscillations are ubiquitous in the solar atmosphere over a wide range of periods, occurring in open structures such as solar plumes and closed magnetic structures such as coronal loops and prominences. Their observation and use in coronal magnetoseismology is a dynamic and growing area of research (Banerjee *et al.*, 2007; De Moortel, 2005; De Moortel and Nakariakov, 2012). Coronal seismology may turn out to be an important new science, given that MHD waves are present in most, if not all, coronal structures and may provide a considerable part of the energy needed to heat the quiet solar corona and drive the solar wind.

A number of different MHD modes are observed in the solar corona. Kink modes are asymmetric standing fast magnetoacoustic waves (azimuthal wave number $m = 1$) that cause bulk transverse motions of coronal flux tubes and are often generated by a nearby flare event. The oscillations have frequency of 3–4 mHz and are heavily damped (Aschwanden *et al.*, 1999; Nakariakov *et al.*, 1999). The phase speed of the fast kink mode is (De Moortel, 2005)

$$c_K = \frac{2L}{T} = V_A \left(\frac{2}{1 + \rho_e/\rho_0} \right)^{1/2}, \tag{9.1}$$

where L is the length of the coronal loop, T is the observed period, $\rho_e/\rho_0 \approx 0.1$ is the ratio of densities outside and inside the loop, and V_A is the usual Alfvén velocity. The observed frequency thus allows the local magnetic field strength in the loop to be estimated, given reasonable assumptions for other better-known quantities (Aschwanden *et al.*, 1999; Nakariakov and Ofman, 2001). Estimated values are of order 5–50 G (10^4 G = 1 T), although comparison with 3D simulations suggests that loop curvature, density, and aspect ratio affect the estimated field strengths (De Moortel and Pascoe, 2009).

Compressible guided fast magnetoacoustic waves called sausage modes ($m = 0$) cause axially symmetric expansion and contraction of flux tubes, varying the magnetic field and hence the orbits of trapped particles within the structure. They are likely responsible for the periodic modulation of microwave and X-ray emissions (Aschwanden *et al.*, 2004; Roberts, Edwin, and Benz, 1984).

Longitudinal, slow acoustic modes are compressive guided waves and both propagating and standing slow modes have been observed, the latter in connection with 1–2 mHz oscillations in UV emissions from hot coronal loops (Wang *et al.*, 2003). Propagating slow modes are discussed further in the next section.

The detection of incompressible, transverse torsional Alfvén waves in the corona is challenging. There is indirect evidence of their existence in the corona, but to date torsional waves have only been detected in the lower solar atmosphere (Jess *et al.*, 2009; Verth, Erdélyi, and Goossens, 2010). One set of such observations was obtained using the Swedish Solar Telescope to examine a 10×10 arc-sec region of a bright point group at 6562.8 nm with spatial resolution of about 110 km at the solar surface. Wavelet analysis revealed coherent full-width half-maximum oscillations with a 180° phase shift across the region of interest, produced by Alfvén waves causing torsional twist of the field lines of $\pm 22°$ (Jess *et al.*, 2009). The energy flux of such waves averaged across the entire solar surface is expected to be around $240\,\mathrm{W\,m^{-2}}$, sufficient to heat the entire corona. In a separate study, high-resolution observations of chromospheric spicules with the Solar Optical Telescope on the Hinode satellite, supported by 3D MHD simulations, have revealed that these exhibit transverse oscillatory displacements indicative of Alfvén waves in the chromosphere (De Pontieu *et al.*, 2007). The energy flux in the chromosphere was estimated at $4\text{–}7\,\mathrm{kW\,m^{-2}}$, resulting in $\sim 120\,\mathrm{W\,m^{-2}}$ in the corona.

Prompted by such observations, Verth, Erdélyi, and Goossens (2010) examined the use of torsional Alfvén eigenmodes to remote sense the corona. Deriving the relevant equations for the two cases of isolated, vertically aligned thin and expanding flux tubes embedded in an exponentially stratified atmosphere, they showed using both a WKB approximation and numerical solution how the fundamental mode frequency varies with plasma density, β, scale height, and flux tube radius. For isolated thin flux tubes, the analysis of eigenmodes can determine the temperature difference between the internal and external plasma, while in the finite width case the wave modes will provide information on the radial structure of the flux tube.

In summary, observational and modeling results show that magnetoseismology of the solar atmosphere is an exciting field that is likely to yield important new understanding of the properties of this region. Furthermore, the MHD waves may play an important role in coronal heating and hence formation of the solar wind (De Pontieu *et al.*, 2007).

9.3 Introduction to Helioseismology and Asteroseismology

Propagating longitudinal, slow acoustic mode waves were mentioned in the previous section. These modes, along with propagating transverse mode waves, occur throughout the corona, often at the footpoints of large, quiescent coronal loop structures, close to active regions. Observed periods are around 170 s (5.9 mHz, $\sim$3 min) and 320 s (3.1 mHz, $\sim$5 min) for loops above sunspots and plage regions, respectively. These periods are similar to the well-established 3 min oscillations in sunspots and 5 min oscillations in the photosphere (De Moortel, 2005). There is accumulating evidence that 2–4 mHz (5 min) acoustic modes of the solar interior couple to the corona, producing features such as oscillations in soft X-ray fluxes (Didkovsky *et al.*, 2011), outward propagating slow mode disturbances in coronal loops (De Moortel *et al.*, 2002), and the fast kink mode in coronal loops (Aschwanden *et al.*, 1999).

The processes by which the internal 5 min modes are generated have been known for some decades. Periodic 300 s oscillations of local regions of the solar surface were reported by Leighton *et al.* in 1962 (Leighton, Noyes, and Simon, 1962), who noted the potential for using the observed period to probe the properties of the solar atmosphere. Since then there have been many studies of these features, including using dedicated multipoint ground observatories such as GONG (Global Oscillation

Network Group) and BiSON (Birmingham Solar Oscillation Network). Several reviews exist (e.g. Christensen-Dalsgaard, 2002; Milone and Wilson, 2008).

The oscillations are manifested as small line-of-sight Doppler shifts of spectral lines, corresponding to velocity variations for each mode of order $10\,\mathrm{cm\,s^{-1}}$ superimposed upon the turbulent background. The oscillations are observable because of their high spatial and temporal coherence in long data sets integrated across the solar disk. Ulrich (1970) proposed that these oscillations are due to standing acoustic modes in the solar interior. Subsequently, Claverie et al. (1979) determined the global mode structure of these oscillations, and a dispersion law was developed by Duvall (1982). Resonant cavities within the Sun result in normal modes formed by acoustic waves propagating around the Sun between upper and lower boundaries and arriving in phase at the original point. These are called *p*-modes, characterized by the quantities l, m, and n, where l is the spherical harmonic around the stellar circumference, m is the azimuthal order, and n is the radial order. Thus, purely radial modes, $l = 0$, penetrate the center of the Sun, while modes of the highest degree, $l \geq 1000$, are trapped in the outermost part of the solar radius.

The dispersion relation for plane acoustic waves is

$$\omega^2 = c^2 |k^2|, \tag{9.2}$$

where $k = k_r \boldsymbol{a}_r + \boldsymbol{k}_h$ is the wave vector and $\boldsymbol{k}_h$ is the tangential wave number. The mode properties are controlled by the variation of adiabatic sound speed $c(r)$. The radial variation of the *p*-mode is given by

$$k_r^2 = \frac{\omega^2}{c^2}\left(1 - \frac{S_l^2}{\omega^2}\right), \tag{9.3}$$

where

$$S_l^2 = \frac{l(l+1)c^2}{r^2} \approx k_h^2 c^2 \tag{9.4}$$

is the Lamb frequency (Christensen-Dalsgaard, 2002).

Figure 9.3 schematically illustrates the propagation of acoustic modes in a solar cross section. Temperature and hence sound speed c increase with depth (decreasing r), while from Equation 9.4 $\boldsymbol{k}_h$ decreases with increasing c. This causes refraction, forming a boundary at the inner turning point (dotted circle in the figure) where $r = r_t$, $\omega = S_l$, and $k_r = 0$. For $r = r_t$, k_r is imaginary and the wave decays exponentially. The outer cavity boundary is defined by the sharp decrease in density near the surface.

The formation of normal modes requires that an integer number of wave cycles are accommodated in the waveguide around the Sun with boundaries at $r = r_t$ and $r = R$. Equation 9.3 can therefore be rewritten as

$$(n + \alpha)\pi \approx \int_{r_t}^{R} k_r \mathrm{d}r \approx \int_{r_t}^{R} \frac{\omega}{c}\left(1 - \frac{S_l^2}{\omega^2}\right)^{1/2} \mathrm{d}r, \tag{9.5}$$

where α is a phase term. The result is a spectrum of very narrow closely-spaced peaks centered near $3300\,\mu\mathrm{Hz}$. A typical disk-averaged oscillation spectrum is

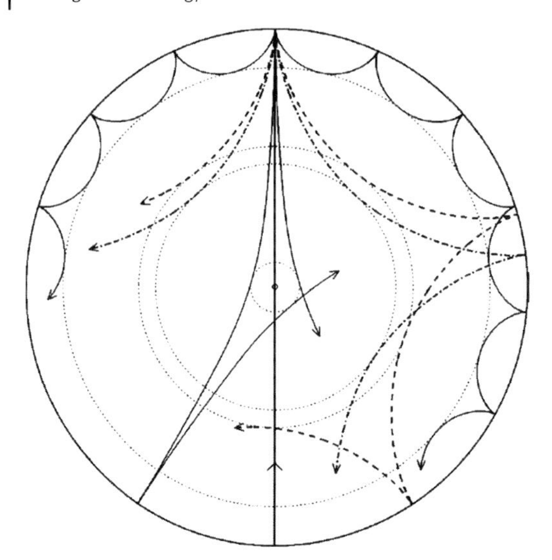

Figure 9.3 Schematic representation of p-mode acoustic waves propagating within the solar interior. From Christensen-Dalsgaard (2002).

shown in Figure 9.4. The narrowness of the peaks points to low damping rates, while the varying height of the peaks arises from the asymptotic nature of the low-degree p-modes.

The oscillation frequencies of different modes can be measured to precisions of a few parts in 10^5 and are used to diagnostically probe the internal structure of the

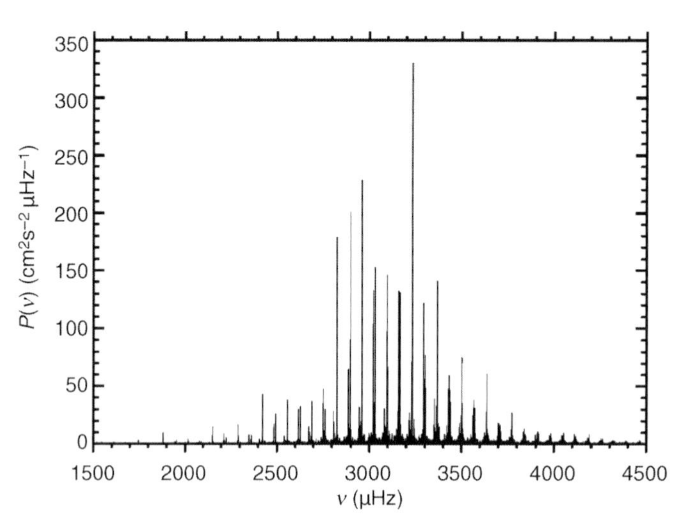

Figure 9.4 Power spectrum of solar oscillations based on 4 months data from the BiSON observatory array. From Christensen-Dalsgaard (2002).

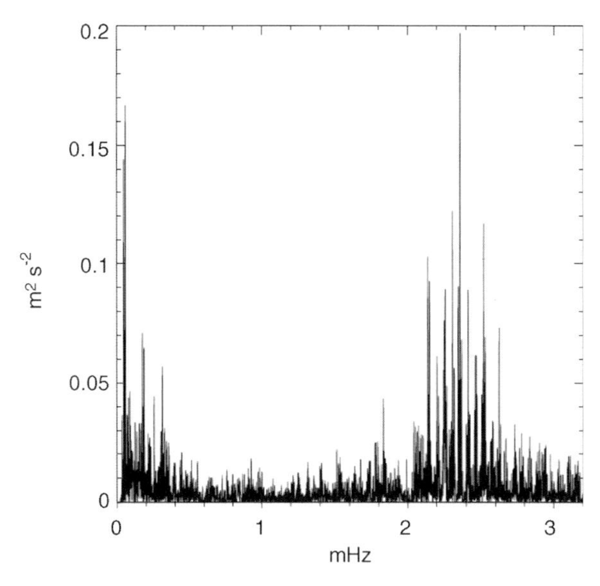

Figure 9.5 Power spectrum of radial velocity oscillations on the star α Centaurus A, based on observations from the 1.2 m Swiss telescope at the ESO La Silla Observatory. From Bouchy and Carrier (2001).

Sun. Studies extending over more than 30 years have shown that the frequencies of low *p*-modes across the 2500–3700 μHz spectral range vary with solar cycle. Frequencies increase by about 0.5 μHz from sunspot minimum to maximum, although the shift for the highest-frequency modes is larger than for the lowest-frequency modes. The frequency shifts correlate with a range of solar activity proxies, in particular the $F_{10.7}$ flux (Chaplin *et al.*, 2007).

Helioseismology has proven to be a remarkable tool for probing solar structure and properties, providing a better understanding of the Sun and improving constraints on solar models (Serenelli, 2010). It is now possible to use helioseismic holography to monitor active regions on the far side of the Sun, potentially providing 1–2 weeks warning of space weather events (González-Hernández *et al.*, 2010). This is possible because the acoustic waves experience a phase shift when interacting with an active region. Full-hemisphere far side maps are calculated twice daily and accessible at the GONG Web site: http://gong.nso.edu/data/farside/. At present seismic holography can detect about 40% of the total active regions that appear on the east limb of the Sun with about 60% confidence.

Even more remarkable has been the extension of this science to study the interior of other stars, called asteroseismology (Aerts, Christensen-Dalsgaard, and Kurtz, 2010; Suárez *et al.*, 2012). An early result from ground-based observations of radial velocity oscillations on the star α Centaurus A, based on 1260 spectra obtained over five nights, is shown in Figure 9.5. Average amplitude of the oscillations is 35 cm s^{-1}. A series of peaks between 1.7 and 3 mHz are evident, while frequency resolution is 2.6 μHz and peak separation is 106 μHz.

The launch of the Kepler spacecraft in March 2009 has permitted asteroseismic study of other solar-type stars, providing new information on stellar structure and evolution. A study (by up to 11 teams) of two Sun-like stars with the first 8 months of Kepler data has revealed *p*-mode global modes with frequency of maximum amplitude of 830 and 675 μHz, frequency separations of 48 and 42 μHz, yielding estimates of the stellar rotation periods, mean density, mass (1.25 and $1.33M_S$), and radius (2.15 and $2.4R_S$) (Mathur *et al.*, 2011). A further asteroseismic study of 22 solar-type stars observed for 1 month each during the first year of the mission used an automated analysis of oscillation frequencies to determine stellar radii and masses to about 1% precision and ages to about 2.5% precision (Mathur *et al.*, 2012).

It is clear that helioseismology and asteroseismology have and will continue to provide astonishing remote sensing capability on the properties of our Sun and other solar-like stars. Detailed analysis of Kepler observations and refinement of helioseismic holography will likely usher in new ways of thinking about the stellar space environment within which our planet swims.

9.4
Field Line Resonances at Other Stars

Remarkably coherent quasi-periodic oscillations (QPOs) occur in the X-ray flux from accreting neutron stars, with frequencies typically in the kilohertz range. Most sources exhibit twin spectral peaks that vary in frequency simultaneously. The characteristic timescale for material orbiting near the accreting star is in the same frequency range, and beat-frequency models have been invoked to explain the QPOs. One alternative explanation supposes that Alfvén oscillations of the accretion disk occur at the quasi-sonic point radius where the Alfvén velocity matches the orbital Keplerian velocity, with different mass densities accounting for the different frequencies.

A more satisfactory model based on the excitation of field line resonances was proposed by Rezania and Samson (2005). Plasma is accelerated supersonically into a neutron star from the accretion disk and produces two effects: (i) the excitation of MHD waves due to compressive action of the accreting plasma on the star's magnetosphere, and (ii) distortion inward of the stellar magnetosphere in the disk plane and outward bulging of flux tubes away from that plane. Resonant coupling between compressional and shear Alfvén waves in the enhanced density regions of these flux tubes results in two MHD modes with frequencies $\omega^{\pm} = k_z(v_p \pm V_A)$ in the kilohertz range, where v_p is the field-aligned plasma velocity. This is represented in Figure 9.6, which compares the predicted variation in separation of the two spectral peaks, with the predicted upper frequency and observations for different stars. The model calculations assumed a stellar dipole moment $\mu = 0.32 \times 10^{23}\,\mathrm{G\,cm^3}$, accretion rate of $10^{17}\,\mathrm{g\,s^{-1}}$, stellar radius of 10 km, and mass of 1 solar mass. Flux tube plasma oscillates at these frequencies resulting in modulation of the accreting plasma flow and hence modulated X-ray flux. This model may explain observed QPO features and also allows estimation of the mass

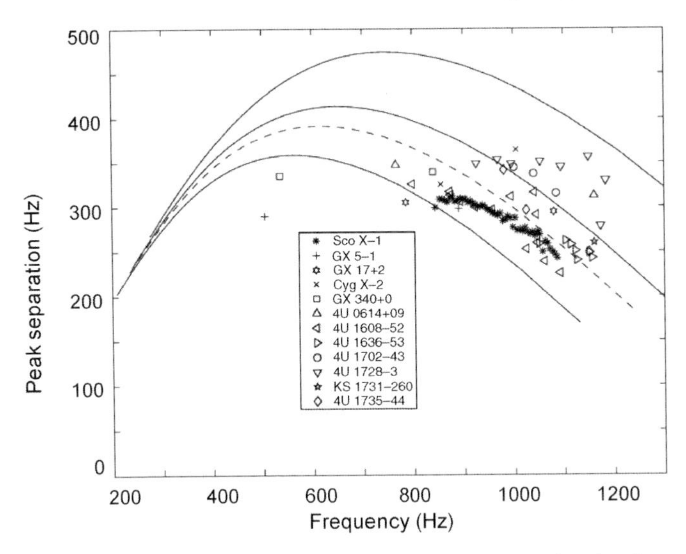

Figure 9.6 Modeled variation in separation of the two spectral peaks of neutron star X-ray oscillations, with the predicted upper frequency and observations for different stars. From Rezania and Samson (2005).

density, which is $\sim 10^{14}\,\mathrm{g\,cm^{-3}}$ at $x = 10$ stellar radii, compatible with realistic values. This model may also explain the $\sim$10 Hz QPOs observed in rapid bursters.

Thus, we see that the concept of field line resonance may be used to remote sense properties of very distant stars as well as other planets and the magnetosphere of Earth.

Appendix A
Computer Codes

Three computer codes are provided. Two of these allow the reader to explore the effects of the ionosphere on ultralow-frequency (ULF) wave propagation from the magnetosphere to the ground. The third code solves the toroidal mode FLR wave equation derived in Appendix B and discussed in Section 3.4. This is the fundamental equation used in magnetoseismology.

The source codes are provided in the Interactive Data Language (IDL). The code to solve the toroidal mode equation is quite short, and so has been reproduced herein. The source codes for the calculation of the reflection and transmission coefficients in addition to the ULF wave fields interacting with the ionosphere may be obtained from the following Web site: http://plasma.newcastle.edu.au. A brief description of these three codes is given in the following sections.

A.1
Reflection and ULF Wave Mode Coefficient Calculation: rcoeff_ionos.pro

The program rcoeff_ionos.pro solves for the reflection and ULF wave mode conversion coefficients as a function of the geomagnetic field dip angle, using the formulas discussed in Section 8.3. The parameters are set to reproduce Figure 8.2. The model assumes a thin sheet ionosphere at altitude d. The input parameters are the ULF frequency (mHz), height of the thin sheet ionosphere, d (m), Alfvén wave speed just above the ionosphere sheet (ms^{-1}), horizontal wave numbers k_x, k_y (m^{-1}), and the Pedersen, Hall, and direct conductances (S).

The conditions for ideal magnetohydrodynamics (MHD) waves in an infinite half-space with uniform Alfvén speed are assumed above the ionosphere sheet. Lines 66–121 in the code define the fields and wave numbers for this ideal MHD region. The mode conversion matrix involves the hyperbolic sine and cosine functions since the solutions assume evanescent waves in the atmosphere. The program prints the values for $k_{z,a}$, the vertical wave number in the atmosphere. For a valid solution, these should be purely imaginary. Line 61 of the code checks this condition.

Magnetoseismology: Ground-based remote sensing of Earth's magnetosphere, First Edition. Frederick W. Menk and Colin L. Waters.

A.2
ULF Wave Fields from a Thin Sheet Ionosphere: ulf_updown_fields.pro

The program ulf_updown_fields.pro uses the wave mode conversion matrix to compute the up- and down-going ULF wave electric and magnetic fields reflected from a thin sheet ionosphere. The code is set up to reproduce Figure 8.3. Inputs are the same as the rcoeff_ionos.pro code, with additional parameters being the ULF wave mode mix and phasing for the incident wave and the geomagnetic field dip angle. The outputs are the ULF electric and magnetic fields with altitude for the up- and down-going shear and fast modes. Above the ionosphere sheet, uniform, ideal MHD conditions are assumed along with an evanescent solution in the atmosphere. The first half of the code reproduces the calculation of the wave mode conversion matrix used in rcoeff_ionos.pro, while the second half of the code computes the various fields.

A.3
Field Line Resonant Frequencies: toroidal_eig_dipole.pro

This short IDL code uses a matrix method to solve the "toroidal" mode differential equation for field line resonant frequencies in a dipole magnetic field, as discussed in Chapter 3. The input parameters are the number of harmonics to print (harm), the *L*-value of the resonant field line (*L*), plasma mass density in the equatorial plane (n_eq) and the *L*-value of this (L_neq), and the plasma mass density exponent (alph) assuming an $r^{-\mathrm{alph}}$ density model.

The code is set up to solve for the half-wave solutions by setting the plasma displacement to zero at both ends of the field line. It is trivial to alter the code to solve for quarter-wave mode oscillations by changing one of the boundary conditions so that the derivative of the plasma displacement is zero at one end of the field line. The output eigenvalues give the harmonic frequencies in millihertz. The eigenfunctions can easily be obtained by uncommenting the la_eigenproblem command to place the eigenfunctions into the "evec" variable.

```
;###########################################################################
; Calculate the toroidal FLR harmonic frequencies given:
; (a) Dipole magnetic field
; (b) r^m plasma mass density profile
;
; C.L. Waters and I.A. Price
; Centre for Space Physics
; University of Newcastle, Australia
;;##########################################################################
Pro toroidal_eig_dipole
  sizen=401              ; number of spatial grid points
  harm = 6               ; number of FLR harmonics to calculate
  L = 6.6                ; L value of field line to solve along
  n_eq = 1.0             ; H+/cm^3 at equatorial plane
  L_neq = 6.6            ; L Value where density is n_eq
  alph = 3               ; exponent on r^m density model
```

```
  Re = 6371.d3          ; Earth radius in m
  Anchor = 150.0d3      ; Altitude of Ionosphere anchor point
  lat0 = acos(sqrt((Re + anchor)/(Re*L))) ; Latitude for this L value (at
                                              ionosphere height)

  Print, 'L_Value = ',L
  Print, 'Latitude=',Lat0*180./!pi
  Print, 'Proton Number density at Equator = ',n_eq
;
  h = 2.0*lat0/double(sizen-1)   ; step size of lambda (latitude)
  mu = 1.2566371d-6              ; Permeability of free space
  LRe=L*Re
  K0=8.0d15                      ; magnetic moment of Earth
  Mp=1.6725d-27
  m_den=n_eq*1.0d6*Mp            ; mass density in kg/m^3
;
; Calculate r^m plasma mass density model
  dens_a = dblarr(sizen)
  f_len = 0.0                    ; Initialise field line length variable
  For ii=0,sizen-1 do begin
    lda = -lat0 + float(ii)*h
    r = LRe*cos(lda)*cos(lda)
    dens_a(ii) = m_den*(L_neq*Re/r)^alph
    f_len = f_len + LRe*cos(lda)*sqrt(4.0-3.0*cos(lda)^2)*h
  end

Print,'Field length = ',f_len/Re,' Re'
;
;   Solve for FLRs
  B = DBLARR(sizen,sizen)          ; finite difference matrix
  C = DBLARR(3)                    ; space for calc matrix elements
  EVAL = COMPLEXARR(sizen)         ; eigenvalue array
;   EVEC = COMPLEXARR(sn,sn)       ; eigenvector array - if required
  ORDER = INTARR(sizen)            ; array used to sort eigenvalues
  const = double(((K0/(LRe^4))^2)/mu)
;
; Populate finite difference (central) matrix
; Do 1st row, including boundary condition -> for zero at both ends - i.e. 1/2 wave mode
; For 1/4 mode wave, change b.c. to set 1st derivation = 0
  lda = -lat0                       ; Start in Sth hemisphere
  C = [-2.0/h, 1.0/h+tan(lda)/2.0]
  B(0:1,0) = C*const/(h*dens_a(0)*(cos(lda)^14))

; Do the internal matrix elements
 For i = 1, sizen-2 do begin
   lda = -lat0 + h*i                ; increment latitude
   C = [1.0/h-tan(lda)/2.0, -2.0/h, 1.0/h+tan(lda)/2.0]
   B(i-1,i) = C * const/(h*dens_a(i)*(cos(lda)^14))
  end

; Do last row, including boundary condition
  lda = lat0
  C = [1.0/h-tan(lda)/2.0, -2.0/h]
  B(sizen-2:sizen-1,sizen-1) = C*const/(h*dens_a(sizen-1)*(cos(lda)^14))
```

```
; Solve eigenvalue problem
; res=la_eigenproblem(B, /double, eigenvectors=evec)  ; Use this if eigenvectors
                                                        are required
  res=la_eigenproblem(B, /double)
  res_s = sort(res)                          ; Sort the eigen values small to large
  flr_a=dblarr(harm)
  For i = 0,harm-1 do flr_a(i) = sqrt(-1.0*double(res(res_s(i))))*500/!pi ; freqs in mHz
  Print,'FLR freqs :'
  Print,flr_a,' mHz'
  Print,'Finished'
End
```

Appendix B
The Transverse MHD Wave Equation for General Magnetic Field Models

Calculation of the resonant period of toroidal mode geomagnetic pulsations requires the magnetohydrodynamic (MHD) wave equation be solved with suitable magnetic field and plasma density models. Formulations that embed dipole magnetic field and R^{α} plasma density models in the differential equations are common, as discussed in Section 3.4. A formulation for a general magnetic field model was given by Singer *et al.* (1981), and a derivation of their toroidal mode wave equation is provided here. A derivation of the general linear MHD wave equation is given in Section 3.3 (Equation 3.33).

Consider two adjacent field lines separated at some point by a distance δ_{α}. At any other point along the field, the separation scales by $h_{\alpha}\delta_{\alpha}$. For the perpendicular unit vector between field lines, $\hat{\alpha}$,

$$\nabla\alpha = \hat{\boldsymbol{\alpha}}/h_{\alpha}. \tag{B.1}$$

A plasma displacement ξ_{α} in the α-direction causes a magnetic perturbation (Equation 3.29):

$$\mathbf{b} = \nabla \times (\xi_{\alpha}\hat{\boldsymbol{\alpha}} \times \mathbf{B}_0), \tag{B.2}$$

and so

$$b \bullet \nabla\alpha = \nabla\alpha \bullet [\nabla \times (\nabla\alpha \times \xi_{\alpha}h_{\alpha}\mathbf{B}_0)]. \tag{B.3}$$

Using the vector identity for the curl of the cross-product of two vectors, we obtain

$$b \bullet \nabla\alpha = \nabla\alpha \bullet [\nabla\alpha(\nabla \bullet \xi_{\alpha}h_{\alpha}\mathbf{B}_0) - \xi_{\alpha}h_{\alpha}\mathbf{B}_0(\nabla \bullet \nabla\alpha) + \xi_{\alpha}h_{\alpha}(\mathbf{B}_0 \bullet \nabla)\nabla\alpha \\ -(\nabla\alpha \bullet \nabla)\xi_{\alpha}h_{\alpha}\mathbf{B}_0]. \tag{B.4}$$

Since $\nabla\alpha$ is perpendicular to $\boldsymbol{B}_0$,

$$\nabla\alpha \bullet \mathbf{B}_0 = 0, \tag{B.5}$$

and so

$$b \bullet \nabla\alpha = |\nabla\alpha|^2(\nabla \bullet \xi_{\alpha}h_{\alpha}\mathbf{B}_0) + \xi_{\alpha}h_{\alpha}\nabla\alpha \bullet [(\mathbf{B}_0 \bullet \nabla)\nabla\alpha] - \nabla\alpha \bullet [(\nabla\alpha \bullet \nabla)\xi_{\alpha}h_{\alpha}\mathbf{B}_0]. \tag{B.6}$$

Magnetoseismology: Ground-based remote sensing of Earth's magnetosphere, First Edition. Frederick W. Menk and Colin L. Waters.

Equation B.5 also implies that

$$\nabla(\nabla\alpha \bullet \xi_\alpha h_\alpha \mathbf{B}_0) = 0. \tag{B.7}$$

The vector identity,

$$\nabla(\mathbf{C} \bullet \mathbf{D}) = (\mathbf{D} \bullet \nabla)\mathbf{C} + (\mathbf{C} \bullet \nabla)\mathbf{D} + \mathbf{D} \times (\nabla \times \mathbf{C}) + \mathbf{C} \times (\nabla \times \mathbf{D}), \tag{B.8}$$

applied to Equation B.7 gives

$$(\xi_\alpha h_\alpha \mathbf{B}_0 \bullet \nabla)\nabla\alpha + (\nabla\alpha \bullet \nabla)\xi_\alpha h_\alpha \mathbf{B}_0 + \xi_\alpha h_\alpha \mathbf{B}_0 \times (\nabla \times \nabla\alpha) + \nabla\alpha \times (\nabla \times \xi_\alpha h_\alpha \mathbf{B}_0) = 0. \tag{B.9}$$

However, the curl of the gradient of α is zero. Therefore, Equation B.9 becomes

$$(\xi_\alpha h_\alpha \mathbf{B}_0 \bullet \nabla)\nabla\alpha + (\nabla\alpha \bullet \nabla)\xi_\alpha h_\alpha \mathbf{B}_0 + \nabla\alpha \times (\nabla \times \xi_\alpha h_\alpha \mathbf{B}_0) = 0. \tag{B.10}$$

Applying $\nabla\alpha$ to both sides of Equation B.10 gives

$$\nabla\alpha \bullet [\nabla\alpha \times (\nabla \times \xi_\alpha h_\alpha \mathbf{B}_0)] + \nabla\alpha \bullet [(\xi_\alpha h_\alpha \mathbf{B}_0 \bullet \nabla)\nabla\alpha] + \nabla\alpha \bullet [(\nabla\alpha \bullet \nabla)\xi_\alpha h_\alpha \mathbf{B}_0] = 0. \tag{B.11}$$

Since the divergence of a curl equals zero, Equation B.11 is

$$\nabla\alpha \bullet (\xi_\alpha h_\alpha \mathbf{B}_0 \bullet \nabla)\nabla\alpha = -\nabla\alpha \bullet (\nabla\alpha \bullet \nabla)\xi_\alpha h_\alpha \mathbf{B}_0. \tag{B.12}$$

Equation B.12 is now substituted for the last term in Equation B.6 to give

$$b \bullet \nabla\alpha = |\nabla\alpha|^2 \nabla \bullet (\xi_\alpha h_\alpha \mathbf{B}_0) + 2\xi_\alpha h_\alpha \mathbf{B}_0 \bullet (\nabla\alpha \bullet \nabla)\nabla\alpha. \tag{B.13}$$

Using the vector identity of Equation B.8, this becomes

$$\nabla(\nabla\alpha \bullet \nabla\alpha) = 2(\nabla\alpha \bullet \nabla) + 2\nabla\alpha \times (\nabla \times \nabla\alpha). \tag{B.14}$$

Therefore,

$$2(\nabla\alpha \bullet \nabla)\nabla\alpha = \nabla|\nabla\alpha|^2. \tag{B.15}$$

Substituting Equation B.15 into Equation B.13 yields

$$b \bullet \nabla\alpha = |\nabla\alpha|^2 \nabla \bullet (\xi_\alpha h_\alpha \mathbf{B}_0) + \xi_\alpha h_\alpha \mathbf{B}_0 \nabla|\nabla\alpha|^2 = \nabla \bullet (|\nabla\alpha|^2 \xi_\alpha h_\alpha \mathbf{B}_0), \tag{B.16}$$

so that

$$b \bullet \nabla\alpha = \nabla \bullet (\xi_\alpha h_\alpha |\nabla\alpha|^2 \mathbf{B}_0). \tag{B.17}$$

From Equation B.1,

$$|\nabla\alpha|^2 = \frac{1}{h_\alpha^2}. \tag{B.18}$$

Equation B.17 then becomes

$$b \bullet \nabla\alpha = \frac{b_\alpha}{h_\alpha} = \mathbf{B}_0 \bullet \nabla\left(\frac{\xi_\alpha}{h_\alpha}\right). \tag{B.19}$$

For plasma displacement with negligible magnetic field compression, combining Ampère's law and the momentum equation gives

$$\mu_0\rho\frac{\partial^2}{\partial t^2}(\xi_\alpha\hat{\boldsymbol{\alpha}}) = (\nabla\times b_\alpha\hat{\boldsymbol{\alpha}})\times\mathbf{B}_0. \tag{B.20}$$

Using Equation B.1, we have

$$\mu_0\rho\frac{\partial^2}{\partial t^2}(\xi_\alpha\hat{\boldsymbol{\alpha}}) = (\nabla b_\alpha h_\alpha\times\nabla\alpha)\times\mathbf{B}_0. \tag{B.21}$$

We make use of the vector identity

$$(\mathbf{A}\times\mathbf{B})\times\mathbf{C} = \mathbf{B}(\mathbf{A}\bullet\mathbf{C}) - \mathbf{A}(\mathbf{B}\bullet\mathbf{C}), \tag{B.22}$$

so that Equation B.21 becomes

$$\mu_0\rho\frac{\partial^2}{\partial t^2}(\xi_\alpha\hat{\boldsymbol{\alpha}}) = \nabla\alpha(\nabla b_\alpha h_\alpha\bullet\mathbf{B}_0) - \nabla b_\alpha h_\alpha(\nabla\alpha\bullet\mathbf{B}_0). \tag{B.23}$$

Using Equations B.1 and B.5, we get

$$\mu_0\rho\frac{\partial^2\xi_\alpha}{\partial t^2} = \frac{1}{h_\alpha}\mathbf{B}_0\bullet\nabla b_\alpha h_\alpha. \tag{B.24}$$

Substituting Equation B.19 into Equation B.24, we obtain

$$\mu_0\rho\frac{\partial^2}{\partial t^2}\left(\frac{\xi_\alpha}{h_\alpha}\right) = \frac{1}{h_\alpha^2}\mathbf{B}_0\bullet\nabla\left[h_\alpha^2\left(\mathbf{B}_0\bullet\nabla\left(\frac{\xi_\alpha}{h_\alpha}\right)\right)\right]. \tag{B.25}$$

This must be in a form that provides a suitable numerical solution. To do this, assume a time variation of the form $e^{i\omega t}$ so that

$$-\omega^2\mu_0\rho\left(\frac{\xi_\alpha}{h_\alpha}\right) = \frac{1}{h_\alpha^2}\mathbf{B}_0\bullet\nabla\left[h_\alpha^2\left(\mathbf{B}_0\bullet\nabla\left(\frac{\xi_\alpha}{h_\alpha}\right)\right)\right]. \tag{B.26}$$

For *ds*, the increment of length along the magnetic field direction at any point, Equation B.26 becomes

$$-\omega^2\mu_0\rho\left(\frac{\xi_\alpha}{h_\alpha}\right) = \frac{1}{h_\alpha^2}B_0\frac{\partial}{\partial s}\left[h_\alpha^2\left(B_0\frac{\partial}{\partial s}\left(\frac{\xi_\alpha}{h_\alpha}\right)\right)\right], \tag{B.27}$$

giving

$$-\omega^2\mu_0\rho\left(\frac{\xi_\alpha}{h_\alpha}\right) = \frac{1}{h_\alpha^2}B_0\left[h_\alpha^2 B_0\frac{\partial^2}{\partial s^2}\left(\frac{\xi_\alpha}{h_\alpha}\right) + h_\alpha^2\frac{\partial}{\partial s}\left(\frac{\xi_\alpha}{h_\alpha}\right)\frac{\partial B_0}{\partial s} + B_0\frac{\partial}{\partial s}\left(\frac{\xi_\alpha}{h_\alpha}\right)\frac{\partial h_\alpha^2}{\partial s}\right], \tag{B.28}$$

and so

$$\frac{-\omega^2\mu_0\rho}{B_0^2}\left(\frac{\xi_\alpha}{h_\alpha}\right) = \frac{\partial^2}{\partial s^2}\left(\frac{\xi_\alpha}{h_\alpha}\right) + \frac{1}{B_0}\frac{\partial}{\partial s}\left(\frac{\xi_\alpha}{h_\alpha}\right)\frac{\partial B_0}{\partial s} + \frac{1}{h_\alpha^2}\frac{\partial}{\partial s}\left(\frac{\xi_\alpha}{h_\alpha}\right)\frac{\partial h_\alpha^2}{\partial s}, \tag{B.29}$$

that is,

$$\frac{-\omega^2\mu_0\rho}{B_0^2}\left(\frac{\xi_\alpha}{h_\alpha}\right) = \frac{\partial^2}{\partial s^2}\left(\frac{\xi_\alpha}{h_\alpha}\right) + \frac{\partial}{\partial s}\left(\frac{\xi_\alpha}{h_\alpha}\right)\left[\frac{\partial}{\partial s}\ln\left(h_\alpha^2 B_0\right)\right], \tag{B.30}$$

giving

$$\frac{\partial^2}{\partial s^2}\left(\frac{\xi_\alpha}{h_\alpha}\right) + \frac{\partial}{\partial s}\left(\frac{\xi_\alpha}{h_\alpha}\right)\frac{\partial}{\partial s}\ln\left(h_\alpha^2 B_0\right) + \frac{\omega^2}{V_A^2}\left(\frac{\xi_\alpha}{h_\alpha}\right) = 0. \tag{B.31}$$

References

Ables, S.T. and Fraser, B.J. (2005) Observing the open-closed boundary using cusp-latitude magnetometers. *Geophys. Res. Lett.*, **32** (L10104). doi: 10.1029/2005GL022824

Acuña, M.H., Neubauer, F.M., and Ness, N.F. (1981) Standing Alfvén wave current system at Io: Voyager 1 observations. *J. Geophys. Res.*, **86** (A10), 8513–8521. doi: 10.1029/JA086iA10p08513

Aerts, C., Christensen-Dalsgaard, J., and Kurtz, D.W. (2010) *Asteroseismology*, Springer, Heidelberg.

Akasofu, S.-I. (1965) Attenuation of hydromagnetic waves in the ionosphere. *Radio Sci.*, **69**D (3), 361–366.

Alfvén, H. (1942) Existence of electromagnetic–hydrodynamic waves. *Nature*, **150** (3805), 405–406. doi: 10.1038/150405d0

Alfvén, H. (1948) *Cosmical Electrodynamics*, Oxford University Press, Oxford.

Alfvén, H. and Fälthammar, C.-G. (1963) *Cosmical Electrodynamics Fundamental Principles*, 2nd edn, Clarendon Press, Oxford.

Allan, W. and Knox, F.B. (1979a) A dipole field model for axisymmetric Alfvén waves with finite ionosphere conductivities. *Planet Space Sci.*, **27**, 79–85. doi:10.1016/0032-0633(79)90149-1

Allan, W. and Knox, F.B. (1979b) The effect of finite ionosphere conductivities on axisymmetric toroidal Alfvén wave resonances, *Planet.Space Sci.*, **27**, 939–950. doi: 10.1016/0032-0633(79)90024-2

Allan, W. and Poulter, E.M. (1992) ULF waves: their relationship to the structure of the Earth's magnetosphere. *Rep. Prog. Phys.*, **55**, 533–598.

Allan, W., Poulter, E.M., and White, S.P. (1986) Hydromagnetic wave coupling in the magnetosphere: plasmapause effects on impulse-excited resonances. *Planet Space Sci.*, **34** (12), 1189–1200. doi: 10.1016/0032-0633(86)90056-5

Allan, W., White, S.P., and Poulter, E.M. (1985) Magnetospheric coupling of hydromagnetic waves: initial results. *Geophys. Res. Lett.*, **12** (5), 287–290. doi: 10.1029/GL012i005p00287

Alperovich, L.S. and Federov, E.N. (2007) *Hydromagnetic Waves in the Magnetosphere*, Springer, New York.

Al'pert, Y. (1980) 40 years of whistlers. *J. Atmos. Terr. Phys.*, **42**, 1–20. doi: 10.1016/0021-9169(80)90117-8

Andrews, M.K. (1977) Magnetic pulsation behaviour in the magnetosphere inferred from whistler mode signals. *Planet Space Sci.*, **25** (10), 957–966. doi: 10.1016/0032-0633(77)90008-3

Andrews, D.J., Coates, A.J., Cowley, S.W.H., Dougherty, M.K., Lamy, L., Provan, G., and Zarka, P. (2010) Magnetospheric period oscillations at Saturn: comparison of equatorial and high-latitude magnetic field periods with north and south Saturn kilometric radiation periods. *J. Geophys. Res.*, **115** (A12252). doi: 10.1029/2010JA015666

Andrews, M.K., Knox, F.B., and Thomson, N. R. (1978) Magnetospheric electric fields and protonspheric coupling fluxes inferred from simultaneous phase and group path measurements on whistler-mode signals. *Planet Space Sci.*, **26** (2), 171–183.

Magnetoseismology: Ground-based remote sensing of Earth's magnetosphere, First Edition. Frederick W. Menk and Colin L. Waters.

Angerami, J.J. (1970) Whistler duct properties deduced from VLF observations made with the Ogo 3 satellite near the magnetic equator. *J. Geophys. Res.*, **75** (31), 6115–6135. doi: 10.1029/JA075i031p06115

Archer, M., Horbury, T.S., Lucek, E.A., Mazelle, C., Balough, A., and Dandouras, I. (2005) Size and shape of ULF waves in the terrestrial foreshock. *J. Geophys. Res.*, **110** (A05208). doi: 10.1029/2004JA010791

Arnoldy, R.L., Cahill, L.J., Engebretson, M.J., Lanzerotti, L.J., and Wolfe, A. (1988) Review of hydromagnetic wave studies in the Antarctic. *Rev. Geophys.*, **26** (1), 181–207. doi: 10.1029/RG026i001p00181

Arnoldy, R.L. *et al.* (2005) Pc1 waves and associated unstable distributions of magnetospheric protons observed during a solar wind pressure pulse. *J. Geophys. Res.*, **110** (A07229). doi: 10.1029/2005JA011041

Arthur, C.W. and McPherron, R.L. (1977) Interplanetary magnetic field conditions associated with synchronous orbit: observations of Pc3 magnetic pulsations. *J. Geophys. Res.*, **82** (32), 5138–5142. doi: 10.1029/JA082i032p05138

Aschwanden, M. (2005) *Physics of the Solar Corona: An Introduction with Problems and Solutions*, Springer-Praxis, Berlin.

Aschwanden, M.J., Fletcher, L., Schrijver, C.J., and Alexander, D. (1999) Coronal loop oscillations observed with the Transition Region Coronal Explorer. *Astrophys. J.*, **520**, 880–894. doi: 10.1086/307502

Aschwanden, M.J., Nakariakov, V.M., and Melnikov, V.F. (2004) Magnetohydrodynamic sausage-mode oscillations in coronal loops. *Astrophys. J.*, **600**, 458–463. doi: 10.1086/379789

Baddeley, L.J., Yeoman, T.K., Wright, D.M., Trattner, K.J., and Kellet, B.J. (2005) On the coupling between unstable magnetospheric particle populations and resonant high m ULF wave signatures in the ionosphere. *Ann. Geophys.*, **23**, 567–577. doi: 10.5194/angeo-23-567-2005

Bailey, G.J. and Sellek, R. (1990) A mathematical model of the Earth's plasmasphere and its application in a study of He^+ at $L = 3$. *Ann. Geophys.*, **8**, 171–190.

Baker, D.N., Kanekal, S.G., Horne, R.B., Meredith, N.P., and Glauert, S.A. (2007) Low-altitude measurements of 2–6 MeV electron trapping lifetimes at $1.5 \leq L \leq 2.5$. *Geophys. Res. Lett.*, **34** (L20110), 5. doi: 10.1029/2007GL031007

Baker, D.N., Turner, N.E., and Pulkinnen, T.I. (2001) Energy transport and dissipation in the magnetosphere during geomagnetic storms. *J. Atmos. Solar-Terr. Phys.*, **63** (6), 421–429. doi: 10.1016/S1364-6826(00)00169-3

Baker, D.N. *et al.* (2008) *Severe Space Weather Events: Understanding Societal and Economic Impacts*, The National Academies Press, Washington, DC, pp. 144.

Banerjee, D., Erdélyi, R., Oliver, R., O'Shea, E. (2007) Present and future observing trends in atmospheric magnetoseismology. *Solar Phys.*, **246** (1), 3–29. doi: 10.1007/s11207-007-9029-z

Baransky, L.N., Belokris, S.P., Borovkov, U.E., and Green, C.A. (1990) Two simple methods for the determination of the resonance frequencies of magnetic field lines. *Planet Space Sci.*, **38** (12), 1573–1576. doi: 10.1016/0032-0633(90)90163-K

Baransky, L.N., Borovkov, J.E., Gokhberg, M. B., Krylov, S.M., and Troitskaya, V.A. (1985) High resolution method of direct measurement of the magnetic field lines' eigen frequencies. *Planet Space Sci.*, **33** (12), 1369–1374.

Barbiera, L.P. and Mahmot, R.E. (2004) October–November 2003's space weather and operations lessons learned. *Space Weather*, **2** (S09002). doi: 10.1029/2004SW000064

Barnhart, B.L., Kurth, W.S., Groene, J.B., Faden, J.B., Santolik, O., and Gurnett, D.A. (2009) Electron densities in Jupiter's outer magnetosphere determined from Voyager 1 and 2 plasma wave spectra. *J. Geophys. Res.*, **114** (A05218). doi: 10.1029/2009JA014069

Barton, C.E. (1989) Geomagnetic secular variation: direction and intensity, in *Encyclopedia of Solid Earth Geophysics* (ed. D. E. James), Van Nostrand Reinhold, New York, pp. 560–577.

Barton, C.E. (2002) Survey tracks current position of South Magnetic Pole. *EoS*, **83** (27), 291. doi: 10.1029/2002EO000210

Bauer, S. (1969) Diffusive equilibrium in the topside ionosphere. *Proc. IEEE*, **57** (6), 1114–1118.

Beamish, D., Hanson, H.W., and Webb, D.C. (1979) Complex demodulation applied to Pi2 geomagnetic pulsations. *Geophys. J. R. Astron. Soc.*, **58**, 471–493.

Bedingfield, K.L., Leach, R.D., and Alexander, M.B. (1996) *Spacecraft system failures and anomalies attributed to the natural space environment*. NASA Reference Publication 1390, Huntsville, AL.

Belcher, J.W. and Davis, L. Jr. (1971) Large-amplitude Alfvén waves in the interplanetary medium. *J. Geophys. Res.*, **76** (16), 3534–3563. doi: 10.1029/JA076i016p03534

Bencze, P. and Lemperger, I. (2011) Characteristics of field line resonance type geomagnetic pulsations and variations of plasmaspheric plasma density composition. *Adv. Space Res.*, **47** (9), 1568–1577. doi: 10.1016/j.asr.2010.12.016

Bennett, J.A. (1967) The calculation of Doppler shifts due to a changing ionosphere. *J. Atmos. Terr. Phys.*, **29** (7), 887–891.

Berube, D., Moldwin, M.B., Fung, S.F., and Green, J.L. (2005) A plasmaspheric mass density model and constraints on its heavy ion concentration. *J. Geophys. Res.*, **110** (A04212). doi: 10.1029/2004JA010684

Berube, D., Moldwin, M.B., and Weygand, J.M. (2003) An automated method for the detection of field line resonance frequencies using ground magnetometer techniques. *J. Geophys. Res.*, **108** (A9). doi: 10.1029/2002JA009737

Biermann, L. (1951) Kometenschweife und solare Korpuskularstrahlung. *Z. Astrophys.*, **29**, 274–286.

Bilitza, D. and Reinisch, B.W. (2008) International Reference Ionosphere 2007: improvements and new parameters. *Adv. Space Res.*, **42** (4), 599–609. doi: 10.1016/j.asr.2007.07.048

Blanco-Cano, X., Omidi, N., and Russell, C.T. (2009) Global hybrid simulations: foreshock waves and cavitons under radial interplanetary magnetic field geometry. *J. Geophys. Res.*, **114** (A01216). doi: 10.1029/2008JA013406

Blelly, P.-L. and Alcaydé, D. (2007) Ionosphere, in *Handbook of the solar-terrestrial environment*, (eds Y. Kamide, and A.C.-L. Chian), Springer, Berlin, p. 202. doi: 10.1007/b104478

Boardsen, S.A., Anderson, B.J., Acuña, M.H., Slavin, J.A., Korth, H., and Solomon, S.C. (2009) Narrow-band ultra-low frequency wave observations by MESSENGER during its January 2008 flyby through Mercury's magnetosphere. *Geophys. Res. Lett.*, **36** (L01104). doi: 10.1029/2008GL036034

Bolton, S.J., Janssen, M., Thorne, R., Levin, S., Klein, M., Gulkis, S., Bastian, T., Sault, R., Elachi, C., Hofstadter, M., Bunker, A., Dulk, G., Gudmin, E., Hamilton, G., Johnson, W. T.K., Leblanc, Y., Liepack, O., McLeod, R., Roller, J., Roth, L., and West, R. (2002) Ultra-relativistic electrons in Jupiter's radiation belts. *Nature*, **415**, 987–991. doi: 10.1038/415987a

Bonfond, B., Grodent, D., Gerard, J.-C., Stallard, T., Clarke, J.T., Yoneda, M., Radioti, A., and Gustin, J. (2012) Auroral evidence of Io's control over the magnetosphere of Jupiter. *Geophys. Res. Lett.*, **39** (L01105). doi: 10.1029/2011GL050253

Bouchy, F. and Carrier, F. (2001) P-mode observations on α Cen A. *Astron. Astrophys.*, **374** (1), L5–L8. doi: 10.1051/0004-6361:20010792

Boudouridis, A. and Zesta, E. (2007) Comparison of Fourier and wavelet techniques in the determination of geomagnetic field line resonances. *J. Geophys. Res.*, **112** (A08205). doi: 10.1029/2006JA011922

Campbell, W.H. (2003) *Introduction to Geomagnetic Fields*, 2nd edn, Cambridge University Press, Cambridge.

Carovillano, R.L. and Siscoe, G.L. (1973) Energy and momentum theorems in magnetospheric dynamics. *Rev. Geophys.*, **11** (2), 289–353. doi: 10.1029/RG011i002p00289

Carpenter, D.L. (1962) Electron-density variations in the magnetosphere deduced from whistler data. *J. Geophys. Res.*, **67** (9), 3345–3360. doi: 10.1029/JZ067i009p03345

Carpenter, D.L. (1963) Whistler evidence of a 'knee' in the magnetospheric ionization density profile. *J. Geophys. Res.*, **68** (6), 1675–1682. doi: 10.1029/JZ068i006p01675

Carpenter, D.L. (1970) Whistler evidence of the dynamic behavior of the duskside bulge in the plasmasphere. *J. Geophys. Res.*, **75** (19), 3837–3847. doi: 10.1029/JA075i019p03837

Carpenter, D.L. (1983) Some aspects of plasmapause probing by whistlers. *Radio Sci.*, **18** (6), 917–925. doi: 10.1029/RS018i006p00917

Carpenter, D.L. (1988) Remote sensing of the magnetospheric plasma by means of whistler mode signals. *Rev. Geophys.*, **26** (3), 535–549. doi: 10.1029/RG026i003p00535

Carpenter, D.L. and Anderson, R.R. (1992) An ISEE/whistler model of equatorial electron density in the magnetosphere. *J. Geophys. Res.*, **97** (A2), 1097–1108. doi: 10.1029/91JA01548

Carpenter, D.L., Anderson, R.R., Bell, T.F., and Miller, T.R. (1981) A comparison of equatorial electron densities measured by whistlers and by a satellite radio technique. *Geophys. Res. Lett.*, **8** (10), 1107–1110. doi: 10.1029/GL008i010p01107

Carpenter, D.L., Anderson, R.R., Calvert, W., and Moldwin, M.B. (2000) CRRES observations of density cavities inside the plasmasphere. *J. Geophys. Res.*, **105** (A10), 23323–23338. doi: 10.1029/2000JA000013

Carpenter, D.L. and Park, C.G. (1973) On what ionospheric workers should know about the plasmapause–plasmasphere. *Rev. Geophys.*, **11** (1), 133–154. doi: 10.1029/RG011i001p00133

Carpenter, D.L. and Smith, R.L. (1964) Whistler measurements of electron density in the magnetosphere. *Rev. Geophys.*, **2** (3), 415–441. doi: 10.1029/RG002i003p00415

Chan, K.L., Kanellakos, D.P., and Villard, O.G. (1962) Correlation of short-period fluctuations of the Earth's magnetic field and instantaneous frequency measurements. *J. Geophys. Res.*, **67** (5), 2066–2072. doi: 10.1029/JZ067i005p02066

Chaplin, W.J., Elsworth, Y., Miller, B.A., Verner, G.A., and New, R. (2007) Solar p-mode frequencies over three solar cycles. *Astrophys. J.*, **659**, 1749–1760. doi: 10.1086/512543

Chapman, S. (1931) The absorption and dissociative or ionizing effect of monochromatic radiation in an atmosphere on a rotating Earth. *Proc. Phys. Soc.*, **43**, 26–45. doi: 10.1088/0959-5309/43/1/305

Chapman, S. and Bartels, J. (1940) *Geomagnetism*, vol. 1, Oxford University Press, Oxford.

Chapman, S. and Ferraro, C.A. (1930) A new theory of magnetic storms. *Nature*, **126**, 129–130. doi: 10.1038/126129a0

Chappell, C.R., Harris, K.K., and Sharp, G.W. (1970a) The morphology of the bulge region of the plasmasphere. *J. Geophys. Res.*, **75** (19), 3848–3861. doi: 10.1029/JA075i019p03848

Chappell, C.R., Harris, K.K., and Sharp, G.W. (1970b) A study of the influence of magnetic activity on the location of the plasmapause as measured by OGO 5. *J. Geophys. Res.*, **75** (1), 50–56. doi: 10.1029/JA075i001p00050

Chappell, C.R., Harris, K.K., and Sharp, G.W. (1971) The dayside plasmasphere. *J. Geophys. Res.*, **76** (31), 7632–7647. doi: 10.1029/JA076i031p07632

Chen, F.F. (1984) *Introduction to Plasma Physics and Controlled Fusion*, 2nd edn, Plenum Press, New York.

Chen, L. and Hasegawa, A. (1974a) A theory of long-period magnetic pulsations: 1. Steady state excitation of field line resonance. *J. Geophys. Res.*, **79** (7), 1024–1032. doi: 10.1029/JA079i007p01024

Chen, L. and Hasegawa, A. (1974b) A theory of long-period magnetic pulsations: 2. Impulse excitation of surface eigenmode. *J. Geophys. Res.*, **79** (7), 1033–1037. doi: 10.1029/JA079i007p01033

Chi, P.J., Russell, C.T., Musman, S., Peterson, W.K., Le, G., Angelopoulos, V., Reeves, G.D., Moldwin, M.B., and Chun, F.K. (2000) Plasmaspheric depletion and refilling associated with the September 25, 1998 magnetic storm observed by ground magnetometers at L = 2. *Geophys. Res. Lett.*, **27** (5). doi: 10.1029/1999GL010722

Chi, P.J., Lee, D.H., and Russell, C.T. (2006) Tamao travel time of sudden impulses and its relationship to ionospheric convection vortices. *J. Geophys. Res.*, **111** (A08205). doi: 10.1029/2005JA011578

Chi, P.J. and Russell, C.T. (2005) Travel-time magnetoseismology: magnetospheric sounding by timing the tremors in space. *Geophys. Res. Lett.*, **32** (L18108). doi: 10.1029/2005GL023441

Chi, P.J., Russell, C.T., Foster, J.C., Moldwin, M.B., Engebretson, M.J., and Mann, I.R. (2005) Density enhancement in plasmasphere–ionosphere plasma during the 2003 Halloween superstorm: observations along the 330th magnetic meridian in North America. *Geophys. Res. Lett.*, **32** (L03S07). doi: 10.1029/2004GL021722

Chisham, G. (1996) Giant pulsations: an explanation for their rarity and occurrence during geomagnetically quiet times. *J. Geophys. Res.*, **101** (A11), 24755–24763. doi: 10.1029/96JA02540

Chisham, G., Lester, M., Milan, S.E., Freeman, M.P., Bristow, W.A., Grocott, A., McWilliams, K.A., Ruohoniemi, J.M., Yeoman, T.K., Dyson, P.L., Greenwald, R.A., Kikuchi, T., Pinnock, M., Rash, J.P.S., Sato, N., Sofko, G.J., Villain, J.-P., and Walker, A.D.M. (2007) A decade of the Super Dual Auroral Radar Network (SuperDARN): scientific achievements, new techniques and future directions. *Surv. Geophys.*, **28**, 33–109. doi: 10.1007/s10712-007-9017-8

Christensen-Dalsgaard, J. (2002) Helioseismology. *Rev. Mod. Phys.*, **74**, 1073–1129. doi: 10.1103/RevModPhys.74.1073

Clarke, J.T., Moos, H.W., Atreya, S.K., and Lane, A.L. (1980) Observations from Earth orbit and variability of the polar aurora on Jupiter. *Astrophys. J. Lett.*, **241**, L179–L182. doi: 10.1086/183386

Claudepierre, S.G., Elkington, S.R., and Wiltberger, M. (2008) Solar wind driving of magnetospheric ULF waves: pulsations driven by solar wind velocity shear at the magnetopause. *J. Geophys. Res.*, **113** (A05218). doi: 10.1029/2007/JA012890

Clausen, L.B.N. and Yeoman, T.K. (2009) Comprehensive survey of Pc4 and Pc5 band spectral content in Cluster magnetic field data. *Ann. Geophys.*, **27**, 3237–3248. doi: 10.5194/angeo-27-3237-2009

Clausen, L.B.N., Yeoman, T.K., Behlke, R., and Lucek, E.A. (2008) Multi-instrument observations of a large scale Pc4 pulsation. *Ann. Geophys.*, **26**, 185–199. doi: 10.5194/angeo-26-185-2008

Clausen, L.B.N., Yeoman, T.K., Fear, R.C., Behlke, R., Lucek, E.A., and Engebretson, M.J. (2009) First simultaneous measurements of waves generated at the bow shock in the solar wind, the magnetosphere and on the ground. *Ann. Geophys.*, **27**, 357–371. doi: 10.5194/angeo-27-357-2009

Claverie, A., Isaak, G.R., McLeod, C.P., van der Raay, H.B., and Roca Cortes, T. (1979) Solar structure from global studies of the 5-minute oscillation. *Nature*, **282** (5739), 591–594. doi: 10.1038/282591a0

Clilverd, M.A., Clark, T.D.G., Clarke, E., Rishbeth, H., and Ulich, T. (2002) The causes of long-term changes in the *aa* index. *J. Geophys. Res.*, **107** (A12), SSH 4-1–SSH 4-7. doi: 10.1029/2001JA000501

Clilverd, M.A., Clarke, E., Ulich, T., Linthe, J., and Rishbeth, H. (2005) Reconstructing the long-term *aa* index. *J. Geophys. Res.*, **110** (A7025). doi: 10.1029/2004JA010762

Clilverd, M.A., Menk, F.W., Milinevski, G., Sandel, B.R., Goldstein, J., Reinisch, B.W., Wilford, C.R., Rose, M.C., Thomson, N.R., Yearby, K.H., Bailey, G.J., Mann, I.R., and Carpenter, D.L. (2003) *In situ* and ground-based intercalibration measurements of plasma density at $L = 2.5$. *J. Geophys. Res.*, **108** (A10). doi: 10.1029/2003JA009866

Clilverd, M.A., Meredith, N.P., Horne, R.B., Glauert, S.A., Anderson, R.R., Thomson, N. R., Menk, F.W., and Sandel, B.R. (2007) Longitudinal and seasonal variations in plasmaspheric electron density: implications for electron precipitation. *J. Geophys. Res.*, **112** (A11210). doi: 10.2019/2007JA012416

Clilverd, M.A., Rodger, C.J., Moffatt-Griffin, T., Spanswick, E., Breen, P., Menk, F.W., Grew, R.S., Hayashi, K., and Mann, I.R. (2010) Energetic outer radiation belt electron precipitation during recurrent solar activity. *J. Geophys. Res.*, **115** (A08323). doi: 10.1029/2009JA015204

Clilverd, M.A., Rodger, C.J., Thomson, N.R., Brundell, J.B., Ulich, Th., Lichtenberger, J., Cobbett, N., Collier, A.B., Menk, F.W., Seppälä, A., Verronen, P.T., and Turunen, E. (2009) Remote sensing space weather events: the AARDDVARK network. *Space Weather*, **7** (S04001). doi: 10.1029/2008SW000412

Cole, K.D. (1971) Formation of field-aligned irregularities in the magnetosphere. *J. Atmos. Terr. Phys.*, **33** (5), 741–750.

Collett, E. (1993) *Polarized Light: Fundamentals and Applications*, Marcel Dekker, Inc., New York.

Constantinescu, O.D., Glassmeier, K.-H., Decreau, P.M.E., Franz, M., and Fornaçon, K.-H. (2007) Low frequency wave sources in the outer magnetosphere, magnetosheath, and near Earth solar wind. *Ann. Geophys.*, **25**, 2217–2228. doi: 10.5194/angeo-25-2217-2007

Constantinescu, O.D., Glassmeier, K.-H., Plaschke, F., Auster, U., Angelopoulos, V., Baumjohann, W., Fornaçon, K.-H., Georgescu, E., Larson, D., Magnes, W., McFadden, J.P., Nakamura, R., and Narita, Y. (2009) THEMIS observations of duskside compressional Pc5 waves. *J. Geophys. Res.*, **114** (A00C25). doi: 10.1029/2008JA013519

Coster, A. and Komjathy, A. (2008) Space weather and the Global Positioning System. *Space Weather*, **6** (S06D04). doi: 10.1029/2008SW00040

Cowley, S.W.H. (1995) The Earth's magnetosphere: a brief beginner's guide. *EOS*, **76** (51), 525–529. doi: 10.1029/95EO00322

Cramm, R., Glassmeier, K.-H., Stellmacher, M., and Othmer, C. (1998) Evidence for resonance mode coupling in Saturn's magnetosphere. *J. Geophys. Res.*, **103** (A6), 11951–11960. doi: 10.1029/98JA00629

Cummings, W.D., O'Sullivan, R.J., and Coleman, P.J., Jr. (1969) Standing Alfvén waves in the magnetosphere. *J. Geophys. Res.*, **74** (3), 778–793. doi: 10.1029/JA074i003p00778

Daglis, I.A. (2004) *Effects of Space Weather on Technology Infrastructure*, Kluwer Academic Press, Dordrecht, The Netherlands.

De Moortel, I. (2005) An overview of coronal seismology. *Philos. Trans. R. Soc. Lond. A*, **363** (1837), 2743–2760. doi: 10.1098/rsta.2005.1665

De Moortel, I., Ireland, J., Hood, A.W., and Walsh, R.W. (2002) The detection of 3 & 5 min period oscillations in coronal loops. *Astron. Astrophys. Lett.*, **387** (1), L13–L16. doi: 10.1051/0004-6361:20020436

De Moortel, I. and Nakariakov, V.M. (2012) Magnetohydrodynamic waves and coronal seismology: an overview of recent results. *Philos. Trans. R. Soc. Lond. A*, **370** (1970), 3193–3216. doi: 10.1098/rsta.2011.0640

De Moortel, I. and Pascoe, D.J. (2009) Putting coronal seismology estimates of the magnetic field strength to the test. *Astrophys. J. Lett.*, **699** (L72). doi: 10.1088/0004-637X/699/2/L72

De Pontieu, B., McIntosh, S.W., Carlsson, M., Hansteen, V.H., Tarbell, T.D., Schrijver, C.J., Title, A.M., Shine, R.A., Tsuneta, S., Katsukawa, Y., Ichimoto, K., Suematsu, Y., Shimizu, T., and Nagata, S. (2007) Chromospheric Alfvén waves strong enough to power the solar wind. *Science*, **318** (5856), 1574–1577. doi: 10.1126/science.1151747

Degeling, A.W., Rankin, R., Kabin, K., Rae, I. J., and Fenrich, F.R. (2010) Modeling ULF waves in a compressed dipole magnetic field. *J. Geophys. Res.*, **115** (A10212). doi: 10.1029/2010JA015410

Demekhov, A.G. (2007) Recent progress in understanding Pc1 pearl formation. *J. Atmos. Solar-Terr. Phys.*, **69** (14), 1609–1622. doi: 10.1016/j.jastp.2007.01.014

Demetrescu, C. and Dobrica, V. (2008) Signature of Hale and Gleissberg solar cycles in the geomagnetic activity. *J. Geophys. Res.*, **113** (A02103). doi: 10.1029/2001JA000501

Dent, Z.C., Mann, I.R., Goldstein, J., Menk, F. W., and Ozeke, L.G. (2006) Plasmaspheric depletion, refilling, and plasmapause dynamics: a coordinated ground-based and IMAGE satellite study. *J. Geophys. Res.*, **111** (A03205). doi: 10.1029/2005JA011046

Dent, Z.C., Mann, I.R., Menk, F.W., Goldstein, J., Wilford, C.R., Clilverd, M.A., and Ozeke, L.G. (2003) A coordinated ground-based and IMAGE satellite study of quiet-time plasmaspheric density profiles. *Geophys. Res. Lett.*, **30** (12). doi: 10.1029/2003GL016946

Denton, R.E., Lessard, M.R., and Kistler, L.M. (2003) Radial localization of magnetospheric guided poloidal Pc4-5 waves. *J. Geophys. Res.*, **108** (A3). doi: 10.1029/2002JA009679

Denton, R.E., Takahashi, K., Galkin, I.A., Nsumei, P.A., Huang, X., Reinisch, B.W., Anderson, R.R., Sleeper, M.K., and Hughes, W.J. (2006) Distribution of density along geomagnetic field lines. *J. Geophys. Res.*, **111** (A04213). doi: 10.1029/2005JA011414

Denton, R.E. and Vetoulis, G. (1998) Global poloidal mode. *J. Geophys. Res.*, **103** (A4), 6729–6739. doi: 29/97JA03594

Denton, R.E., Wang, Y., Webb, P.A., Tengdin, P.M., Goldstein, J., Redfern, J.A., and Reinisch, B.W. (2012) Magnetospheric electron density long term (>1 day) refilling rates inferred from passive radio emissions

measured by IMAGE RPI during geomagnetically quiet times. *J. Geophys. Res.*, **117** (A03221). doi: 10.1029/2011JA017274
Dessler, A.J. (1959) Ionospheric heating by hydromagnetic waves. *J. Geophys. Res.*, **64** (4), 397–401. doi: 10.1029/JZ064i004p00397
D'haeseleer, W.D., Hitchon, W.N.G., Callen, J. D., and Shohet, J.L. (1991) *Flux Coordinates and Magnetic Field Structure: A Guide to a Fundamental Tool of the Plasma Theory*, Springer, Berlin.
Didkovsky, L., Judge, D., Kosovichev, A.G., Wieman, S., and Woods, T. (2011) Observations of five-minute solar oscillations in the corona using the Extreme Ultraviolet Spectrophotometer (ESP) on board the Solar Dynamics Observatory Extreme Ultraviolet Variability Experiment (SDO/EVE). *Astrophys. J. Lett.*, **738** (1). doi: 10.1088/2041-8205/738/L7
Duffus, H.J. and Boyd, G.M. (1968) The association between ULF geomagnetic fluctuations and Doppler ionospheric observations. *J. Atmos. Terr. Phys.*, **30** (4), 481–496.
Duffus, H.J. and Shand, J.A. (1958) Some observations of geomagnetic micropulsations. *Can. J. Phys.*, **36** (4), 508–526. doi: 10.1139/p58-052
Duncan, R.A. (1961) Some studies of geomagnetic micropulsations. *J. Geophys. Res.*, **66** (7), 2087–2094. doi: 10.1029/JZ066i007p02087
Dungey, J.W. (1954) Electrodynamics of the outer atmosphere, *Scientific Rept. No. 69*, Ionospheric Research Lab., Penn. State Univ.
Dungey, J.W. (1955) Electrodynamics of the outer atmosphere, *The Physics of the Ionosphere*, The Physical Society, London, pp. 229–236.
Dungey, J.W. (1961) Interplanetary field and the auroral zones. *Phys. Rev. Lett.*, **6**, 47–48.
Dungey, J.W. (1963) Hydromagnetic waves and the ionosphere, *Proceedings of the International Conference*, The Institute of Physics, London, p. 230.
Dunlop, I.S., Menk, F.W., Hansen, H.J., Fraser, B.J., and Morris, R.J. (1994) A multistation study of long period geomagnetic pulsations at cusp and boundary layer latitudes. *J. Atmos. Terr. Phys.*, **56** (5), 667–671.
Duvall, T.L., Jr. (1982) A dispersion law for solar oscillations. *Nature*, **300** (5889), 242–243. doi: 10.1038/300242a0
Dyer, C. and Rodgers, D. (1998) Effects on spacecraft & aircraft electronics. *ESA Workshop on Space Weather*, ESTEC, Noordwijk, The Netherlands, ESA.
Dyrud, L.P., Engebretson, M.J., Posch, J.L., Hughes, W.J., Fukunishi, H., Arnoldy, R.L., Newell, P.T., and Horne, R.B. (1997) Ground observations and possible source regions of two types of Pc1-2 micropulsations at very high latitudes. *J. Geophys. Res.*, **102** (A12), 27011–27027. doi: 10.1029/97JA02191
Eastwood, J.P., Balogh, A., Lucek, E.A., Mazelle, C., and Dandouras, I. (2005a) Quasi-monochromatic ULF foreshock waves as observed by the four-satellite Cluster mission: 2. Oblique propagation. *J. Geophys. Res.*, **110** (A11220). doi: 10.1029/2004JA010618
Eastwood, J.P., Balogh, A., Lucek, E.A., Mazelle, C., and Dandouras, I. (2005b) Quasi-monochromatic ULF foreshock waves as observed by the four-spacecraft Cluster mission: 1. Statistical properties. *J. Geophys. Res.*, **110** (A11219). doi: 10.1029/2004JA010617
Elkington, S.R. (2006) A review of ULF interactions with radiation belt electrons, in *Magnetospheric ULF Waves: Synthesis and New Directions* (eds K. Takahashi, P.J. Chi, and R.L. Lysak), American Geophysical Union, Washington, DC, pp. 177–193.
Ellis, G.R.A. (1960) Geomagnetic micropulsations. *Aust. J. Phys.*, **13**, 625–632.
Engebretson, M.J., Glassmeier, K.-H., Stellmacher, M., Hughes, W.J., and Lühr, H. (1998) The dependence of high-latitude Pc5 wave power on solar wind velocity and on the phase of high speed solar wind streams. *J. Geophys. Res.*, **103** (A11), 26271–26283. doi: 10.1029/97JA03143
Engebretson, M.J., Lessard, M.R., Bortnik, J., Green, J.C., Horne, R.B., Detrick, D.L., Weatherwax, A.T., Manninen, J., Petit, N.J., Posch, J.L., and Rose, M.C. (2008) Pc1–Pc2 waves and energetic particle precipitation during and after magnetic storms: superposed epoch analysis

and case studies. *J. Geophys. Res.*, **113** (A01211). doi: 10.1029/2007JA012362

Engebretson, M.J., Zanetti, L., Potemra, T., Baumjohann, W., Lühr, H., and Acuña, M. (1987) Simultaneous observations of Pc3–4 pulsations in the solar wind and in the Earth's magnetosphere. *J. Geophys. Res.*, **92** (A9), 10053–10062. doi: 10.1029/JA092iA09p10053

Engwall, E. (2004) Numerical studies of spacecraft–plasma interaction: simulations of wake effects on the Cluster Electric Field Instrument EFW. IRF Scientific Report 284, Institutet for Rymdfysik (IRF).

Eriksson, P.T.I., Blomberg, L.G., Schaefer, S., and Glassmeier, K.-H. (2008) Sunward propagating Pc5 waves observed on the post-midnight magnetospheric flank. *Ann. Geophys.*, **26**, 1567–1579. doi: 10.5194/angeo-26-1567-2008

Eriksson, P.T.I., Blomberg, L.G., Walker, A.D. M., and Glassmeier, K.-H. (2005) Poloidal ULF oscillations in the dayside magnetosphere: a Cluster study. *Ann. Geophys.*, **23**, 2679–2686. doi: 10.5194/angeo-23-2679-2005

Eriksson, P.T.I., Walker, A.D.M., and Stephenson, J.A.E. (2006) A statistical correlation of Pc5 pulsations and solar wind pressure oscillations. *Adv. Space Res.*, **38**, 1763–1771. doi: 10.1016/j.asr.2005.08.023

Eschenhagen, M. (1897) Über schnelle periodische Veränderungen des Erdmagnetismus von sehr kleiner Amplitude. *Sitzungberichte der Akademie der Wissenschaften zu Berlin*, pp. 678–686 with plate.

Evans, J.V. and Holt, J.M. (1978) Nighttime proton fluxes at Millstone Hill. *Planet. Space Sci.*, **26** (8), 727–744.

Fairfield, D.H. (1969) Bow shock associated waves observed in the far upstream interplanetary medium. *J. Geophys. Res.*, **74** (14), 3541–3553. doi: 10.1029/JA074i014p03541

Fenrich, F.R. and Samson, J.C. (1997) Growth and decay of field line resonances. *J. Geophys. Res.*, **102** (A9), 20031–20039. doi: 10.1029/97JA01376

Fenrich, F.R., Samson, J.C., Sofko, G., and Greenwald, R.A. (1995) ULF high- and low-m field line resonances observed with the Super Dual Auroral Radar Network. *J. Geophys. Res.*, **100** (A11), 21535–21547. doi: 10.1029/95JA02024

Fenrich, F.R. and Waters, C.L. (2008) Phase coherence analysis of a field line resonance and solar wind oscillation. *Geophys. Res. Lett.*, **35** (L20102). doi: 10.1029/2008GL035430

Finlay, C.C., Maus, S., Beggan, C.D., Bondar, T.N., Chambodut, A., Chernova, T.A., Chulliat, A., Golovkov, V.P., Hamilton, B., Hamoudi, M., Holme, R., Hulot, G., Kuang, W., Langlais, B., Lesur, V., Lowes, F.J., Luhr, H., Macmillan, S., Mandea, M., McLean, S., Manoj, C., Menviellw, M., Michaelis, I., Olsen, N., Rauberg, J., Rother, M., Sabaka, T.J., Tangborn, A., Toffner-Clausen, L., Thebault, E., Thomson, A.W.P., Wardinski, I., Wei, Z., and Zvereva, T.I. (2010) International Geomagnetic Reference Field: the eleventh generation. *Geophys. J. Int.*, **183** (3), 1216–1230. doi: 10.1111/j.1365-246X.2010.04804.x

Fitzenreiter, R.J. (1995) The electron foreshock. *Adv. Space Res.*, **15** (8-9), 9–27.

Foster, J.C., Erickson, P.J., Coster, A.J., Goldstein, J., and Rich, F.J. (2002) Ionospheric signatures of plasmaspheric tails. *Geophys. Res. Lett.*, **29** (13). doi: 10.1029/2002GL015067

Francia, P. and Villante, U. (1997) Some evidence of ground power enhancements at frequencies of global magnetospheric modes at low latitude. *Ann. Geophys.*, **15**, 17–23. doi: 10.1007/s00585-997-0017-2

Fraser, B.J. (1975) Ionospheric duct propagation and Pc1 pulsation sources. *J. Geophys. Res.*, **80** (19), 2790–2796. doi: 10.1029/JA080i019p02790

Fraser, B.J. (1976) Pc1 geomagnetic pulsation source regions and ionospheric waveguide propagation. *J. Atmos. Terr. Phys.*, **38** (11), 1141–1146.

Fraser, B.J., Horwitz, J.L., Slavin, J.A., Dent, Z. C., and Mann, I.R. (2005) Heavy ion mass loading of the geomagnetic field near the plasmapause and ULF wave implications. *Geophys. Res. Lett.*, **32** (L04102). doi: 10.1029/2004GL021315

Fraser, B.J. and McPherron, R.L. (1982) Pc1–2 magnetic pulsation spectra and heavy ion effects at synchronous orbit: ATS 6 results. *J. Geophys. Res.*, **87** (A6), 4560–4566. doi: 10.1029/JA087iA06p04560

Fraser, B.J., Samson, J.C., Hu, Y.D., McPherron, R.L., and Russell, C.T. (1992) Electromagnetic ion cyclotron waves observed near the oxygen cyclotron frequency by ISEE 1 and 2. *J. Geophys. Res.*, **97** (A3), 3063–3074. doi: 10.1029/91JA02447

Fraser, B.J., Singer, H.J., Hughes, W.J., Wygant, J.R., Anderson, R.R., and Hu, Y.D. (1996) CRRES Poynting vector observations of electromagnetic ion cyclotron waves near the plasmapause. *J. Geophys. Res.*, **101** (A7), 15331. doi: 10.1029/95JA03480

Fraser-Smith, A.C. (1987) Centered and eccentric geomagnetic dipoles and their poles, 1600–1985. *Rev. Geophys.*, **23** (1), 1–16. doi: 10.1029/RG025i001p00001

Fukunishi, H., Toya, T., Koike, K., and Kuwashima, M. (1981) Classification of hydromagnetic emissions based on frequency–time spectra. *J. Geophys. Res.*, **86** (A11), 9029–9039. doi: 1029/JA086iA11p09029

Fukushima, N. (1969) Equivalence in ground geomagnetic effect of Chapman–Vestine's and Bireland–Alfvén's current systems for polar magnetic storms. *Rep. Ionos. Space Phys. Jpn.*, **23**, 219–227.

Gallagher, D.L., Adrian, M.L., and Liemohn, M.W. (2005) Origin and evolution of deep plasmaspheric notches. *J. Geophys. Res.*, **110** (A09201). doi: 10.1029/2004JA010906

Garcia, L.N., Fung, S.F., Green, J.L., Boardsen, S.A., Sandel, B.R., and Reinisch, B.W. (2003) Observations of the latitudinal structure of plasmaspheric convection plumes by IMAGE-RPI and EUV. *J. Geophys. Res.*, **108** (A8). doi: 10.1029/2002JA009496

Getley, I.L. (2004) Observation of solar particle event on board a commercial flight from Los Angeles to New York on 29 October 2003. *Space Weather*, **2** (S05002). doi: 10.1029/2003SW00058

Giampieri, G., Dougherty, M.K., Smith, E.J., and Russell, C.T. (2006) A regular period for Saturn's magnetic field that may track its internal rotation. *Nature*, **441**, 62–64. doi: 10.1038/nature04750

Glassmeier, K.-H. (1980) Magnetometer array observations of a giant pulsation event. *J. Geophys.*, **48**, 127–138.

Glassmeier, K.-H., Ness, N.F., Acuña, M.H., and Neubauer, F.M. (1989) Standing hydromagnetic waves in the Io plasma torus: Voyager 1 observations. *J. Geophys. Res.*, **94** (A11), 15063–15076. doi: 10.1029/JA094iA11p15063

Glassmeier, K.-H., Soffel, H., and Negendank, J.F.W. (2009) *Geomagnetic field variations*, Springer, Berlin.

Goldstein, M.L., Eastwood, J.P., Treumann, R. A., Lucek, E.A., Pickett, J., and Décreau, P. (2005) The near-Earth solar wind. *Space Sci. Rev.*, **118** (1-4), 7–39. doi:10.1007/s11214-005-3823-4

Goldstein, J., Spiro, R.W., Reiff, P.H., Wolf, R. A., and Sandel, B.R. (2002) IMF-driven overshielding electric field and the origin of the plasmaspheric shoulder of March 24, 2000. *Geophys. Res. Lett.*, **29** (16). doi: 10.1029/2001GL014534

González-Hernández, I., Hill, F., Scherrer, P. H., Lindsey, C., and Braun, D.C. (2010) On the success rate of the farside seismic imaging of active regions. *Space Weather*, **8**, S0600210.1029/2009SW000560.

Gopalswamy, N., Barbieri, L., Cliver, E.W., Lu, G., Plunkett, S.P., and Skoug, R.M. (2005) Introduction to violent Sun–Earth connection events of October–November 2003. *J. Geophys. Res.*, **110** (A09S00). doi: 10.1029/2005JA011268

Gosling, J.T. and Pizzo, V.J. (1999) Formation and evolution of corotating interaction regions and their three dimensional structure. *Space Sci. Rev.*, **89**, 21–52.

Grebowsky, J.M. (1970) Model study of plasmapause motion. *J. Geophys. Res.*, **75** (22), 4329–4333. doi: 10.1029/JA075i022p04329

Grec, G., Fossat, E., and Pomerantz, M. (1980) Solar oscillations: full disk observations from the geographic South Pole. *Nature*, **288**, 541–544. doi: 10.1038/288541a0

Green, C.A. (1985) Giant pulsations in the plasmasphere, *Planet. Space Sci.*, **33** (10), 1155–1168. doi: 10.1016/0032-0633(85)90073-X

Green, J.L., Sandel, B.R., Fung, S.F., Gallagher, D.L., and Reinisch, B.W. (2002) On the origin of kilometric radiation. *J. Geophys. Res.*, **107** (A7). doi: 10.1029/2001JA000193

Green, A.W., Worthington, E.W., Baransky, L. N., Fedorov, E.N., Kurneva, N.A., Pilipenko, V.A., Shvetzov, D.N., Bektemirov, A.A., and Philipov, G.V. (1993) Alfvén field line

resonances at low latitudes ($L = 1.5$). *J. Geophys. Res.*, **98** (A9), 15693–15699. doi: 10.1029/93JA00644
Greenstadt, E.W., Le, G., and Strangeway, R.J. (1995) ULF waves in the foreshock. *Adv. Space Res.*, **15** (8-9), 71–84.
Greenstadt, E.W., Mellott, M.M., McPherron, R.L., Singer, H.J., and Knecht, D.J. (1983) Transfer of pulsation-related wave activity across the magnetopause: observations of corresponding spectra by ISEE-1 and ISEE-2. *Geophys. Res. Lett.*, **10** (8), 659–662. doi: 10.1029/GL010i008p00659
Greenstadt, E.W. and Russell, C.T. (1994) Stimulation of exogenic, daytime geomagnetic pulsations: a global perspective, in *Solar Wind Sources of Magnetospheric Ultra-Low Frequency Waves* (eds M. Engebretson, K. Takahashi, and M. Scholer), American Geophysical Union, Washington, DC, pp. 13–23.
Greenwald, R.A. and Walker, A.D.M. (1980) Energetics of long period resonant hydromagnetic waves. *Geophys. Res. Lett.*, **7** (10), 745–748. doi: 10.1029/GL007i010p00745
Grew, R.S., Menk, F.W., Clilverd, M.A., and Sandel, B.R. (2007) Mass and electron densities in the inner magnetosphere during a prolonged disturbed interval. *Geophys. Res. Lett.*, **34** (L02108). doi: 10.1029/2006GL028254
Gul'elmi, A.V. (1966) Plasma concentration at great heights according to data on toroidal fluctuations of the magnetosphere. *Geomagn. Aeron.*, **6**, 98.
Gurnett, D.A. (2001) Solar wind plasma waves, in *Encyclopedia of Astronomy and Astrophysics* (ed. P. Murdin), Institute of Physics and Macmillan Publishing, Bristol, pp. 2805–2813.
Gurnett, D.A. and Kurth, W.S. (2008) Intense plasma waves at and near the solar wind termination shock. *Nature*, **454**, 78–80. doi: 10.1038/nature07023
Gurnett, D.A., Scarf, F.L., Kurth, W.S., Shaw, R.R., and Poynter, R.L. (1981) Determination of Jupiter's electron density profile from plasma wave observations. *J. Geophys. Res.*, **86** (A10), 8199–8212. doi: 10.1029/JA086iA10p08199
Haerendel, G., Baumjohann, W., Georgescu, E., Nakamura, R., Kistler, L.M., Klecker, B., Kucharek, H., Vaivads, A., Mukai, T., and Kokubun, S. (1999) High-beta plasma blobs in the morningside plasma sheet. *Ann. Geophys.*, **17**, 1592–1601. doi: 10.1007/s00585-999-1592-1
Halford, A.J., Fraser, B.J., and Morley, S.K. (2010) EMIC wave activity during geomagnetic storm and nonstorm periods: CRRES results. *J. Geophys. Res.*, **115** (A12248). doi: 10.1029/2010JA015716
Halley, E. (1702) A new and correct sea chart of the whole world shewing the variations of the compass as they were found in the year MDCC. R. Mount and T. Page, London.
Hamlin, D.A., Karplus, R., Vik, R.C., and Watson, K.M. (1961) Mirror and azimuthal drift frequencies for geomagnetically trapped particles. *J. Geophys. Res.*, **66** (1), 1–4. doi: 10.1029/JZ066i001p00001
Hansen, H.J., Fraser, B.J., Menk, F.W., Hu, Y.D., Newell, P.T., and Meng, C.-I. (1991) High latitude unstructured Pc1 emissions generated in the vicinity of the dayside auroral oval. *Planet Space Sci.*, **39** (5), 709–719.
Hansen, H.J., Fraser, B.J., Menk, F.W., Hu, Y. D., Newell, P.T., Meng, C.-I., and Morris, R. J. (1992) High-latitude Pc1 bursts arising in the dayside boundary layer region. *J. Geophys. Res.*, **97** (A4), 3993–4008. doi: 10.1029/91JA01456
Hansen, H.J. and Scourfield, M.W.J. (1990) Associated ground-based observations of optical aurorae and discrete whistler waves. *J. Geophys. Res.*, **95** (A1), 233–239. doi: 10.1029/JA095iA01p00233
Hapgood, M. (2010) Space weather: its impact on Earth and implications for business, in *Lloyd's Risk Insight*, Lloyd's and RAL, London, pp. 34.
Harang, L. (1939) Pulsations in an ionized region at height of 650–800 km during the appearance of giant pulsations in the geomagnetic records. *Terr. Magn. Atmos. Elect.*, **44**, 17–19.
Harris, K.K. (1974) The measurement of cold ion densities in the plasma trough. *J. Geophys. Res.*, **79** (31), 4654–4660. doi: 10.1029/JA079i031p04654
Harrold, B.G. and Samson, J.C. (1992) Standing ULF modes of the magnetosphere: a theory. *Geophys. Res. Lett.*, **19** (18), 1811–1814. doi: 10.1029/92GL01802

Hasegawa, A. and Chen, L. (1974) Theory of magnetic pulsations. *Space Sci. Rev.*, **16**, 347–359. doi: 10.1007/BF00171563

Hasegawa, H., Retinò, A., Vaivads, A., Khotyaintsev, Y., André, M., Nakamura, T.K. M., Teh, W.-L., Sonnerup, B.U.Ö., Schwartz, S.J., Seki, Y., Fujimoto, M., Saito, Y., Rème, H., and Canu, P. (2009) Kelvin–Helmholtz waves at the Earth's magnetopause: multiscale development and associated reconnection. *J. Geophys. Res.*, **114** (A12207). doi: 10.1029/2009JA014042

Hastings, D. and Garrett, H. (1996) *Spacecraft–Environment Interactions*, Cambridge University Press, Cambridge.

Hattingh, S. and Sutcliffe, P.R. (1987) Pc3 pulsation eigenperiod determination at low latitudes. *J. Geophys. Res.*, **92** (A11), 12433–12436. doi: 10.1029/JA092iA11p12433

Heilig, B., Lotz, S., Verö, J., Sutcliffe, P., Reda, J., Pajunpää, K., and Raita, T. (2010) Empirically modelled Pc3 activity based on solar wind parameters. *Ann. Geophys.*, **28**, 1703–1722. doi: 10.5194/angeo-28-1703-2010

Heilig, B., Lühr, H., and Rother, M. (2007) Comprehensive study of ULF upstream waves observed in the topside ionosphere by CHAMP and on the ground. *Ann. Geophys.*, **25**, 737–754. doi: 10.5194/angeo-25-737-2007

Ho, D. and Bernard, L.C. (1973) A fast method to determine the nose frequency and minimum group delay of a whistler when the causative spheric is unknown. *J. Atmos. Terr. Phys.*, **35** (5), 881–887.

Ho, D. and Carpenter, D.L. (1976) Outlying plasmasphere structure detected by whistlers. *Planet Space Sci.*, **24** (10), 987–994. doi: 10.1016/0032-0633(76)90010-6

Horita, R.E., Barfield, J.N., Heacock, R.R., and Kangas, J. (1979) IPDP source regions and resonant proton energies. *J. Atmos. Terr. Phys.*, **41** (3), 293–309.

Howard, T.H. (2011) *Coronal Mass Ejections*, Springer, New York.

Howard, T.A. and Menk, F.W. (2001) Propagation of 10–50 mHz ULF waves with high spatial coherence at high latitudes. *Geophys. Res. Lett.*, **28** (2), 231–234. doi: 10.1029/2000GL011993

Howard, T.A. and Menk, F.W. (2005) Ground observations of high-latitude Pc3-4 ULF waves. *J. Geophys. Res.*, **110** (A04205). doi: 10.1029/2004JA010417

Hughes, W.J. (1974) The effect of the atmosphere and ionosphere on long period magnetospheric micropulsations. *Planet Space Sci.*, **22** (8), 1157–1172. doi: 10.1016/0032-0633(74)90001-4

Hughes, W.J., McPherron, R.L., and Barfield, J.N. (1978a) Geomagnetic pulsations observed simultaneously on three geostationary satellites. *J. Geophys. Res.*, **83** (A3), 1109–1116. doi: 10.1029/JA083iA03p01109

Hughes, W.J. and Southwood, D.J. (1976) The screening of micropulsation signals by the atmosphere and ionosphere. *J. Geophys. Res.*, **81** (19), 3234–3240. doi: 10.1029/JA081i019p03234

Hughes, W.J., Southwood, D.J., Mauk, B., McPherron, R.L., and Barfield, J.N. (1978b) Alfvén waves generated by an inverted plasma energy distribution. *Nature*, **275**, 43–45. doi: 10.1038/275043a0

Hundhausen, A.J. (1995) The solar wind, in *Introduction to Space Physics* (eds M.G. Kivelson and C.T., Russell), Cambridge University Press, Cambridge, pp. 91–128.

Iijima, T. and Potemra, T.A. (1976) The amplitude distribution of field-aligned currents at northern high latitudes observed by Triad. *J. Geophys. Res.*, **81** (13), 2165–2174. doi: 10.1029/JA081i013p02165

Jackson, A., Jonkers, A.R.T., and Walker, M.R. (2000) Four centuries of geomagnetic secular variation from historical records. *Philos. Trans. R. Soc. Lond. A*, **358** (1768), 957–990. doi: 10.1098/rsta.2000.05969

Jacobs, J.A. (1987). *Geomagnetism vols. 1, 2 and 3*, Academic Press, London.

Jacobs, J.A. (1991). *Geomagnetism vol. 4*, Academic Press, London.

Jacobs, J.A., Kato, Y., Matsushita, S., and Troitskaya, V.A. (1964) Classification of geomagnetic micropulsations. *J. Geophys. Res.*, **69** (1), 180–181. doi: 10.1029/JZ069i001p00180

Jacobs, J.A. and Sinno, K. (1960) World-wide characteristics of geomagnetic micropulsations. *Geophys. J. R. Astron. Soc.*, **3**, 333–353. doi: 10.1111/j.1365-246X.1960.tb01707.x

Janardhan, P., Fujiki, K., Sawant, H.S., Kojima, M., Hakamada, K., and Krishnan, R. (2008) Source regions of solar wind disappearance events. *J. Geophys. Res.*, **113** (A03102). doi: 10.1029/2007JA012608

Jess, D.B., Mathioudakis, M., Erdélyi, R., Crockett, P.J., Keenan, F.P., and Christian, D.J. (2009) Alfvén waves in the lower solar atmosphere. *Science*, **323** (5921), 1582–1585. doi: 10.1126/science.1168680

Jian, L.K., Russell, C.T., Luhmann, J.G., Anderson, B.J., Boardsen, S.A., Strangeway, R.J., Cowee, M.M., and Wennmacher, A. (2010) Observations of ion cyclotron waves in the solar wind near 0.3 AU. *J. Geophys. Res.*, **115** (A12115). doi: 10.1029/2010JA015737

Jonkers, A.R.T., Jackson, A., and Murray, A. (2003) Four centuries of geomagnetic data from historical records. *Rev. Geophys.*, **41** (2), 957–990. doi: 10.1029/2002RG000115

Kabin, K., Rankin, R., Mann, I.R., Degeling, A.W., and Marchand, R. (2007a) Polarization properties of standing shear Alfvén waves in non-axisymmetric background magnetic fields. *Ann. Geophys.*, **25**, 815–822. doi: 10.5194/angeo-25-815-2007

Kabin, K., Rankin, R., Waters, C.L., Marchand, R., Donovan, E.F., and Samson, J.C. (2007b) Different eigenproblem models for field line resonances in cold plasma: effect on magnetospheric density estimates. *Planet Space Sci.*, **55** (6), 820–828. doi: 10.1016/j.pss.2006.03.014

Kale, Z.C., Mann, I.R., Waters, C.L., Goldstein, J., Menk, F.W., and Ozeke, L.G. (2007) Ground magnetometer observation of a cross-phase reversal at a steep plasmapause. *J. Geophys. Res.*, **112** (A10222). doi: 10.1029/2007JA012367

Kale, Z.C., Mann, I.R., Waters, C.L., Vellante, M., Zhang, T.L., and Honary, F. (2009) Plasmaspheric dynamics resulting from the Hallowe'en 2003 geomagnetic storms. *J. Geophys. Res.*, **114** (A08204). doi: 10.1029/2009JA014194

Kato, Y. and Tamao, T. (1956) Hydromagnetic oscillations in a conducting medium with Hall conductivity under the uniform magnetic field. *Sci. Rep. Tohoku Imper. Univ., Ser. 5, Geophys.* **7** (3), 147.

Kato, Y. and Watanabe, T. (1957) A survey of observational knowledge of the geomagnetic pulsation. *Sci. Rep. Tohoku Imper. Univ., Ser. 5, Geophys.* **8**, 157–185.

Kawano, H., Yumoto, K., Pilipenko, V.A., Tanaka, Y.-M., Takasaki, S., Iizima, M., and Seto, M. (2002) Using two ground stations to identify magnetospheric field line eigenfrequency as a continuous function of ground latitude. *J. Geophys. Res.*, **107** (A8). doi: 10.1029/2001JA000274

Kelley, M.C. (2009) *The Earth's Ionosphere: Plasma Physics & Electrodynamics*, Elsevier, Amsterdam.

Kepko, L. and Spence, H.E. (2003) Observations of discrete, global magnetospheric oscillations directly driven by solar wind density variations. *J. Geophys. Res.*, **108** (A6). doi: 10.1029/2002JA009676

Kessel, R.L. (2008) Solar wind excitation of Pc5 fluctuations in the magnetosphere and on the ground. *J. Geophys. Res.*, **113** (A04202). doi: 10.1029/2007JA012255

Kessel, R.L., Mann, I.R., Fung, S.F., Milling, D.K., and O'Connell, N. (2004) Correlation of Pc5 wave power inside and outside the magnetosphere during high speed streams. *Ann. Geophys.*, **21**, 629–641. doi: 10.5194/angeo-22-629-2004

Khurana, K.K. and Kivelson, M.G. (1989) Ultralow frequency MHD waves in Jupiter's middle magnetosphere. *J. Geophys. Res.*, **94** (A5), 5241–5254. doi: 10.1029/JA094iA05p05241

Kikuchi, T. and Araki, T. (2002) Comment on "Propagation of the preliminary reverse impulse of sudden commencements to low latitudes" by P.J. Chi *et al. J. Geophys. Res.*, **107** (A12). doi: 10.1029/2001JA009220

Kim, K.-H., Kim, K.C., Lee, D.-H., and Rostoker, G. (2006) Origin of geosynchronous relativistic electron events. *J. Geophys. Res.*, **111** (A03208). doi: 10.1029/2005JA011469

Kim, K.-H., Lee, D.-H., Takahashi, K., Russell, C.T., Moon, Y.-J., and Yumoto, K. (2005) Pi2 pulsations observed from the Polar satellite outside the plasmapause. *Geophys. Res. Lett.*, **32** (L18102). doi: 10.1029/2005GL023872

Kivelson, M.G. and Russell, C.T. (1995). *Introduction to space physics*, Cambridge University Press, New York.

Kivelson, M.G. and Southwood, D.J. (1985) Resonant ULF waves: a new interpretation.

Geophys. Res. Lett., **12** (1), 49–52. doi: 10.1029/GL012i001p00049

Kivelson, M.G. and Southwood, D.J. (1986) Coupling of global magnetospheric MHD eigenmodes to field line resonances. *J. Geophys. Res.*, **91** (A4), 4345–4351. doi: 10.1029/JA091iA04p04345

Kleindienst, G., Glassmeier, K.-H., Simon, S., Dougherty, M.K., and Krupp, N. (2009) Quasiperiodic ULF pulsations in Saturn's magnetosphere. *Ann. Geophys.*, **27** (2), 885–894. doi: 10.5194/angeo-27-885-2009

Klimushkin, D.Y. (1998) Resonators for hydromagnetic waves in the magnetosphere. *J. Geophys. Res.*, **103** (A2), 2369–2375. doi: 10.1029/97JA02193

Klimushkin, D.Y., Mager, P.N., and Glassmeier, K.-H. (2004) Toroidal and poloidal Alfvén waves with arbitrary azimuthal wavenumbers in a finite pressure plasma in the Earth's magnetosphere. *Ann. Geophys.*, **22**, 267–287. doi: 10.5194/angeo-22-267-2004

Korotova, G.I., Sibeck, D.G., Kondratovich, V., Angelopolous, V., and Constantinescu, O.D. (2009) THEMIS observations of compressional pulsations in the dawn-side magnetosphere: a case study. *Ann. Geophys.*, **27**, 3725–3735. doi: 10.5194/angeo-27-3725-2009

Korotova, G.I., Sibeck, D.G., Weatherwax, A., Angelopolous, V., and Styazhkin, V. (2011) THEMIS observations of a transient event at the magnetopause. *J. Geophys. Res.*, **116** A07224. doi: 10.1029/2011JA016606

Koskinen, H.E.J. and Tanskanen, E. (2002) Magnetospheric energy budget and the epsilon parameter. *J. Geophys. Res.*, **107** (A11). doi: 10.1029/2002JA009283

Koskinen, H., Tanskanen, E., Pirjola, R., Pulkkinen, A., Dyer, C., Rodgers, D., Cannon, P., Mandeville, J.-C., Boscher, D., and Hilgers, A. (2001) Space weather effects catalogue. ESA Space Weather Study, ESWS-FMI-RP-0001.

Kozlov, D.A. and Leonovich, A.S. (2008) The structure of field line resonances in a dipole magnetosphere with moving plasma. *Ann. Geophys.*, **26**, 689–698. doi: 10.5194/angeo-26-689-2008

Krauss-Varban, D. (1994) Bow shock and magnetosheath simulations: wave transport and kinetic properties, in *Solar Wind Sources of Magnetospheric Ultra-Low Frequency Waves* (eds M. Engebretson, K. Takahashi, and M. Scholer), American Geophysical Union, Washington, DC, pp. 121–134.

Kurchashov, Y.P., Nikomarov, Y.S., Pilipenko, V.A., and Best, A. (1987) Field line resonance effects in local meridional structure of mid-latitude geomagnetic pulsations. *Ann. Geophys.*, **5**, 147–154.

Kurth, W.S., Strayer, B.D., Gurnett, D.A., and Scarf, F.L. (1985) A summary of whistlers observed by Voyager 1 at Jupiter. *Icarus*, **61** (3), 497–507. doi: 10.101/0019-1035(85)90138-1

Langmuir, I. (1928) Oscillations in ionized gases. *Proc. Natl. Acad. Sci. USA*, **14** (8), 627–637. doi. 10.1073/pnas.14.8.627

Lanzerotti, L.J., Fukunishi, H., and Chen, L. (1974) ULF pulsation evidence of the plasmapause 3. Interpretation of polarization and spectral amplitude studies of Pc3 and Pc4 pulsations near $L = 4$. *J. Geophys. Res.*, **79** (31), 4648–4653. doi: 10.1029/JA079i031p04648

Le, G. and Russell, C.T. (1992) A study of ULF wave foreshock morphology – I: ULF foreshock boundary. *Planet Space Sci.*, **40** (9), 1203–1213. doi: 10.1016/0032-0633(92)90077-2

Le, G. and Russell, C.T. (1996) Solar wind control of upstream wave frequency. *J. Geophys. Res.*, **101** (A2), 2571–2575. doi: 10.1029/95JA03151

Le, G., Russell, C.T., and Petrinec, S.M. (2000) The magnetosphere on May 11, 1999, the day the solar wind almost disappeared: I. Current systems. *Geophys. Res. Lett.*, **27** (13), 1827–1830. doi: 10.1029/1999GL010774

Lee, L.C., Albano, R.K., and Kan, J.R. (1981) Kelvin–Helmholtz instability at the magnetopause-boundary layer region. *J. Geophys. Res.*, **86** (A1), 54–58. doi: 10.1029/JA086iA01p00054

Lee, D.-H. and Lysak, R.L. (1991) Monochromatic ULF wave excitation in the dipole magnetosphere. *J. Geophys. Res.*, **96** (A4), 5811–5817. doi: 10.1029/90JA01592

Lee, D.-H. and Lysak, R.L. (1999) MHD waves in a three-dimensional dipolar magnetic field: a search for Pi2 pulsations. *J. Geophys. Res.*, **104** (A12), 28691–28699. doi: 10.1029/1999JA9000377

Lee, L.C. and Olson, J.V. (1980) Kelvin–Helmholtz instability and the variation of geomagnetic pulsation activity. *Geophys. Res. Lett.*, **7** (10), 777–780. doi: 10.1029/GL007i010p00777

Lee, D.-H. and Takahashi, K. (2006) MHD eigenmodes in the inner magnetosphere, in *Magnetospheric ULF Waves: Synthesis and New Directions* (eds K. Takahashi, P.J. Chi, R.E. Denton, and R.L. Lysak), American Geophysical Union, Washington, DC, pp. 73–90.

Leighton, R.B., Noyes, R.W., and Simon, G.W. (1962) Velocity fields in the solar atmosphere: I. Preliminary report. *Astrophys. J.*, **135**, 474–499.

Lemaire, J.F. and Gringauz, K.I. (1998) *The Earth's Plasmasphere*, Cambridge University Press, Cambridge.

Li, X., Baker, D.N., Kanekal, S.G., Looper, M., and Temerin, M. (2001) Long term measurements of radiation belts by SAMPEX and their variations. *Geophys. Res. Lett.*, **28** (20), 3827–3830.

Li, H., Wang, C., and Richardson, J.D. (2008) Properties of the termination shock observed by Voyager 2. *Geophys. Res. Lett.*, **35** (L19107). doi: 10.1029/2008GL034869

Lichtenberger, J., Ferencz, C., Bodnar, L., Hamar, D., and Steinbach, P. (2008) Automatic whistler detector and analyzer system: automatic whistler detector. *J. Geophys. Res.*, **113** (A12201). doi: 10.1029/2008JA013467

Lichtenberger, J., Ferencz, C., Hamar, D., Steinbach, P., Rodger, C.J., Clilverd, M.A., and Collier, A.B. (2010) Automatic whistler detector and analyzer system: implementation of the analyzer algorithm. *J. Geophys. Res.*, **115** (A12214). doi: 10.1029/2010JA015931

Liou, K., Takahashi, K., Newell, P.T., and Yumoto, K. (2008) Polar ultraviolet imager observations of solar wind-driven ULF auroral pulsations. *Geophys. Res. Lett.*, **35** (L16101). doi: 10.1029/2008GL034953

Liu, Y.H., Fraser, B.J., Ables, S.T., Dunlop, M. W., Zhang, B.C., Liu, R.Y., and Zong, Q.G. (2008) Phase structure of Pc3 waves observed by Cluster and ground stations near the cusp. *J. Geophys. Res.*, **113** (A07S37). doi: 10.1029/2007JA012754

Liu, Y.H., Fraser, B.J., Ables, S.T., Zhang, B.C., Liu, R.Y., Dunlop, M.W., and Waterman, J. (2009a) Transverse-scale size of Pc3 ULF waves near the exterior cusp. *J. Geophys. Res.*, **114** (A08208). doi: 10.1029/2008JA013971

Liu, W., Sarris, T.E., Li, X., Elkington, S.R., Ergun, R., Angelopolous, V., Bonnell, J., and Glassmeier, K.-H. (2009b) Electric and magnetic field observations of Pc4 and Pc5 pulsations in the inner magnetosphere: a statistical study. *J. Geophys. Res.*, **114** (A12206). doi: 10.1029/2009JA014243

Lockwood, M., Stamper, R., and Wild, M.N. (1999) A doubling of the Sun's coronal magnetic field during the past 100 years. *Nature*, **399**, 437–439. doi: 10.1038/20867

Loto'aniu, T.M., Fraser, B.J., and Waters, C.L. (2005) Propagation of electromagnetic ion cyclotron wave energy in the magnetosphere. *J. Geophys. Res.*, **110** (A07214). doi: 10.1029/2004JA010816

Loto'aniu, T.M., Fraser, B.J., and Waters, C.L. (2009) The modulation of electromagnetic ion cyclotron waves by Pc5 ULF waves. *Ann. Geophys.*, **27**, 121–130. doi: 10.5194/angeo-27-121-2009

Loto'aniu, T.M., Mann, I.R., Ozeke, L.G., Chan, A.A., Dent, Z.C., and Milling, D.K. (2006) Radial diffusion of relativistic electrons into the radiation belt slot region during the 2003 Halloween geomagnetic storms. *J. Geophys. Res.*, **111** (A04218). doi: 10.1029/2005JA011355

Loto'aniu, T.M., Waters, C.L., Fraser, B.J., and Samson, J.C. (1999) Plasma mass density in the plasmatrough: comparison using ULF waves and CRRES. *Geophys. Res. Lett.*, **26** (21), 3277–3280. doi: 10.1029/1999GL003641

Luhmann, J.G. (1995) Ionospheres, in *Introduction to Space Physics* (eds M.G. Kivelson and C.T. Russell), Cambridge University Press, Cambridge, pp. 183–202.

Lysak, R.L. (1997) Propagation of Alfvén waves through the ionosphere. *Phys. Chem. Earth*, **22** (7–8), 757–766. doi: 10.1016/S0079-1946(97)00208-5

Lysak, R.L. (2004) Magnetosphere–ionosphere coupling by Alfvén waves at midlatitudes. *J. Geophys. Res.*, **109** (A07201). doi: 10.1029/2004JA010454

Maeda, N., Takasaki, S., Kawano, H., Ohtani, S., Décréau, P.M.E., Trotignon, J.G., Solovyev, S.I., Baishev, D.G., and Yumoto, K. (2009) Simultaneous observations of the plasma density on the same field line by the CPMN ground magnetometers and the Cluster satellites. *Adv. Space Res.*, **43**, 265–272. doi: 10.1016/j.asr.2008.04.016

Mann, I.R., O'Brien, T.P., and Milling, D.K. (2004) Correlations between ULF wave power, solar wind speed, and relativistic electron flux in the magnetosphere: solar cycle dependence. *J. Atmos. Solar-Terr. Phys.*, **66** (2), 187–198. doi: 10.1016/j.jastp.2003.10.002

Mann, I.R., Wright, A.N., Mills, K.J., and Nakariakov, V.M. (1999) Excitation of magnetospheric waveguide modes by magnetosheath flows. *J. Geophys. Res.*, **104** (A1), 333–353. doi: 10.1029/1998JA9000026

Mannucci, A.J., Tsurutani, B.T., Solomon, S. C., Verkhoglyadova, O.P., and Thayer, J.P. (2012) How do coronal hole storms affect the upper atmosphere? *EOS*, **93** (8), 77. doi: 10.1029/2012EO08000

Marshall, R.A. and Menk, F.W. (1999) Observations of Pc3–4 and Pi2 geomagnetic pulsations in the low-latitude ionosphere. *Ann. Geophys.*, **17**, 1397–1410. doi: 10.1007/s00585-999-1397-2

Marshall, R.A., Waters, C.L., and Sciffer, M.D. (2010) Spectral analysis of pipe-to-soil potentials with variations of the Earth's magnetic field in the Australian region. *Space Weather*, **8** (S05002). doi: 10.1029/SW000553

Marshall, R.A., Dalzell, M., Waters, C.L., Goldthorpe, P., and Smith, E.A. (2012) Geomagnetically induced currents in the New Zealand power network. *Space Weather*, **10** (S08003). doi: 10.1029/2012SW000806

Marshall, R.A., Smith, E.A., Francis, M.J., Waters, C.L., and Sciffer, M.D. (2011) A preliminary risk assessment of the Australian region power network to space weather. *Space Weather*, **9** (S100004). doi: 10.1029/2011SW000685

Mathie, R.A. and Mann, I.R. (2000a) A correlation between extended intervals of ULF wave power and storm-time geosynchronous relativistic electron flux enhancements. *Geophys. Res. Lett.*, **27** (20), 3261–3264. doi: 10.1029/2000GL003822

Mathie, R.A. and Mann, I.R. (2000b) Observations of Pc5 field line resonance azimuthal phase speeds: a diagnostic of their excitation mechanism. *J. Geophys. Res.*, **105** (A5), 10713–10728. doi: 10.1029/1999JA000174

Mathie, R.A. and Mann, I.R. (2001) On the solar wind control of Pc5 ULF power at mid-latitudes: implications for MeV electron precipitation in the outer radiation belt. *J. Geophys. Res.*, **106** (A12), 29783–29796. doi: 10.1029/2001JA000002

Mathie, R.A., Mann, I.R., Menk, F.W., and Orr, D. (1999a) Pc5 ULF pulsations associated with waveguide modes observed with the IMAGE magnetometer array. *J. Geophys. Res.*, **104** (A4), 7025–7036. doi: 10.1029/1998JA900150

Mathie, R.A., Menk, F.W., Mann, I.R., and Orr, D. (1999b) Discrete field line resonances and the Alfvén continuum in the outer magnetosphere. *Geophys. Res. Lett.*, **26** (6), 659–662. doi: 10.1029/1999GL900104

Mathur, S., Handburg, R., Campante, T.L., Garcia, R.A. *et al.* (2011) Solar-like oscillations in KIC 11395018 and KIC 11234888 from 8 months of Kepler data. *Astrophys. J.*, **733** (2). doi: 10.1088/0004-637X/733/2/95

Mathur, S., Metcalfe, T.S., Woitaszek, M., Bruntt, H. *et al.* (2012) A uniform asteroseismic analysis of 22 solar-type stars observed by Kepler. *Astrophys. J.*, **749** (2). doi: 10.1088/0004-637X/749/2/152

Maus, S., Macmillan, S., McLean, S., Hamilton, B., Thomson, A., Nair, M., and Rollins, C. (2010) The US/UK World Magnetic Model for 2010–2015. NOAA Technical Report NESDIS/NGDC.

McGregor, S.L., Hughes, W.J., Arge, C.N., Owens, M.J., and Odstrcil, D. (2011) The distribution of solar wind speeds during solar minimum: calibration for numerical solar wind modeling constraints on the source of the slow solar wind. *J. Geophys. Res.*, **116** (A03101). doi: 10.1029/2010JA015881

Mead, G.D. and Beard, D.B. (1964) Shape of the geomagnetic field-solar wind boundary.

J. Geophys. Res., **69** (7), 1169–1179. doi: 10.1029/JZ069i007p01169
Menk, F.W. (1988) Spectral structure of low latitude Pc2–4 geomagnetic pulsations. *J. Geomag. Geolectr.*, **40**, 33–61.
Menk, F.W. (1992) Characterization of ionospheric Doppler oscillations in the Pc3–4 and Pi2 magnetic pulsation frequency range. *Planet Space Sci.*, **40** (4), 495–507. doi: 10.1016/0032-0633(92)90169-O
Menk, F.W. (2007) Geomagnetic field, in *Encyclopaedia of the Antarctic* (ed. B. Riffenburgh), Routledge, New York, pp. 437–441.
Menk, F.W. (2011) Magnetospheric ULF waves: a review, in *The Dynamic Magnetosphere* (eds W. Liu and M. Fujimoto), Springer, Berlin, pp. 223–256.
Menk, F.W., Ables, S.T., Grew, R.S., Clilverd, M.A., and Sandel, B.R. (2012) The annual and longitudinal variations in plasmaspheric ion density. *J. Geophys. Res.*, **117** (A03215). doi: 10.1029/2011JA017071
Menk, F.W., Clilverd, M.A., Yearby, K.H., Milinevski, G., Thomson, N.R., and Rose, M.C. (2006) ULF Doppler oscillations of $L = 2.5$ field lines. *J. Geophys. Res.*, **111** (A07205). doi: 10.1029/2005JA011192
Menk, F.W., Cole, K.D., and Devlin, J.C. (1983) Associated geomagnetic and ionospheric variations. *Planet Space Sci.*, **31** (5), 569–572. doi: 10.1016/0032-0633(83)90045-4
Menk, F.W., Fraser, B.J., Hansen, H.J., Newell, P.T., Meng, C.-I., and Morris, R.J. (1992) Identification of the magnetospheric cusp and cleft using Pc1–2 ULF pulsations. *J. Atmos. Terr. Phys.*, **54** (7-8), 1021–1043.
Menk, F.W., Fraser, B.J., Hansen, H.J., Newell, P.T., Meng, C.-I., and Morris, R.J. (1993) Multistation observations of Pc1–2 pulsations in the vicinity of the polar cusp. *J. Geomag. Geolectr.*, **45**, 1159–1173.
Menk, F.W., Fraser, B.J., Waters, C.L., Ziesolleck, C.W.S., Feng, Q., Lee, S.-H., and McNabb, P.W. (1994) Ground measurements of low latitude magnetospheric field line resonances, in *Solar Wind Sources of Magnetospheric Ultra-Low Frequency Waves* (eds M. Engebretson, K. Takahashi, and M. Scholer), American Geophysical Union, Washington, DC, pp. 299–310.
Menk, F.W., Mann, I.R., Smith, A.J., Waters, C.L., Clilverd, M.A., and Milling, D.K. (2004) Monitoring the plasmapause using geomagnetic field line resonances. *J. Geophys. Res.*, **109** (A04216). doi: 10.1029/2003JA010097
Menk, F.W., Marshall, R.A., McNabb, P.W., and Dunlop, I.S. (1995) An experiment to study the effects of geomagnetic fluctuations on ionospheric HF radio paths. *Aust. J. Elec. Electron. Eng.*, **15**, 325–332.
Menk, F.W., Orr, D., Clilverd, M.A., Smith, A. J., Waters, C.L., Milling, D.K., and Fraser, B. J. (1999) Monitoring spatial and temporal variations in the dayside plasmasphere using geomagnetic field line resonances. *J. Geophys. Res.*, **104** (A9), 19955–19969. doi: 10.1029/1999JA900205
Menk, F.W., Waters, C.L., and Dunlop, I.S. (2007) ULF Doppler oscillations in the low latitude ionosphere. *Geophys. Res. Lett.*, **34** (L10104). doi: 10.1029/2007GL029300
Menk, F.W., Waters, C.L., and Fraser, B.J. (2000) Field line resonances and waveguide modes at low latitudes: 1. Observations. *J. Geophys. Res.*, **105** (A4), 7747–7761. doi: 10.1029/1999JA900268
Menk, F.W., Yeoman, T.K., Wright, D.M., Lester, M., and Honary, F. (2003) High-latitude observations of impulse-driven ULF pulsations in the ionosphere and on the ground. *Ann. Geophys.*, **21**, 559–576. doi: 10.5194/angeo-21-559-2003
Meredith, N.P., Horne, R.B., Lam, M.M., Denton, M.H., Borovsky, J.E., and Green, J.C. (2011) Energetic electron precipitation during high-speed solar wind stream driven storms. *J. Geophys. Res.*, **116** (A05223). doi: 10.1029/2010JA016293
Merrill, R.T., McElhinny, M.W., and McFadden, P.L. (1998) *The Magnetic Field of the Earth*, vol. **63**, Academic Press, San Diego, CA.
Meziane, K., Hamza, A.M., Wilber, M., Mazelle, C., and Lee, M.A. (2011) Anomalous foreshock field-aligned beams observed by Cluster. *Ann. Geophys.*, **29**, 1967–1975. doi: 10.5194/angeo-29-1967-2011
Mier-Jedrzejowicz, W.A.C. and Southwood, D. J. (1979) The East–West structure of mid-latitude geomagnetic pulsations in the

8–25 mHz band. *Planet Space Sci.*, **27** (5), 617–630.

Milan, S.E., Provan, G., and Hubert, B. (2007) Magnetic flux transport in the Dungey cycle: a survey of dayside and nightside reconnection events. *J. Geophys. Res.*, **12** (A01209). doi: 10.1029/2006JA011642

Miletits, J.C., Verö, J., Szendröi, J., Ivanova, P., Best, A., and Kivinen, M. (1990) Pulsation periods at mid-latitudes: a seven-station study. *Planet Space Sci.*, **38**, 85–95.

Milling, D.K., Mann, I.R., and Menk, F.W. (2001) Diagnosing the plasmapause with a network of closely spaced ground-based magnetometers. *Geophys. Res. Lett.*, **28** (1), 115–118. doi: 10.1029/2000GL011935

Milone, E.F. and Wilson, W.J.F. (2008) *Solar System Astrophysics: Background Science and the Inner Solar System*, Springer, New York.

Morley, S.K., Ables, S.T., Sciffer, M.D., and Fraser, B.J. (2009) Multipoint observations of Pc1–2 waves in the afternoon sector. *J. Geophys. Res.*, **114** (A09205). doi: 10.1029/2009JA014162

Nagata, T., Hirasawa, T., Fukunishi, H., Ayukawa, M., Sato, M., and Fujii, R. (1980) Classification of Pc1 and Pi1 waves observed in high latitudes, in *IMS in Antarctica* (ed. T. Hirasawa), National Institute of Polar Research, Tokyo, pp. 56–71.

Nagata, T., Kokubun, S., and Iijima, T. (1963) Geomagnetically conjugate relationships of giant pulsations at Syowa Base, Antarctica, and Reykjavik, Iceland. *J. Geophys. Res.*, **68** (15), 4621–4625. doi: 10.1029/JZ068i015p04621

Nakariakov, V.M. and Ofman, L. (2001) Determination of the coronal magnetic field by coronal loop oscillations. *Astron. Astrophys. Lett.*, **372** (3), L53–L56. doi: 10.1051/0004-6361:20010607

Nakariakov, V.M., Ofman, L., De Luca, E.E., Roberts, B., and Davila, J.M. (1999) TRACE observation of damped coronal loop oscillations: implications for coronal heating. *Science*, **285** (5429), 862–864. doi: 10.1126/science.285.5429.862

Narita, Y., Glassmeier, K.-H., Fornaçon, K.-H., Richter, I., Schäfer, S., Motschmann, U., Dandouras, I., Rème, H., and Georgescu, E. (2006) Low-frequency wave characteristics in the upstream and downstream region of the terrestrial bow shock. *J. Geophys. Res.*, **111** (A01203). doi: 10.1029/2005JA011231

Narita, Y., Glassmeier, K.-H., Schäfer, S., Motschmann, U., Franz, M., Dandouras, I., Fornaçon, K.-H., Georgescu, E., Korth, A., Rème, H., and Richter, I. (2004) Alfvén waves in the foreshock propagating upstream in the plasma rest frame: statistics from Cluster observations. *Ann. Geophys.*, **22**, 2315–2323. doi: 10.5194/angeo-22-2315-2004

Ndiitwani, D.C. and Sutcliffe, P.R. (2009) The structure of low-latitude Pc3 pulsations observed by CHAMP and on the ground. *Ann. Geophys.*, **27**, 1267–1277. doi: 10.5194/angeo-27-1267-2009

Ndiitwani, D.C. and Sutcliffe, P.R. (2010) A study of L-dependent Pc3 pulsations observed by low Earth orbiting CHAMP satellite. *Ann. Geophys.*, **28**, 407–414. doi: 10.5194/angeo-28-407-2010

Neudegg, D.A., Fraser, B.J., Menk, F.W., Hansen, H.J., Burns, G.B., Morris, R.J., and Underwood, M.J. (1995) Sources and velocities of Pc1-2 ULF waves at high latitudes. *Geophys. Res. Lett.*, **22** (21), 2965–2968. doi: 10.1029/95GL02939

Newell, P.T. and Meng, C.-I. (1988) The cusp and cleft/boundary layer: low altitude identification and statistical local time variation. *J. Geophys. Res.*, **93** (A12), 14549–14556. doi: 10.1029/JA093iA12p14549

Newell, P.T., Sotirelis, T., Liou, K., Meng, C.-I., and Rich, F.J. (2007) A nearly universal solar wind–magnetosphere coupling function inferred from 10 magnetospheric state variables. *J. Geophys. Res.*, **112** (A01206). doi: 10.1029/2006JA012015

Newnham, D.A., Espy, P.J., Clilverd, M.A., Rodger, C.J., Seppälä, A., Maxfield, D.J., Hartogh, P., Holmén, K., and Horne, R.B. (2011) Direct observations of nitric oxide produced by energetic electron precipitation into the Antarctic middle atmosphere. *Geophys. Res. Lett.*, **38** (L20104). doi: 10.1029/2011GL048666

Nishida, A. (1964) Ionospheric screening effect and storm sudden commencement. *J. Geophys. Res.*, **69** (9), 1861–1874. doi: 10.1029/JZ069i009p01861

Nishida, A. (1966) Formation of plasmapause, or magnetospheric plasma knee, by the combined action of magnetospheric

convection and plasma escape from the tail. *J. Geophys. Res.*, **71**, (23) 5669–5679. doi: 10.1029/JZ071i023p05669
Nosé, M., Takahashi, K., Anderson, R.R., and Singer, H.J. (2011) Oxygen torus in the deep inner magnetosphere and its contribution to recurrent process of O^+-rich ring current formation. *J. Geophys. Res.*, **116** (A10224). doi: 10.1029/2011JA016651
Obana, Y., Menk, F.W., Sciffer, M.D., and Waters, C.L. (2008) Quarter-wave modes of standing Alfvén waves detected by cross-phase analysis. *J. Geophys. Res.*, **113** (A08203). doi: 10.1029/2007JA012917
Obana, Y., Menk, F.W., and Yoshikawa, I. (2010) Plasma refilling rates for $L = 2.3$–3.8 flux tubes. *J. Geophys. Res.*, **115** (A03204). doi: 10.1029/2009JA014191
Obayashi, T. (1965) Hydromagnetic whistlers. *J. Geophys. Res.*, **70** (5), 1069–1078. doi: 10.1029/JZ070i005p01069
Obayashi, T. and Jacobs, J.A. (1958) Geomagnetic pulsations and the Earth's outer atmosphere. *Geophys. J. R. Astron. Soc.*, **1**, 53–63.
O'Brien, T.P., Looper, M.D., and Blake, J.B. (2004) Quantification of relativistic electron microburst losses during the GEM storms. *Geophys. Res. Lett.*, **31** (L04802). doi: 10.1029/2003GL018621
O'Brien, T.P. and Moldwin, M.B. (2003) Empirical plasmapause models from magnetic indices. *Geophys. Res. Lett.*, **30** (4), 1152. doi: 10.1029/2002GL016007
Odera, T.J. (1986) Solar wind controlled pulsations: a review. *Rev. Geophys.*, **24** (1), 55–74. doi: 10.1029/RG024i001p00055
Olson, W.P. (1969) The shape of the tilted magnetopause. *J. Geophys. Res.*, **74** (24), 5642–5651. doi: 10.1029/JA074i024p05642
Olson, W.P. (1970) Variations in the Earth's surface magnetic field from the magnetopause current system. *Planet Space Sci.*, **18** (10), 1471–1484. doi: 10.1016/0032-0633(70)90119-4
Olson, W.P. (1982) The geomagnetic field and its extension into space. *Adv. Space Res.*, **2** (1), 13–17.
Olson, J.V. (1987) Diurnal characteristics of field line resonances at College, Alaska. *J. Geophys. Res.*, **92** (A8), 8805–8811. doi: 10.1029/JA092iA08p08805
Olson, J.V. and Rostoker, G. (1978) Longitudinal phase variations of Pc4-5 micropulsations. *J. Geophys. Res.*, **83** (A6), 2481–2488. doi: 10.1029/JA083iA06p02481
Olson, J.V. and Samson, J.C. (1979) On the detection of the polarization states of Pc micropulsations. *Geophys. Res. Lett.*, **6** (5), 413–416. doi: 10.1029/GL006i005p00413
Orr, D. (1973) Magnetic pulsations within the magnetosphere: a review. *J. Atmos. Terr. Phys.*, **35**, 1–50.
Orr, D. (1984) Magnetospheric hydromagnetic waves: their eigenperiods, amplitudes and phase variations – a tutorial introduction. *J. Geophys.*, **55**, 76–84.
Orr, D. and Hanson, H.W. (1981) Geomagnetic pulsation phase patterns over an extended latitudinal array. *J. Atmos. Terr. Phys.*, **43** (9), 899–910.
Orr, D. and Matthew, J.A.D. (1971) The variation of geomagnetic micropulsation periods with latitude and the plasmapause. *Planet Space Sci.*, **19** (8), 897–905.
Orr, D. and Webb, D.C. (1975) Statistical studies of geomagnetic pulsations with periods between 10 and 70 sec and their relationship to the plasmapause region. *Planet Space Sci.*, **23**, 1169–1178.
Ostwald, P.M., Menk, F.W., Fraser, B.J., and McNabb, P.W. (1993) Spatial and temporal characteristics of 10–100 mHz ULF waves recorded across a low latitude azimuthal array. *Ann. Geophys.*, **11**, 742–752.
Ozeke, L.G. and Mann, I.R. (2001) Modeling the properties of high-*m* Alfvén waves driven by the drift-bounce resonance mechanism. *J. Geophys. Res.*, **106** (A8), 15583–15597. doi: 10.1029/2000JA000393
Ozeke, L.G. and Mann, I.R. (2008) Energization of radiation belt electrons by ring current ion driven ULF waves. *J. Geophys. Res.*, **113** (A02201). doi: 10.1029/2007JA012468
Ozeke, L.G., Mann, I.R., and Rae, I.J. (2009) Mapping guided Alfvén wave magnetic field amplitudes observed on the ground to equatorial electric field amplitudes in space. *J. Geophys. Res.*, **114** (A01214). doi: 10.1029/2008JA013041
Pahud, D.M., Rae, I.J., Mann, I.R., Murphy, K. R., and Amalraj, V. (2009) Ground-based Pc5 ULF wave power: solar wind speed and MLT dependence. *J. Atmos. Solar-Terr. Phys.*, **71** (10–11), 1082–1092. doi: 10.1016/jastp.2006.12.004

Park, C.G. (1970) Whistler observations of the interchange of ionization between the ionosphere and protonsphere. *J. Geophys. Res.*, **75** (22), 4249–4260. doi: 10.1029/JA075i022p04249

Park, C.G. (1974) Some features of plasma distribution in the plasmasphere deduced from Antarctic whistlers. *J. Geophys. Res.*, **79** (1), 169–173. doi: 10.1029/JA079i001p00169

Park, C.G., Carpenter, D.L., and Wiggin, D.B. (1978) Electron density in the plasmasphere: whistler data on solar cycle, annual and diurnal variations. *J. Geophys. Res.*, **83** (A7), 3137–3144. doi: 10.1029/JA083iA07p03137

Parker, E.N. (1959) Dynamics of the interplanetary gas and magnetic fields. *Astrophys. J.*, **128**, 664–676.

Paschmann, G., Schwartz, S.J., Escoubet, C.P., and Haaland, S. (2005) Outer magnetospheric boundaries: Cluster results. *Space Sci. Rev.*, **118** (1), 231–424.

Peristykh, A.N. and Damon, P.E. (2003) Persistence of the Gleissberg 88-year solar cycle over the last 12,000 years: evidence from cosmogenic isotopes. *J. Geophys. Res.*, **108** (A1). doi: 10.1029/2002JA009390

Perrault, P. and Akasofu, S.-I. (1978) A study of geomagnetic storms. *Geophys. J. R. Astron. Soc.*, **54** (3), 547–573. doi: 10.1111/j.1365-246X.1978.tb05494.x

Pilipenko, V.A. and Fedorov, E.N. (1994) Magnetotelluric sounding of the crust and hydromagnetic monitoring of the magnetosphere with the use of ULF waves, in *Solar Wind Sources of Magnetosphere Ultra-Low Frequency Waves* (eds M.J. Engebreston, K. Takahashi, and M. Scholer,), American Geophysical Union, Washington, DC, pp. 283–292.

Pilipenko, V.A., Yumoto, K., Fedorov, E., and Yagova, N. (1999) Hydromagnetic spectroscopy of the magnetosphere with Pc3 geomagnetic pulsations along the 210° meridian. *Ann. Geophys.*, **17**, 53–65. doi: 10.1007/s00585-999-0053-1

Pirjola, R. (2002) Review on the calculation of surface electric and magnetic fields of geomagnetically induced currents in ground-based technological systems. *Surv. Geophys.*, **23**, 71–90.

Pitteway, M.L.V. (1965) The numerical calculation of wave-fields, reflexion coefficients and polarizations for long radio waves in the lower ionosphere. *Philos. Trans. R. Soc. Lond. A*, **257**, 219–241.

Plaschke, F., Glassmeier, K.-H., Auster, H.U., Constaninescu, O.D., Magnes, W., Angelopoulos, V., Sibeck, D.G., and McFadden, J.P. (2009a) Standing Alfvén waves at the magnetopause. *Geophys. Res. Lett.*, **36** (L02104). doi: 10.1029/2008GL036411

Plaschke, F., Glassmeier, K.-H., Constaninescu, Mann, I.R., Milling, D.K., Motschmann, U., and Rae, I.J. (2008) Statistical analysis of ground based magnetic field measurements with the field line resonance detector. *Ann. Geophys.*, **26**, 3477–3489. doi: 10.5194/angeo-26-3477-2008

Plaschke, F., Glassmeier, K.-H., Sibeck, D.G., Auster, H.U., Constaninescu, O.D., Angelopoulos, V., and Magnes, W. (2009b) Magnetopause surface oscillation frequencies at different solar wind conditions. *Ann. Geophys.*, **27**, 4521–4532. doi: 10.5194/angeo-27-4521-2009

Ponomarenko, P.V., Fraser, B.J., Menk, F.W., Ables, S.T., and Morris, R.J. (2002) Cusp-latitude Pc3 spectra: band-limited and power-law components. *Ann. Geophys.*, **20**, 1539–1551. doi: 10.5194/angeo-20-1539-2002

Ponomarenko, P.V., Menk, F.W., and Waters, C.L. (2003) Visualization of ULF waves in SuperDARN data. *Geophys. Res. Lett.*, **30** (18), 1926. doi: 10.1029/2003GL017757

Ponomarenko, P.V., Menk, F.W., Waters, C.L., and Sciffer, M.D. (2005) Pc3-4 ULF waves observed by the SuperDARN TIGER radar. *Ann. Geophys.*, **23** (1271–1280). doi: 10.5194/angeo-23-1271-2005

Ponomarenko, P.V., Waters, C.L., Sciffer, M.D., Fraser, B.J., and Samson, J.C. (2001) Spatial structure of ULF waves: comparison of magnetometer and Super Dual Auroral Radar Network data. *J. Geophys. Res.*, **106** (A6), 10509–10517. doi: 10.1029/2000JA000281

Ponomarenko, P.V., Waters, C.L., and St-Maurice, J.-P. (2010) Upstream Pc3-4 waves: experimental evidence of propagation to the nightside plasmapause/plasmatrough. *Geophys. Res. Lett.*, **37** (L22102). doi: 10.1029/2010GL045416

Poole, A.W.V., Sutcliffe, P.R., and Walker, A.D.M. (1988) The relationship between ULF geomagnetic pulsations and ionospheric Doppler oscillations: derivation of a model. *J. Geophys. Res.*, **93** (A12), 14656–14664. doi: 10.1029/JA093iA12p14656

Potemra, T., Lühr, H., Zanetti, L.J., Takahashi, K., Erlandson, R.E., Block, L.P., Blomberg, L.G., and Lepping, R.P. (1989) Multisatellite and ground-based observations of transient ULF waves. *J. Geophys. Res.*, **94** (A3), 2543–2554. doi: 10.1029/JA094iA03p02543

Poulter, E.M., Allan, W., and Bailey, G.J. (1988) ULF pulsation eigenperiods within the plasmasphere. *Planet Space Sci.*, **36** (2), 185–196. doi: 10.1016/0032-0633(88)90054-2

Poulter, E.M., Allan, W., Bailey, G.J., and Moffett, R.J. (1984) On the diurnal period variation of mid-latitude ULF pulsations. *Planet Space Sci.*, **32** (6), 727–734. doi: 10.1016/0032-0633(84)90096-5

Preece, W.H. (1894) Earth currents. *Nature*, **49** (554). doi: 10.1038/049554b0

Price, I.A., Waters, C.L., Menk, F.W., Bailey, G.J., and Fraser, B.J. (1999) A technique to investigate plasma mass density in the topside ionosphere using ULF waves. *J. Geophys. Res.*, **104** (A6), 12723–12732. doi: 10.1029/1999JA900042

Pu, Z.-Y. and Kivelson, M.G. (1983) Kelvin–Helmholtz instability at the magnetopause: solution for compressible plasmas. *J. Geophys. Res.*, **88** (A2), 841–852. doi: 10.1029/JA088iA02p00841

Radoski, H.R. (1967) A note on oscillating field lines. *J. Geophys. Res.*, **72** (1), 418–419. doi: 10.1029/JZ072i001p00418

Radoski, H.R. (1971) A note on the problem of hydromagnetic resonances in the magnetosphere. *Planet. Space Sci.*, **19**, 1012–1013.

Radoski, H.R. (1972) The effect of asymmetry on toroidal hydromagnetic waves in a dipole field. *Planet Space Sci.*, **20**, 1015–1023.

Radoski, H.R. (1974) A theory of latitude dependent geomagnetic micropulsations: the asymptotic fields. *J. Geophys. Res.*, **79** (4), 595–603. doi: 10.1029/JA079i004p00595

Rae, I.J., Mann, I.R., Watt, C.E.J., Kistler, L.M., and Baumjohann, W. (2007a) Equator-S observations of drift mirror mode waves in the dawnside magnetosphere. *J. Geophys. Res.*, **112** (A11203). doi: 10.1029/2006JA012064

Rae, I.J., Watt, C.E.J., Fenrich, F.R., Mann, I. R., Ozeke, L.G., and Kale, A. (2007b) Energy deposition in the ionosphere through a global field line resonance. *Ann. Geophys.*, **25**, 2529–2539. doi: 10.5194/angeo-25-2529-2007

Rankin, R., Kabin, K., Lu, J.Y., Mann, I.R., Marchand, R., Rae, I.J., Tikhonchuk, V.T., and Donovan, E.F. (2005) Magnetospheric field-line resonances: ground-based observations and modeling. *J. Geophys. Res.*, **110** (A10809). doi: 10.1029/2004JA010919

Rankin, R., Kabin, K., and Marchand, R. (2006) Alfvénic field line resonances in arbitrary magnetic field topology. *Adv. Space Res.*, **38** (8), 1720–1729. doi: 10.1016/j.asr.2005.09.034

Ratcliffe, J.A. (1972) *An Introduction to the Ionosphere and Magnetosphere*, Cambridge University Press, Cambridge.

Rees, M.H. (1989) *Physics and Chemistry of the Upper Atmosphere*, Cambridge University Press, Cambridge.

Reinisch, B.W., Haines, D.M., Bibl, K., Cheney, G., Galkin, I.A., Huang, X., Myers, S.H., Sales, G.S., Benson, R.F., Fung, S.F. *et al.* (2000) The radio plasma imager investigation on the IMAGE spacecraft. *Space Sci. Rev.*, **91** (1–2), 319–359. doi: 10.1023/A:1005252602159.

Reinisch, B.W., Huang, X., Song, P., Green, J. L., Fung, S.F., Vasyliunas, V.M., Gallagher, D.L., and Sandel, B.R. (2004) Plasmaspheric mass loss and refilling as a result of a magnetic storm. *J. Geophys. Res.*, **109** (A01202). doi: 10.1029/2003JA009948

Reinisch, B.W., Moldwin, M.B., Denton, R.E., Gallagher, D.L., Matsui, H., Pierrard, V., and Tu, J. (2009) Augmented empirical models of plasmaspheric density and electric field using IMAGE and CLUSTER data. *Space Sci. Rev.*, **145**, 231–261. doi: 10.1007/s11214-008-9481-6

Rezania, V. and Samson, J.C. (2005) Quasi-periodic oscillations: resonant shear Alfvén waves in neutron star magnetospheres. *Astron. Astrophys.*, **436** (3), 999–1008. doi: 10.1051/0004-6361:20041796

Richards, P.G. (2002) Ion and neutral density variations during ionospheric storms in September 1974: comparison of measurement and models. *J. Geophys. Res.*, **107** (1361). doi: 10.1029/2002JA009278

Rishbeth, H. (1973) Physics and chemistry of the ionosphere. *Contemp. Phys.*, **14** (3), 229–249. doi: 10.1080/00107517308210752

Rishbeth, H. and Müller-Wodarg, I.C.F. (2006) Why is there more ionosphere in January than in July? The annual asymmetry in the F2-layer. *Ann. Geophys.*, **24**, 3293–3311. doi: 10.5194/angeo-24-3293-2006

Roberts, B., Edwin, P.M., and Benz, A.O. (1984) On coronal oscillations. *Astrophys. J.*, **279**, 857–865. doi: 10.1086/161956

Rodger, C.J., Clilverd, M.A., Seppälä, A., Thomson, N.R., Gamble, R.J., Parrot, M., Sauvaud, J.-A., and Ulich, T. (2010) Radiation belt electron precipitation due to geomagnetic storms: significance to middle atmosphere ozone chemistry. *J. Geophys. Res.*, **115** (A11320). doi: 10.1029/2010JA015599

Rostoker, G., Skone, S., and Baker, D.N. (1998) On the origin of relativistic electrons in the magnetosphere associated with some geomagnetic storms. *Geophys. Res. Lett.*, **25** (19), 3701–3704. doi: 10.1029/98GL02801

Russell, C.T. (1989) ULF waves in the Mercury magnetosphere. *Geophys. Res. Lett.*, **16** (11), 1253–1256. doi: 10.1029/GL016i011p01253

Russell, C.T., Luhmann, J.G., Odera, T.J., and Stuart, W.F. (1983) The rate of occurrence of dayside Pc3,4 pulsations: the *L*-value dependence of the IMF cone angle effect. *Geophys. Res. Lett.*, **10** (8), 663–666. doi: 10.1029/GL010i008p00663

Russell, C.T. and McPherron, R.L. (1973) Semiannual variation of geomagnetic activity. *J. Geophys. Res.*, **78** (1), 92–108. doi: 10.1029/JA078i001p00092

Saito, T. (1964) A new index of geomagnetic pulsation and its relation to solar M-regions, Part I. *Rep. Ionos. Space Phys. Jpn.*, **18**, 260–274.

Saito, T. (1969) Geomagnetic pulsations. *Space Sci. Rev.*, **10** (3), 319–412. doi: 10.1007/BF00203620

Sakurai, T., Tonegawa, Y., Kitagawa, T., Yumoto, K., Yamamoto, T., Kokubun, S., Mukai, T., and Tsuruda, K. (1999) Dayside magnetopause Pc3 and Pc5 ULF waves observed by the GEOTAIL satellite. *Earth Planets Space*, **51**, 965–978.

Samson, J.C. (1973) Descriptions of the polarization states of vector processes: applications to ULF magnetic fields. *Geophys. J. R. Astron. Soc.*, **34**, 403–419.

Samson, J.C. (1983) The spectral matrix, eigenvalues, and principal components in the analysis of multichannel geophysical data. *Ann. Geophys.*, **1** (3), 115–119.

Samson, J.C., Greenwald, R.A., Ruohoniemi, J.M., Hughes, T.J., and Wallis, D.D. (1991) Magnetometer and radar observations of magnetohydrodynamic cavity modes in the earth's magnetosphere. *Can. J. Phys.*, **69**, 929–939.

Samson, J.C., Harrold, B.G., Ruohoniemi, J. M., Greenwald, R.A., and Walker, A.D.M. (1992) Field line resonances associated with MHD waveguides in the magnetosphere. *Geophys. Res. Lett.*, **19** (5), 441–444. doi: 10.1029/92GL00116

Samson, J.C., Jacobs, J.A., and Rostoker, G. (1971) Latitude-dependent characteristics of long-period micropulsations. *J. Geophys. Res.*, **76** (16), 3675–3683. doi: 10.1029/JA076i016p03675

Samson, J.C. and Olson, J.V. (1981) Data-adaptive polarization filters for multichannel geophysical data. *Geophysics*, **46** (10), 1423–1431. doi: 10.1190/1.1441149

Samson, J.C., Waters, C.L., Menk, F.W., and Fraser, B.J. (1995) Fine structure in the spectra of low latitude field line resonances. *Geophys. Res. Lett.*, **22** (16), 2111–2114. doi: 10.1029/95GL01770

Sandel, B.R. and Denton, M.H. (2007) Global view of refilling of the plasmasphere. *Geophys. Res. Lett.*, **34** (L17102). doi: 10.1029/2007GL030669

Sandel, B.R., Goldstein, J., Gallagher, D.L., and Spasojevic, M. (2003) Extreme ultraviolet imager observations of the structure and dynamics of the plasmasphere. *Space Sci. Rev.*, **109** (1–4), 25–46.

Sarris, T.E., Li, X., and Singer, H.J. (2009a) A long-duration narrowband Pc5 pulsation. *J. Geophys. Res.*, **114** (A01213). doi: 10.1029/2007JA012660

Sarris, T.E., Liu, W., Kabin, K., Li, X., Elkington, S.R., Ergun, R., Rankin, R., Angelopoulos, V., Bonnell, J., Glassmeier, K.-H., and Auster, U. (2009b) Characterization of ULF pulsations by THEMIS. *Geophys. Res. Lett.*, **36** (L04104). doi: 10.1029/2008GL036732

Sarris, T.E., Loto'aniu, T.M., Li, X., and Singer, H.J. (2007) Observations at geosynchronous orbit of a persistent Pc5 geomagnetic pulsation and energetic electron flux modulations. *Ann. Geophys.*, **25**, 1653–1667. doi: 10.5194/angeo-25-1653-2007

Saxton, J.M. and Smith, A.J. (1989) Quiet time plasmaspheric electric fields and plasmasphere–ionosphere coupling fluxes at $L = 2.5$. *Planet Space Sci.*, **37** (3), 283–293.

Schäfer, S., Glassmeier, K.-H., Eriksson, P.T.I., Mager, P.N., Pierrard, V., Fornaçon, K.H., and Blomberg, L.G. (2008) Spatio-temporal structure of a poloidal Alfvén wave detected by Cluster adjacent to the dayside plasmapause. *Ann. Geophys.*, **26**, 1805–1817. doi: 10.5194/angeo-26-1805-2008

Schäfer, S., Glassmeier, K.-H., Eriksson, P.T.I., Pierrard, V., Fornaçon, K.H., and Blomberg, L.G. (2007) Spatial and temporal characteristics of poloidal waves in the terrestrial plasmasphere: a Cluster study. *Ann. Geophys.*, **25**, 1011–1024. doi: 10.5194/angeo-25-1011-2007

Schield, M.A., Freeman, J.W., and Dessler, A.J. (1969) A source for field-aligned currents at auroral latitudes. *J. Geophys. Res.*, **74** (1), 247–256. doi: 10.1029/JA074i001p00247

Scholer, M. (1970) On the motion of artificial ion clouds in the magnetosphere. *Planet Space Sci.*, **18** (7), 977–1004. doi: 10.1016/0032-0633(70)90101-7

Schröder, W. and Wiederkehr, W.S. (2000) A history of the early recording of geomagnetic variations. *J. Atmos. Solar-Terr. Phys.*, **62**, 323–334.

Schulz, M. (1996) Eigenfrequencies of geomagnetic field lines and implications for plasma density modeling. *J. Geophys. Res.*, **101** (A8), 17385–17397. doi: 10.1029/95JA03727

Schulz, M., Blake, J.B., Mazuk, S.M., Balogh, A., Dougherty, M.K., Forsyth, R.J., Keppler, E., Phillips, J.L., and Bame, S.J. (1993) Energetic particle, plasma and magnetic field signatures of a poloidal pulsation in Jupiter's magnetosphere. *Planet Space Sci.*, **41** (11-12), 967–975.

Schulz, M. and Lanzerotti, L.J. (1974) *Particle Diffusion in the Radiation Belts*, vol. **7**, Springer, Berlin.

Schunk, R. and Nagy, A. (2009) *Ionospheres: Physics, Plasma Physics, and Chemistry*, 2nd edn, Cambridge University Press, Cambridge.

Schwartz, S.J., Burgess, D., and Moses, J.J. (1996) Low frequency waves in the Earth's magnetosheath: present status. *Ann. Geophys.*, **14**, 1134–1150. doi: 10.1007/s00585-996-1134-z

Schwartz, S.J., Henley, E., Mitchell, J., and Krasnoselskikh, V. (2011) Electron temperature gradient scale at collisionless shocks. *Phys. Rev. Lett.*, **107** (215002). doi: 10.1103/PhysRevLett.107.215002

Schwenn, R. (2001) Solar wind: global properties, in *Encyclopedia of Astronomy and Astrophysics* (ed. P. Murdin), Institute of Physics and Macmillan Publishing, Bristol, pp. 1–9.

Sciffer, M.D. and Waters, C.L. (2002) Propagation of ULF waves through the ionosphere: analytic solutions for oblique magnetic fields. *J. Geophys. Res.*, **107** (A10), 1297–1311. doi: 10.1029/2001JA000184

Sciffer, M.D. and Waters, C.L. (2011) Relationship between ULF wave mode mix, equatorial electric fields, and ground magnetometer data. *J. Geophys. Res.*, **116** (A06202). doi: 10.1029/2010JA016307

Sciffer, M.D., Waters, C.L., and Menk, F.W. (2004) Propagation of ULF waves through the ionosphere: inductive effect for oblique magnetic fields. *Ann. Geophys.*, **23**, 1155–1169. doi: 10.5194/angeo-22-1155-2004

Sciffer, M.D., Waters, C.L., and Menk, F.W. (2005) A numerical model to investigate the polarisation azimuth of ULF waves through an ionosphere with oblique magnetic fields. *Ann. Geophys.*, **23**, 3457–3471. doi: 10.5194/angeo-23-3457-2005

Sckopke, N. (1966) A general relation between the energy of trapped particles and the disturbance field near the Earth. *J. Geophys. Res.*, **71** (13), 3125–3130. doi: 10.1029/JZ071i013p03125

Seppälä, A., Randall, C.E., Clilverd, M.A., Rozanov, E., and Rodger, C.J. (2009) Geomagnetic activity and polar surface level air temperature variability. *J. Geophys. Res.*, **114** (A10312). doi: 10.1029/2008JA014029

Serenelli, A.M. (2010) New results on standard solar models. *Astrophys. Space Sci.*, **328** (1–2), 13–21. doi: 10.1007/s10509-009-0174-8

Serson, P.H. (1973) Instrumentation for induction studies on land. *Phys. Earth Planet Inter.*, **7**, 313–322.

Sheeley, B.W., Moldwin, M.B., Rassoul, H.K., and Anderson, R.R. (2001) An empirical plasmasphere and trough density model:

CRRES observations. *J. Geophys. Res.*, **106** (A11), 25631–25641. doi: 10.1029/2000JA000286

Shue, J.-H., Chao, J.K., Fu, H.C., Russell, C.T., Song, P., Khurana, K.K., and Singer, H.J. (1997) A new functional form to study the solar wind control of the magnetopause size and shape. *J. Geophys. Res.*, **102** (A), 9497–9511. doi: 10.1029/97JA00196

Siebert, M. (1964) Geomagnetic pulsations with latitude-dependent periods and their relation to the structure of the magnetosphere. *Planet Space Sci.*, **12** (2), 137–147.

Singer, H.J., Southwood, D.J., and Kivelson, M.G. (1981) Alfvén wave resonances in a realistic magnetospheric magnetic field geometry. *J. Geophys. Res.*, **86** (A6), 4589–4596. doi: 10.1029/JA086iA06p04589

Singh, N. and Horwitz, J.L. (1992) Plasmasphere refilling: recent observations and modeling. *J. Geophys. Res.*, **97** (A2), 1049–1079. doi: 10.1029/91JA02602

Sinha, A.K. and Rajaram, R. (1997) An analytic approach to toroidal eigen mode. *J. Geophys. Res.*, **102** (A8), 17649–17657. doi: 10.1029/97JA01039

Slavin, J.A., Acuña, M.H., Anderson, B.J., Baker, D.N., Benna, M., Gloeckler, G., Gold, R.E., Ho, G.C., Killen, R.M., Korth, H., Krimigis, S.M., McNutt, R.L., Jr., Nittler, L. R., Raines, J.M., Schriver, D., Solomon, S. C., Starr, R.D., Trávnőček, P., and Zurbuchen, T.H. (2008) Mercury's magnetosphere after MESSENGER's first flyby. *Science*, **321** (5885), 85–89. doi: 10.1126/science.1159040

Smith, R.L. (1961a) Propagation characteristics of whistlers trapped in field-aligned columns of enhanced ionization. *J. Geophys. Res.*, **66** (11), 3699–3707. doi: 10.1029/JZ066i011p03699

Smith, R.L. (1961b) Properties of the outer ionosphere deduced from nose whistlers. *J. Geophys. Res.*, **66** (11), 3709–3716. doi: 10.1029/JZ066i011p03709

Smith, R.L. and Angerami, J.J. (1968) Magnetospheric properties deduced from OGO 1 observations of ducted and nonducted whistlers. *J. Geophys. Res.*, **73** (1), 1–20. doi: 10.1029/JA073i001p00001

Song, P., Gombosi, T.I., and Ridley, A.J. (2001) Three-fluid Ohm's law. *J. Geophys. Res.*, **106** (A5), 8149–8156. doi: 10.1029/2000JA000423

Southwood, D.J. (1968) The hydromagnetic stability of the magnetospheric boundary. *Planet Space Sci.*, **16** (5), 587–605.

Southwood, D.J. (1974) Some features of field line resonances in the magnetosphere. *Planet Space Sci.*, **22**, 483–491. doi: 10.1016/0032-0633(74)90078-6

Spasojević, M., Goldstein, J., Carpenter, D.L., Inan, U.S., Sandel, B.R., Moldwin, M.B., and Reinisch, B.W. (2003) Global response of the plasmasphere to a geomagnetic disturbance. *J. Geophys. Res.*, **108** (A9). doi: 10.1029/2003JA009987

Stephenson, J.A.E. and Walker, A.D.M. (2002) HF radar observations of Pc5 ULF pulsations driven by the solar wind. *Geophys. Res. Lett.*, **29** (9). doi: 10.1029/2001GL014291

Stewart, B. (1859) Construction of the self-recording magnetographs: at present in operation at the Kew Observatory of the British Association. 29th Report of the British Association for the Advancement of Science, pp. 200–228.

Stewart, B. (1861) On the great magnetic disturbance which extended from August 28 to September 7, 1859, as recorded by photography at the Kew Observatory. *Philos. Trans. R. Soc. Lond. A*, **151**, 423–430.

Storey, L.R.O. (1953) An investigation of whistling atmospherics. *Phil. Trans. R. Soc. Lond. A*, **246** (908), 113–141. doi: 10.1098/rsta.1953.0011

Suárez, J.C., Garrido, R., Balona, L.A., and Christensen-Dalsgaard, J. (eds) (2012) *Stellar Pulsations: Impact of New Instrumentation and New Insights*, Springer, Heidelberg, p. 285.

Sugiura, M. (1961) Evidence of low-frequency hydromagnetic waves in the exosphere. *J. Geophys. Res.*, **66** (12), 4087–4095. doi: 10.1029/JZ066i012p04087

Sugiura, M. and Wilson, C.R. (1964) Oscillation of the geomagnetic field and associated magnetic perturbations at conjugate points. *J. Geophys. Res.*, **69** (7), 1211–1216. doi: 10.1029/JZ069i007p01211

Sutcliffe, P.R., Hattingh, S.K.F., and Boshoff, H.F.V. (1987) Longitudinal effects on the eigenfrequencies of low-latitude Pc3 pulsations. *J. Geophys. Res.*, **92** (A3), 2535–2543. doi: 10.1029/JA092iA03p02535

Takahashi, K., Berube, D., Lee, D.-H., Goldstein, J., Singer, H.J., Honary, F., and Moldwin, M.B. (2009) Possible evidence of virtual resonance in the dayside magnetosphere. *J. Geophys. Res.*, **114** (A05206). doi: 10.1029/2008JA013898

Takahashi, K., Bonnell, J., Glassmeier, K.-H., Angelopolous, V., Singer, H.J., Chi, P.J., Denton, R.E., Nishimura, Y., Lee, D.-H., Nosé, M., and Liu, W. (2010a) Multipoint observation of fast mode waves trapped in the dayside plasmasphere. *J. Geophys. Res.*, **115** (A12247). doi: 10.1029/2010JA015956

Takahashi, K., Denton, R.E., and Singer, H.J. (2010b) Solar cycle variation of geosynchronous plasma mass density derived from the frequency of standing Alfvén waves. *J. Geophys. Res.*, **115** (A07207). doi: 10.1029/2009JA015243

Takahashi, K., Fennell, J.F., Amata, E., and Higbie, P.R. (1987) Field-aligned structure of the storm time Pc5 wave of November 14–15, 1979. *J. Geophys. Res.*, **92** (A6), 5857–5864. doi: 10.1029/JA092iA06p05857

Takahashi, K., Glassmeier, K.-H., Angelopolous, V., Bonnell, J., Nishimura, Y., Singer, H.J., and Russell, C.T. (2011) Multisatellite observations of a giant pulsation event. *J. Geophys. Res.*, **116** (A11223). doi: 10.1029/2011JA016955

Takahashi, K., Liou, K., Yumoto, K., Kitamura, K., Nosé, M., and Honary, F. (2005) Source of Pc4 pulsations observed on the nightside. *J. Geophys. Res.*, **110** (A12207). doi: 10.1029/2005JA011093

Takahashi, K. and McPherrron, R.L. (1982) Harmonic structure of Pc3-4 pulsations. *J. Geophys. Res.*, **87** (A3), 1504–1516. doi: 10.1029/JA087iA03p01504

Takahashi, K., McPherron, R.L., and Terasawa, T. (1984) Dependence of the spectrum of Pc3-4 pulsations on the interplanetary magnetic field. *J. Geophys. Res.*, **89** (A5), 2770–2780. doi: 10.1029/JA089iA05p02770

Takahashi, K. and Ukhorskiy, A.Y. (2007) Solar wind control of Pc5 pulsation power at geosynchronous orbit. *J. Geophys. Res.*, **112** (A11205). doi: 10.1029/2007JA012483

Takahashi, K. and Ukhorskiy, A.Y. (2008) Timing analysis of the relationship between solar wind parameters and geosynchronous Pc5 amplitude. *J. Geophys. Res.*, **113** (A12204). doi: 10.1029/2008JA013327

Tamao, T. (1964) The structure of three-dimensional hydromagnetic waves in a uniform cold plasma. *J. Geomag. Geolectr.*, **16**, 89–114.

Tamao, T. (1965) Transmission and coupling resonance of hydromagnetic disturbances in the non-uniform Earth's magnetosphere. *Sci. Rep. Tohoku Imp. Univ.*, **17** (2), 43–72.

Tan, L.C., Shao, X., Sharma, A.S., and Fung, S. F. (2011) Relativistic electron acceleration by compressional-mode ULF waves: evidence from correlated Cluster, Los Alamos National Laboratory spacecraft, and ground-based magnetometer measurements. *J. Geophys. Res.*, **116** (A07226). doi: 10.1029/2010JA016226

Tanskanen, E.I., Slavin, J.A., Tanskanen, A.J., Viljanen, A., Pulkkinen, T.I., Koskinen, H. E.J., Pulkkinen, A., and Eastwood, J. (2005) Magnetospheric substorms are strongly modulated by interplanetary high-speed streams. *Geophys. Res. Lett.*, **32** (L16104). doi: 10.1029/2005GL023318

Taylor, J.P.H. and Walker, A.D.M. (1984) Accurate approximate formulae for toroidal standing hydromagnetic oscillations in a dipolar geomagnetic field. *Planet Space Sci.*, **32** (9), 1119–1124. doi: 10.1016/0032-0633(84)90138-7

Tepley, L. and Landshoff, R.K. (1966) Waveguide theory for ionospheric propagation of hydromagnetic emissions. *J. Geophys. Res.*, **71** (5), 1499–1504. doi: 10.1029/JZ071i005p01499

Teramoto, M., Nosé, M., and Sutcliffe, P.R. (2008) Statistical analysis of Pi2 pulsations inside and outside the plasmasphere observed by the polar orbiting DE-1 satellite. *J. Geophys. Res.*, **113** (A07203). doi: 10.1029/2007JA012740

Thorne, R.M. (2010) Radiation belt dynamics: the importance of wave-particle interactions. *Geophys. Res. Lett.*, **37** (L22107). doi: 10.1029/2010GL044990

Toth, G., Sokolov, I.V., Gombosi, T.I., Chesney, D.R., Clauer, C.R., De Zeeuw, D.L., Hansen, K.C., Kane, K.J., Manchester, W.B., Oehmke, R.C., Powell, K.G., Ridley, A.J., Roussev, I.I., Stout, Q. F., Volberg, O., Wolf, R.A., Sazykin, S., Chan, A., Yu, B., and Kota, J. (2005)

Space Weather Modeling Framework: a new tool for the space science community. *J. Geophys. Res.*, **110** (A12226). doi: 10.1029/2005JA011126

Trakhtengerts, V.Y. and Demekhov, A.G. (2007) Generation of Pc1 pulsations in the regime of backward wave oscillator. *J. Atmos. Solar-Terr. Phys.*, **69** (14), 1651–1656.

Tribble, A.C. (2003) *The Space Environment; Implications for Spacecraft Design*, Princeton University Press, Princeton, NJ.

Troitskaya, V.A., Plyasova-Bakunina, Y.A., Gul'yel'mi, A.V. (1971) Relationship between Pc2–3 pulsations and the interplanetary magnetic field. *Dokl. Akad. Nauk SSSR*, **197**, 1321.

Tsurutani, B.T., Arballo, J.K., Mok, J., Smith, E.J., Mason, G.M., and Tan, L.C. (1994) Electromagnetic waves with frequencies near the local proton gyrofrequency: ISEE-3 1 AU observations. *Geophys. Res. Lett.*, **21** (7), 633–636. doi: 10.1029/94GL00566

Tsyganenko, N.A. (1990) Quantitative models of the magnetospheric magnetic field: methods and results. *Space Sci. Rev.*, **54** (1-2), 75–186. doi: 10.1007/BF00168021

Tsyganenko, N.A. (1995) Modeling the Earth's magnetospheric magnetic field confined within a realistic magnetopause. *J. Geophys. Res.*, **100** (A4), 5599–5612. doi: 10.1029/94JA03193

Tsyganenko, N.A. and Sitnov, M.I. (2007) Magnetospheric configurations from a high-resolution data-based magnetic field model. *J. Geophys. Res.*, **112** (A06225). doi: 10.1029/94JA03193

Tu, C.-Y. and Marsch, E. (1995) MHD structures, waves and turbulence in the solar wind: observations and theories. *Space Sci. Rev.*, **73** (1-2), 1–210. doi: 10.1007/BF00748891

Turner, R. (2012) National response to a severe space weather event. *Space Weather*, **10** (S03008), 12–16. doi: 10.1029/2011SW000756

Uchida, Y. (1970) Diagnosis of coronal magnetic structure by flare-associated hydromagnetic disturbances. *Publ. Astron. Soc. Jpn*, **22**, 341–364.

Ukhorskiy, A.Y., Takahashi, K., Anderson, B.J., and Korth, H. (2005) Impact of toroidal ULF waves on the outer radiation belt electrons. *J. Geophys. Res.*, **110** (A10202). doi: 10.1029/2005JA011017

Ulrich, R.K. (1970) The five-minute oscillations on the solar surface. *Astrophys. J.*, **162**, 993–1001.

Usanova, M.E., Mann, I.R., Rae, I.J., Kale, Z. C., Angelopoulos, V., Bonnell, J.W., Glassmeier, K.-H., Auster, H.U., and Singer, H.J. (2008) Multipoint observations of magnetospheric compression-related EMIC Pc1 waves by THEMIS and CARISMA. *Geophys. Res. Lett.*, **35** (L17S25). doi: 10.1029/2008GL034458

Vasyliunas, V.M. (2006) Ionospheric and boundary contributions to the Dessler-Parker-Sckopke formula for Dst. *Ann. Geophys.*, **24**, 1085–1097. doi: 10.5194/angeo-24-1085-2006

Vellante, M. and Förster, M. (2006) Inference of the magnetospheric plasma mass density from field line resonances: a test using a plasmasphere model. *J. Geophys. Res.*, **111** (A11204). doi: 10.1029/2005JA011588

Vellante, M., Förster, M., Villante, U., Zhang, T.L., and Magnes, W. (2007) Solar activity dependence of geomagnetic field line resonance frequencies at low latitudes. *J. Geophys. Res.*, **112** (A02205). doi: 10.1029/2006JA011909

Vellante, M., Villante, U., Core, R., Best, A., Lenners, D., and Pilipenko, V.A. (1993a) Simultaneous geomagnetic pulsation observations at two latitudes: resonant mode characteristics. *Ann. Geophys.*, **11**, 734–741.

Vellante, M., Villante, U., De Lauretis, M., and Barchi, G. (1996) Solar cycle variation of the dominant frequencies of Pc3 geomagnetic pulsations at $L = 1.6$. *Geophys. Res. Lett.*, **23** (12), 1505–1508. doi: 10.1029/96GL01399

Vellante, M., Villante, U., De Lauretis, M., Core, R., Best, A., Lenners, D., and Pilipenko, V.A. (1993b) Simultaneous geomagnetic pulsation observations at two latitudes: resonant mode characteristics. *Ann. Geophys.*, **11**, 734–741.

Verö, J. and Holló, L. (1978) Connections between interplanetary magnetic field and geomagnetic pulsations. *J. Atmos. Terr. Phys.*, **40**, 857–865.

Verö, J. and Menk, F.W. (1986) Damping of Pc3-4 pulsations at high F2-region electron concentrations. *J. Atmos. Terr. Phys.*, **48** (3), 231–243.

Verronen, P.T., Rodger, C.J., Clilverd, M.A., and Wang, S. (2011) First evidence of mesospheric hydroxyl response to electron precipitation from the radiation belts. *J. Geophys. Res.*, **116** (D07307). doi: 10.1029/2010JD014965

Veronnen, P.T., Seppälä, A., Clilverd, M.A., Rodger, C.J., Kyrölä, E., Enell, C.-F., Ulich, T., and Turunen, E. (2005) Diurnal variation of ozone depletion during the October–November 2003 solar proton events. *J. Geophys. Res.*, **110** (A09S32). doi: 10.1029/2004JA010932

Verth, G., Erdélyi, R., and Goossens, M. (2010) Magnetoseismology: eigenmodes of torsional Alfvén waves in stratified solar waveguides. *Astrophys. J.*, **714**, 1637–1648. doi: 10.1088/0004-637X/714/2/1637

Viall, N.M., Kepko, L., and Spence, H.E. (2009) Relative occurrence rates and connection of discrete frequency oscillations in the solar wind density and dayside magnetosphere. *J. Geophys. Res.*, **114** (A01201). doi: 10.1029/2008JA013334

Viall, N.M., Spence, H.E., and Kasper, J. (2009) Are periodic solar wind number density structures formed in the solar corona? *Geophys. Res. Lett.*, **36** (L23102). doi: 10.1029/2009GL041191

Villante, U., Francia, P., Vellante, M., Giuseppe, P.D., Nubile, A., and Piersanti, M. (2007) Long-period oscillations at discrete frequencies: a comparative analysis of ground, magnetospheric, and interplanetary observations. *J. Geophys. Res.*, **112** (A04210). doi: 10.1029/2006JA011896

Villante, U., Lepidi, S., Francia, P., Meloni, A., and Palangio, P. (1997) Long period geomagnetic field fluctuations at Terra Nova Bay (Antarctica). *Geophys. Res. Lett.*, **24** (12), 1443–1446. doi: 10.1029/97GL01466

Walker, A.D.M. (1978) Formation of whistler ducts. *Planet Space Sci.*, **26** (4), 375–379.

Walker, A.D.M. (1981) The Kelvin–Helmholtz instability in the low-latitude boundary layer. *Planet Space Sci.*, **29** (10), 1119–1133.

Walker, A.D.M. (2000) Reflection and transmission at the boundary between two counterstreaming MHD plasmas: active boundaries or negative energy waves? *J. Plasma Phys.*, **63** (3), 203–219.

Walker, A.D.M. (2002) Excitation of field line resonances by MHD waves originating in the solar wind. *J. Geophys. Res.*, **107** (A12). doi: 10.1029/2001JA009188

Walker, A.D.M. (2005) *Magnetohydrodynamic Waves in Geospace: The Theory of ULF Waves and Their Interaction with Energetic Particles in the Solar-Terrestrial Environment*, Institute of Physics, Bristol.

Walker, R. and Kivelson, M.G. (1981) Multiply reflected standing Alfvén waves in the Io torus: Pioneer 10 observations. *Geophys. Res. Lett.*, **8** (12), 1281–1284. doi: 10.1029/GL008i012p01281

Walker, A.D.M., Ruoheniemi, J.M., Baker, K. B., Greenwald, R.A., and Samson, J.C. (1992) Spatial and temporal behavior of ULF pulsations observed by the Goose Bay HF radar. *J. Geophys. Res.*, **97** (A8), 12187–12202. doi: 10.1029/92JA00329

Walt, M. (1994) *Introduction to Geomagnetically Trapped Radiation*, Cambridge University Press, Cambridge.

Wang, T.J., Solanki, S.K., Curdt, W., Innes, D. E., Dammasch, I.E., and Kliem, B. (2003) Hot coronal loop oscillations observed with SUMER: examples and statistics. *Astron. Astrophys.*, **406** (3), 1105–1121. doi: 10.1051/0004-6361:20030858

Warner, M.R. and Orr, D. (1979) Time of flight calculations for high latitude geomagnetic pulsations. *Planet Space Sci.*, **27** (5), 679–689. doi: 10.1016/0032-0633(79)90165-X

Watermann, J. (1987) Observations of correlated ULF fluctuations in the geomagnetic field and in the phase path of ionospheric HF soundings. *J. Geophys.*, **61**, 39–45.

Watermann, J., Wintoft, P., Sanahuja, B., Saiz, E., Poedts, S., Palmroth, M., Milillo, A., Metallinou, F.-A., Jacobs, C., Ganushkina, N.Y., Daglis, I.A., Cid, C., Cerrato, G., Aylward, A.D., and Aran, A. (2009) Models of solar wind structures and their interaction with the Earth's space environment. *Space Sci. Rev.*, **147** (3–4), 233–270. doi: 10.1007/s11214-009-9494-9

Waters, C.L. (2000) ULF resonance structure in the magnetosphere. *Adv. Space Res.*, **25** (7–8), 1541–1558.

Waters, C.L., Anderson, B.J., and Liou, K. (2001) Estimation of global field aligned currents using the iridium® System magnetometer data. *Geophys. Res. Lett.*, **28** (11), 2165–2168. doi: 10.1029/2000GL012725

Waters, C.L. and Cox, S.P. (2009) ULF wave effects on high frequency signal propagation through the ionosphere. *Ann. Geophys.*, **27** (7), 2779–2788. doi: 10.5194/angeo-27-2779-2009

Waters, C.L., Harrold, B.G., Menk, F.W., Samson, J.C., and Fraser, B.J. (2000) Field line resonances and waveguide modes at low latitudes: 2. A model. *J. Geophys. Res.*, **105** (A4), 7763–7774. doi: 10.1029/1999JA900267

Waters, C.L., Menk, F.W., and Fraser, B.J. (1991) The resonance structure of low latitude Pc3 geomagnetic pulsations. *Geophys. Res. Lett.*, **18** (12), 2293–2296. doi: 10.1029/91GL02550

Waters, C.L., Menk, F.W., and Fraser, B.J. (1994) Low latitude geomagnetic field line resonance: experiment and modeling. *J. Geophys. Res.*, **99** (A9), 17547–17558. doi: 10.1029/94JA00252

Waters, C.L., Menk, F.W., Thomsen, M.F., Foster, C., and Fenrich, F.R. (2006) Remote sensing the magnetosphere using ground-based observations of ULF waves, in *Magnetospheric ULF Waves: Synthesis and New Directions* (eds K. Takahashi, P.J. Chi, and R.L. Lysak), American Geophysical Union, pp. 319–340.

Waters, C.L., Samson, J.C., and Donovan, E.F. (1995) The temporal variation of the frequency of high latitude field line resonances. *J. Geophys. Res.*, **100** (A5), 7987–7996. doi: 10.1029/94JA02712

Waters, C.L., Samson, J.C., and Donovan, E. F. (1996) Variation of plasmatrough density derived from magnetospheric field line resonances. *J. Geophys. Res.*, **101** (A11), 24737–24745. doi: 10.1029/96JA01083

Waters, C.L. and Sciffer, M.D. (2008) Field line resonant frequencies and ionospheric conductance: results from a 2-D MHD model. *J. Geophys. Res.*, **113** (A05219). doi: 10.1029/2007JA012822

Waters, C.L., Takahashi, K., Lee, D.-H., and Anderson, B.J. (2002) Detection of ultra-low frequency cavity modes using spacecraft data. *J. Geophys. Res.*, **107** (A10). doi: 10.1029/2001JA000224

Webb, D.C., Lanzerotti, L.J., and Park, C.G. (1977) A comparison of ULF and VLF measurements of magnetospheric cold plasma densities. *J. Geophys. Res.*, **82** (32), 5063–5072. doi: 10.1029/JA082i032p05063

Webster, D.J. and Fraser, B.J. (1985) Source regions of low-latitude Pc1 pulsations and their relationship to the plasmapause. *Planet Space Sci.*, **33** (7), 777–793.

Westphal, K.O. and Jacobs, J.A. (1962) Oscillations of the earth's outer atmosphere and micropulsations. *Geophys. J. R. Astron. Soc.*, **6** (3), 360–372.

Wild, J.A., Yeoman, T.K., and Waters, C.L. (2005) Revised time-of-flight calculations for high-latitude geomagnetic pulsations using a realistic magnetospheric magnetic field model. *J. Geophys. Res.*, **110** (A11206). doi: 10.1029/2004JA010964

Wolfe, A., Lanzerotti, L.J., and Maclennan, C. G. (1980) Dependence of hydromagnetic energy spectra on solar wind velocity and interplanetary magnetic field direction. *J. Geophys. Res.*, **85** (A1), 114–118. doi: 10.1029/JA085iA01p00114

Wright, A.N. and Allan, W. (2008) Simulations of Alfvén waves in the geomagnetic tail and their auroral signatures. *J. Geophys. Res.*, **113** (A02206). doi: 10.1029/2007JA012464

Wright, D.M. and Yeoman, T.K. (1999) High-latitude HF Doppler observations of ULF waves: 2. Waves with small spatial scale sizes. *Ann. Geophys.*, **17**, 868–876. doi: doi:10.1007/s00585-999-0868-9

Yearby, K.H. and Clilverd, M.A. (1996) Doppler shift pulsations on whistler mode signals from a VLF transmitter. *J. Atmos. Terr. Phys.*, **58** (13), 1489–1496.

Yeoman, T.K., Wright, D.M., Chapman, P.J., and Stockton-Chalk, A.B. (2000) High-latitude observations of ULF waves with large azimuthal wavenumbers. *J. Geophys. Res.*, **105** (A3), 5453–5462. doi: 10.1029/1999JA005081

Yizengaw, E., Moldwin, M.B., and Galvan, D.A. (2006) Ionospheric signatures of a plasmaspheric plume over Europe. *Geophys. Res. Lett.*, **33** (L17103). doi: 10.1029/2006GL026597

Yoshikawa, A. and Itonaga, M. (1996) Reflection of shear Alfvén waves at the ionosphere and the divergent Hall current.

Geophys. Res. Lett., **23** (1), 101–104. doi: 10.1029/95GL03580

Yoshikawa, A. and Itonaga, M. (2000) The nature of reflection and mode conversion of MHD waves in the inductive ionosphere: multistep mode conversion between divergent and rotational electric fields. *J. Geophys. Res.*, **105** (A5), 10565–10584. doi: 10.1029/1999JA000159

Yuan, Z., Xiong, Y., Pang, Y., Zhou, M., Deng, X., Trotignon, J.G., Lucek, E., and Wang, J. (2012) Wave-particle interaction in a plasmaspheric plume observed by a Cluster satellite. *J. Geophys. Res.*, **117** (A03205). doi: 10.1029/2011JA017152

Yumoto, K., Saito, T., Akasofu, S.-I., Tsurutani, B.T., and Smith, E.J. (1985) Propagation mechanism of daytime Pc3-4 pulsations observed at synchronous orbit and multiple ground stations. *J. Geophys. Res.*, **90** (A7), 6439–6450. doi: 10.1029/JA090iA07p06439

Zhu, X. and Kivelson, M.G. (1989) Global mode ULF pulsations in a magnetosphere with a nonmonotonic Alfvén velocity profile. *J. Geophys. Res.*, **94** (A2), 1479–1485. doi: 10.1029/JA094iA02p01479

Ziesolleck, C.W.S., Fenrich, F.R., Samson, J. C., and McDiarmid, D.R. (1998) Pc5 field line resonance frequencies and structure observed by SuperDARN and CANOPUS. *J. Geophys. Res.*, **103** (A6), 11771–11785. doi: 10.1029/98JA00590

Ziesolleck, C.W.S., Fraser, B.J., Menk, F.W., and McNabb, P.W. (1993) Spatial characteristics of low-latitude Pc3-4 geomagnetic pulsations. *J. Geophys. Res.*, **98** (A1), 197–207. doi: 10.1029/92JA01433

Ziesolleck, C.W.S. and McDiarmid, D.R. (1994) Auroral latitude Pc5 field line resonances: quantized frequencies, spatial characteristics, and diurnal variation. *J. Geophys. Res.*, **99** (A4), 5817–5830. doi: 10.1029/93JA02903

Index

Magnetoseismology: Ground-based remote sensing of Earth's magnetosphere, First Edition. Frederick W. Menk and Colin L. Waters.

n

o